Acknowledgments

Karen Horwath for her kindness, brilliance, and for her exceptional editing skills

Terry LeBarr for designing the great cover on this book

My friends in the following study groups for the wonderful discussions that fueled ideas in this book: Saginaw Valley State University (SVSU) Esoteric Study Group; Saginaw Philosophy Club; Buddhist study group at SVSU.

Rumi quotes flavor this entire book. Poet and translator Coleman Barks has gifted the modern world with a version of Rumi that is powerful, understandable, and emotionally relevant to our complex era.

COVER DESIGN

By Terry LeBarr, Swanton, Ohio.

gonefishn@yahoo.com

Contents

Introduction

Love is the way
messengers from the mystery
tell us things.

~ "Messengers from the Mystery,"
A Year with Rumi,[1] 2006.

At the Third-Eye-Watching Vegan Restaurant

"Welcome back, Dutch."

"Glad to be here, Surge."

"I read the draft of your book, by the way. I see that you re-interpreted physicists Albert Einstein, Werner Heisenberg, and Thomas Young. You think you know better than three of the greatest scientists who ever lived?"

"Not really. I just added a layer of complexity to a long-standing conundrum. I tried to be modest."

"Nice try. However, I detect a hint of flaky speculation."

"That sounds harsh, Surge."

"Really? Well, we have your 3rd grade math scores. Let's review your qualifications, shall we?

"No. Let's not."

"Oh my. Looks like you were having trouble with plus and minus signs—you seem to use them randomly. Your math homework is covered with doodling and bad poetry—plus all the red marks your teacher added. Evidently, you mathematically bottomed-out in the 3rd grade. Shall we look at your 9th grade algebra teacher's comments? Maybe she can shed some light on your ability to re-translate three of mankind's greatest physicists."

"Einstein failed math, too."

"Yeah, but he wasn't spectacular at it, like you. You seem to have exceled at incomprehension of numerical symbols. Mrs. Potter, your 9th grade algebra teacher, says that you were a *spectacular symbolic failure*. Those were the exact words she used at a parent-teacher conference. She says, and I quote: 'Perhaps he should join the Priesthood.' She saw value in your soul, but not so much in your intellect."

"So what's on the dinner menu, Surge? You don't mind if I call you Surge, do you?"

"This is the Hotel California. Nobody cares about names. I'll call you Dr. Einstein today—I'm sure that is okay with someone of your extraordinary intellect."

"There you go again with the sarcasm, Surge."

"For dinner this evening we offer two choices. You can get the Background Broth, or the Foreground Stew."

"They both sound bland. What else have you got?"

"That's all there is, background and foreground. Unless you want to catch a Dark-Matter Shuttle, Number Nine to Nowhere. There's a diner in North

Nowhere where you can order without being weighed down by space and time."

"What's in the Background Broth?"

"Nothing. It's our most mysterious and potential broth. You don't know what you ordered until it shows up. It's a variation on Dark-Matter Stew, except it contains Dim-Matter with photon chunks. It's good if you are on a diet."

"What about the Foreground Stew?"

"Ah, now *there* is something substantial. This stew is made with thick slabs of manifestation—very chewy. Actually, things just appear in the stew without warning. On a good day you might get fried shrimp and glazed carrots. On a bad day, the stew might manifest fish heads and failed lab experiments. It depends on how much love you have in your body. If you are filled with love, the stew might manifest herbal veggies in a delicate almond sauce. If you have a Grinch heart, the stew might manifest cat barf and gorilla lips. Who you are, is what you get."

"You know, I'm not feeling so hungry for a big meal. Maybe I'll just have dessert. What's for dessert?"

"This is your lucky day. We only serve Holodeck Pie during leap year."

"This is leap year?"

"No. This is the Hotel California. If we want it to be leap year, it is. Do you want the pie, or are you going to debate about the dessert menu?"

"Holodeck Pie? What's in it?"

"Whatever you can imagine. I usually suggest that you let your imagination run wild. Dream the impossible dream. You have to eat it quickly, however, because it tends to dematerialize. It has a short half-life."

"The pie spoils quickly?"

"No. It just returns home. It gets lonely for the Mother Pie."

"A dessert with a home?"

"Everything has memory. Of course, you know that, right?"

"Well, sort of. Not really. You mean computers have memory, and grandmothers have memory, and sea squirts have memory?"

"Yeah, that's it. And stars have memory, space has memory, cells have memory, groups of cells have collective memory, your Dog Bruno has memory, and even memory has memory—that's the most important kind of memory: memory of memories. Try to remember that."

"Okay. Bring me the Holodeck Pie and a mug of your famous Third-Eye-Watching coffee."

"So, what are you going to do after your book fails to sell? Not many people are interested in the memory of memory, or the strange behavior of sea squirts."

"Maybe I'll go to Guatemala and watch macaws make out in the trees."

"May I suggest you use the Holodeck Transport System? That way you can control the weather—avoid the rainy season and customs. You can also fill the scenes with birds of paradise—which are really flowers, even though we call them birds. Language is full of ersatz holes and re-mended adverbs. It's a wonder we can communicate at all."

"Are we communicating, Surge? Because, I am not so sure. When are you going to bring me my pie and coffee?"

"I think you need to get back to editing your book. You will embarrass our ancestors unless you make reparations. You can start by apologizing to the quantum physicists."

"I think the book is nearly finished. I just need Karen to do more editing."

"That's sweet. Well, I am sorry to say it's past *personal attention time* here at the famous Third-Eye-Watching Vegan Restaurant and Center for Perceptual Strangeness. I am no longer authorized to pay attention to your mundane problems."

"I don't understand."

"Well then, we *are* making progress!"

"I don't get to eat?"

"Eat your words, Dutch; you seem to have an excess. And the next time you come in for a mental meal, try to order more succinctly and with greater courage. Your wishy-washy nature is mildly irritating. Learn to create with intention. Forget about being modest and editorially correct."

A Small Contribution to the Larger Conundrum

What secret are you hiding with this madness?

~ "Dhu'l-Nun's Instructive Madness,"
A Year with Rumi, 2006.

This is a quirky book (maybe you noticed), although it didn't start out that way. It started as a serious and sincere look at human consciousness—a small contribution to that complex subject. But like the Roman god Janus, twin minds materialized in the foreground broth, each with a different opinion about how a book should be written. Now I find myself with this strange oscillation between a logical explanation for the evolution of consciousness and a strange set of vignettes that take place at the mythological Hotel California.[2]

The vignettes that begin each chapter are dialogues between Surge, a waiter at the Hotel California, and my alter-ego, who Surge insists on calling

Dutch. Use of dialogue is a Socratic method for exploring truth. It is also a method used in early esoteric works to get at mysterious ideas. Obscure statements and emotions bubbled up as I allowed automatic writing to inform my rational mind—frequent editing removed a lot of nonsense but preserved essential insights.

I use these dialogues to lighten the mood and to explore ideas intuitively. I was often surprised as I allowed the dialogue to flow onto a page. For example, in the opening communication between Surge and Dutch, a single sentence contained a summary of the entire book: "That's all there is, background and foreground." Keep that in mind as we explore human consciousness.

Here is the theme of the book: We have two minds—not just two sets of behaviors, not just left-brain/right-brain, or top-brain/bottom-brain, and not just two theories, although I base my ideas on an evolving discipline called dual-process theory. I contend that we have a hardwired, no-getting-around-it, dual-cognition. The god Janus is a visual depiction of an anatomical and physiological reality. This stark split in our cognition, this dual-mind conundrum, has serious consequences for how we manage our earthly domain. It matters on a day-to-day basis that we are torn between one mind and the other. And when one mind dominates, the results can be ugly—when there is no balance we suffer, and so does everyone else who comes into contact with us. We have two amazing cognitive gifts; the trick is using them in a healthy, effective manner.

The dialogue at the Hotel California represents one of our two minds speaking the way it normally communicates: indirectly through metaphor and storytelling, set in a magical location, and containing unexpected twists and turns. The second mind, the one I am using now, speaks with logic and precision, and it seeks to convince through the introduction of evidence. This bouncing from one mind to the other is meant to illustrate our two kinds of cognition. You will find yourself delighted and relieved to revisit Surge and Dutch, especially as you finish each mind-bending chapter of the book.

The Mission

I retired many years ago and now feel the need to record what I learned on my life-journey. How minds work, especially when they are "not normal," is a major interest which developed for me as I taught children in special education. Over the years, I always had more questions than answers. For example, I wondered if we could call any mind "normal" given the trillions of possible connections among neurons. What does a mind do, what is its job? Might a "blind mind" be different from a "deaf mind?" How are the minds of people with autism different from the minds of people without autism? Is the brain the same thing as the mind? [3] Eventually, I wondered how consciousness might play a different role for each of us with our unique, strange minds. I have been thinking about questions like these for over 50 years.

I worked in special education for 33 of those years, as an orientation and mobility specialist. Therefore, my major interests also include two other career-related issues: "navigational disability[4]," a generic term important in special education and rehabilitation; and second, the evolution of my profession, orientation and mobility[5], which benefits greatly from the discussion of the evolution of consciousness. I address navigational disability and the needs of my profession in Chapter Nine. Finally, I find that my books are an unusual kind of memoir. They contain what I enjoy talking about, and they are written in a way that pleases me.[6]

This is my second book.[7] My mission has not changed from that first publication, *Bugs, Blindness, and the Pursuit of Happiness*:

1. My profession, orientation and mobility, has an opportunity to embrace a scientifically sound new perspective based on a fresh understanding of human duality. The new foundation for orientation and mobility is designed around a relatively new field of study called dual-process theory. Therefore, I am hypothetically redefining a profession as I discuss how consciousness evolved.

2. I am making a case about consciousness: that we have two minds and, therefore, two distinct kinds of consciousness. These two minds have two expressive styles, two different ways to learn, to remember, and to pay attention. They are oppositional twins. Dual-process theory is a big deal in the study of consciousness— so says my ego.

3. I will make the case that one of our minds is the source of spirituality. The wisdom tradition of modern religions can be traced to the essence of a so-called *hidden mind*. Failure to recognize and develop the hidden mind results in a lonely barrenness, the feeling we get from being isolated egos living in a "modern" technological society. In Chapter Six, I compare the views of Buddhists with the views of Western neuroscience.

4. I am helping "everyman" understand consciousness. Because we have two minds in conflict, we struggle, both individually and as a collective culture. Understanding dual-process theory can help us better comprehend our biological heritage and our daily challenges.

The circumstantial evidence for our mental duality is found in book after book, in every discipline, throughout recorded history, in philosophy, psychology, and biology.[8] The evidence is overwhelming. *But why would we have two minds?* Why did evolution "decide" to take this route and how did it unfold over the eons? What is there in the anatomy and physiology of the brain and body that provides evidence for our cognitive and behavioral duality? These are the questions that inspired me as I did research for this book.

At first glance, it seems absurd to suppose that one brain, inside one body, could give rise to twin minds—indeed, these sibling minds seem always to be in conflict. Why would nature invent such a strange and paralyzing contraption? I believe there is a logical reason why this came about in evolution. I will argue that my theory, my small contribution, is supported by anatomy, physiology, and evolution. I will even boldly suggest that I have found a plausible explanation for our dual cognition and our paradoxical behaviors.

I am well aware how arrogant that claim might appear. I state my proposition with confidence, even though in my heart I hold many doubts. We humans are massively blind creatures. Therefore, our small viewpoints must be expressed with self-effacing good humor. I speak boldly to drive home my point, but please know that I do so with a wry smile. Ultimately, I do not know whether consciousness is the result of an intelligent, loving universe, or if consciousness is a matter-driven, and often unloving epiphenomenon. I leave that battle to the philosophers and saints. I also write this on a Wednesday afternoon in late April, 2017. From this day forward, as it has done in the past, knowledge will accumulate exponentially in our modern world. Perhaps my perspective will look less radical as time passes.[9]

Ultimately, what has been missing in the discussion of consciousness up until now is the evolutionary reason why brains and nervous systems developed in the first place. There is a flow in evolution that eventually created two minds, out of which came two kinds of consciousness. One of these two minds I call the egocentric mind. The other mind I call the allocentric (hidden) mind. *These mental systems are the result of two ways to pay attention.* I will explain this in detail in the next two chapters.

The Plan

In Chapter One, I explain *the real reason we have brains and nervous systems.* I also explore the key role purposeful movement played in the crafting of our duality. In Chapter Two, I look at the *science of consciousness*, with special attention to dual-process theory, which I discuss in detail.

Chapter Three reviews the perspectives of authors who wrote at a time when hemispheric differentiation was thought to hold the anatomical and physiological key to understanding consciousness. The end of Chapter Three offers other anatomical perspectives beyond hemispheric differentiation for explaining our inherent duality. In this third chapter, I also introduce the proposition that *proprioception is the sensory basis for dual consciousness.*

In Chapter Four, I further define the two core concepts "duality" and "consciousness." A great deal of confusion has resulted after centuries of debate, and much of the problem has centered on a failure to adequately define consciousness. In this chapter, I offer 21 perspectives on the concept of consciousness—what it is not and what it might be.

In Chapter Five, I contend that the egocentric mind gave us our ego, our personality, the familiar entity that we all recognize as who we are. However, the allocentric mind gave us an entity that many have called "self" or "soul." The ego and the self are separate and sophisticated physiological universes that are so starkly different and contrary they are not able to hold center stage alone. The two minds are in perpetual conflict. Indeed, physiologically, they oscillate. Chapter Five is an exploration of the contrasting characteristics of our allocentric and egocentric minds.

Chapter Six is a look at the correspondences between Buddhist practice and science, especially dual-process theory. There is an ongoing and expanding dialogue between Eastern philosophy, as exemplified by Buddhism, and Western science, especially neuroscience.

In Chapter Seven, I discuss quantum theory as the fundamental basis for our duality. Here we find the different perspectives on the theories of Albert Einstein, Werner Heisenberg, and Thomas Young that I mentioned earlier. In Chapter Seven I offer a logical explanation for the dual world we live in.[10]

In Chapter Eight, I explain the anatomical and physiological evidence for dual-process theory. I look at vision, hearing, touch, smell, and the hidden senses: proprioception, kinesthesis, the vestibular system, and photosensitivity. I show how there is a hardwired duality within each sensory system. I contend that as research into neuroscience and neuropsychology moves forward, the evidence for our dual architecture will get stronger until a day comes when dual-process theory is well-established.

Chapter Nine asks the question "Where do we go from here?" I use my experience as an orientation and mobility specialist to demonstrate how we might apply dual-process theory to a very practical field of study. It appears that we can take any kind of disorder, like blindness or deafness, and we can divide our understanding between allocentric and egocentric processing. Therefore, we need to rethink special education categories, and reconsider how we classify disorders. Chapter Nine also contains a look ahead at the next book in this series: *The Confusion Caused by Being Your Own Twin*. There are consequences and responsibilities that arise when we accept and understand our duality. In *The Confusion Caused by Being Your Own Twin*, I look at several broad Western disciplines, for example, religion, education, psychology, poetry, and philosophy, and I show how duality has played a fundamental role in each.

The social sciences use an organizational process called *grounded theory*, which is a blueprint for what I am doing in this book. Grounded theory is a way of thinking about and organizing data. It is a methodology that starts with a collection of clues and then begins to assemble the evidence into relevant categories. Slowly an overarching pattern emerges from all the bits of evidence. That pretty much explains my approach. I shuffle and reshuffle stacks of notes that are everywhere on my desk, tables, chairs, and floors. Every morning I am up early, refreshed and ready anew to rethink concepts, rewrite paragraphs and rearrange whole chapters as the ever-changing puzzle pieces are added or subtracted to the emerging gestalt. I am searching for how things fit together, what has relevance, how various disciplines have looked at the evidence, and whether or not my final dual-reality theory is lovely and useful for humanity—or not. I am also searching for a dynamic gestalt that can adjust to the evidence that flows into our collective knowledge-database over time.

This book has been an ambitious undertaking which I much enjoyed. If nothing else, I have gotten a fine liberal arts education from all the reading and research that has gone into this effort. I hope you find the book easy to read and an enjoyable adventure.

Given all these efforts to understand our duality, and all the evidence that there *actually are* two minds, we need to ask: What is going on? How can we have two minds? Why would nature invent such an unwieldy contraption? We will turn next to these questions and a possible answer in Chapter One.

We cannot rush the story because each step forward depends on the experiences we gained from the past. This is a progression wherein each step must be studied, experienced, and understood before we can move to the next proposition. I like to think of this adventure as an anthropological expedition; join me as we explore our mental makeup and our cognitive history.

Notes

(1) Quotes from the Sufi mystic Rumi. Rumi expresses and exposes, like no one else, the split between our two minds. Rumi poems contain amazing examples of our dual-consciousness. He wrote these poems over 700 years ago. It is clear that the reason we can appreciate and marvel at Rumi's poetry is because of the brilliance of his translator, poet and author Dr. Coleman Barks.

(2) The mythological Hotel California. There is a story behind why I featured the Eagle's song "Welcome to the Hotel California" in this book. At the Millet Learning Center, where I worked for 33 years as an orientation and mobility specialist, I was the unofficial staff photographer. I also took all the videos of events. I self-selected this role because I was uncomfortable in social settings and could hide behind the camera. The Millet Center is a public school for children with disabilities, located in Saginaw, Michigan.

We had yearly proms at Millet, which were elaborate affairs with limousines, a live band, disco lights, and over-the-top themes—giant Scottish castles or Greek Pantheons, for example. The boys wore tuxedos; the girls wore gorgeous prom dresses. However, the planning, running, and cleaning up after prom night was exhausting for the teachers and therapists who did all the work. The prom was also frequently hilarious.

On one prom night, I consoled a young girl who was weeping and babbling—something about her boyfriend, of course. When she calmed down, she was able to tell me that she was pregnant. After a stunned silence and some further probing, it turned out that the pregnancy was caused by a secret kiss that had occurred about ten minutes earlier in a dark corner of the gym.

After the prom was over, and the cleanup had begun, I wandered into the main therapy room with the large video camera still resting on my shoulder; this was in the 1990s. There I found my friend and fellow staff member Emily Browning sitting in the shadows, staring blankly into space. Hold that image while I fill in some back story.

The day before prom night, I had watched a documentary called *Burden of Dreams*, a 1982 film directed by Les Blank about the making of director Werner Herzog's movie *Fitzcarraldo*. The movie and the documentary were both filmed in the jungles of South America. The *Fitzcarraldo* story is about the S.S. Molly Aida, a three-story, 320-ton steamer that was dragged from a river, over a mountain, to another river. To make the movie, Herzog actually dragged a steam ship through the jungle and over a large hill to create a realistic depiction of a true historical event.

There is an unforgettable scene in *Burden of Dreams* in which Herzog goes off on a self-pitying tirade about the horrors of the jungle: blood-sucking bugs, suffocating heat and humidity, and the lack of good showers— everyone was in a murderous mood. The ship was only halfway up the hill, and Herzog was near total collapse. As bad as this might sound, the scene is hilarious. Herzog badly needs happy hour. He needs friends willing to hear his rant, friends who can appreciate his dilemma. That's how I found Emily after prom night—an exhausted young woman stuck in the jungle we call special education—in bad need of happy hour.

Welcome to the Hotel California

I had to go into that background story because when I discovered Emily in the twilight of the therapy room, alone, staring blankly into the darkness, I had a documentary film maker's moment. I lifted the camera to my eye, turned on the recording, and began to talk with Emily. What followed could have been Werner Herzog in the jungle. Emily was totally exhausted from the days of preparation for the prom; she had worked late every evening for a week. She had watched over prom night activities, and then she helped clean up. That was as far as her body would go. She sat in a chair turned toward a bare wall, staring into the darkness like a zombie. With the cameras rolling, Emily gave a speech remonstrant of Herzog's rant:

"I hate proms," she said to the darkness. "I don't want to work in special education. It's too hard. I don't like limousines either. Balloons are irritating. I hate confetti, too. There's too much pain, too much work, too much sadness. I can't do it anymore. There's no money for the kids, no help for the parents, the equipment is always breaking down, the paperwork and rules and regulations are just too much. You have to go to meetings and be yelled at by grieving mothers. The doctors can be rude and on autopilot. The administration has piles of money, but little of it trickles down to the teachers and the kids. And you want to know something else? Look at me. Turn that stupid camera on my face. I want to tell you something. Working in special education is like living at the Hotel California. Once you check in, you can never leave. I can't go another day, I can't do another prom. I hate fruit punch. You and me, Doug, we are stuck at the Hotel California where every night is prom night. We are stuck here because we care, and because our hearts refuse to leave, no matter how many proms await us. The Millet Center is the Hotel California."

Thanks to Emily, I frequently saw myself walking the halls of the Hotel California year after year, in a surreal space that was both magic and emotionally draining. I am glad I checked in. If I had to do it again, I would. I don't regret one second of my 33 years at Millet, and neither did Emily, despite her Herzog moment.

I do have to state the obvious, in case it isn't so obvious to you. The Hotel California is a metaphor. On one layer of understanding, it is life itself, which seems at times to be absurd. The only proper reaction to our stay at the Hotel California is to laugh hilariously—between bouts of weeping. What in the world are we doing in this stage play called life, stuck in this absurd reality? I don't know, to be honest. We just get up each morning, happy to be granted another day, and we do it all again.

There is a level of consciousness, level 5/6 on the Cook-Greuter Scale of ego development—which I discuss in *The Confusion Caused by Being Your Own Twin*—called Post-Modern Man. People who have evolved to this relatively advanced state of awareness go through the day as if all existence was transcendent. In this state of consciousness, every moment is a special event; everything we do is part of some greater reality. One level below this heightened Post-Modern stage is the consciousness of Modern Man, Cook-Grueter's level 4/5, where most of humanity resides. At this lower level of awareness there is no elevated perception. The world is just physics and psychology, bowling alleys and bank accounts, routines and regulations, car repairs, the struggle for career recognition, and root canals.

As for Emily Browning, she taught for over 40 years at the Millet Center before she retired, and she worked on many more proms. Her heart, however, never did leave the Hotel California. Emily and I still have lunch together from time-to-time—we stare into space together and reminisce.

(3) Is the brain the same thing as the mind? No, although it is an integral component. The mind has three integral parts: the brain, the body, and the environment. All three components are necessary for the mind to function. *The elimination of any of the three negates the mind.*

(4) "Navigational disability" is a term that I coined, informally, during my years of teaching. It refers to two generic kinds of disability: *allocentric disability*, and *egocentric disability*. To oversimplify and give an example, blindness can be understood as an allocentric disability. Deafness would be an egocentric disability.

(5) Orientation and mobility was my profession for 33 years. Although I was initially trained as an optometrist, my career revolved around my master's degree in blind rehabilitation. The evolution of orientation and mobility is one of my main concerns in this book. I feel that the discipline of orientation and mobility needs a philosophical foundation that is more sophisticated than the current practice. This book presents a body of research and an emergent philosophy that can become that new foundation.

(6) I write in a way that pleases me. I care what people think, and I care whether or not my ideas have value. However, I self-publish which gives me two powerful gifts: I can say whatever I want, however I want, without editorial oversight from a publishing house; and second, I can issue corrections and editions whenever I want. All I have to do is give Amazon Kindle and CreateSpace 80 dollars each for every set of changes, which I don't mind doing if it makes the book better. There are no more second and third editions in the electronic world; there is, instead, continual editing and publishing on-demand.

(7) This is my second book. The first, called ***Bugs, Blindness, and the Pursuit of Happiness***, was a compilation, a set of vignettes, taken from three other books that I was writing at the time—including this book. I wrote *Bugs, Blindness, and the Pursuit of Happiness* as an experiment in self-publishing, which proved to be highly satisfying.

(8) The circumstantial evidence for our cognitive duality is everywhere we look: Finding duality "everywhere we look" is an oversimplification. Actually, I mean *everywhere that I looked* as I did research I saw authors struggling with duality. My focus was on books and articles about consciousness, about how our minds work, so it makes sense that duality would continually be an issue. In my next book *The Confusion Caused by Being Your Own Twin,* I looked with greater detail at duality in several disciplines, including philosophy, psychology, poetry, and education.

(9) My perspective will look less radical as time passes. Actually, it is hardly radical at all; it is rather obvious and well-documented. All that is different is my insistence that we don't need any more evidence—we have dual minds.

(10) Here I offer a logical explanation for our dual world. Many people are working on the conundrum of consciousness, so there is a steady influx of new perspectives. Therefore, please read with caution. What I recorded in this book are thoughts—and a line of reasoning—that came at a certain time and place. As you read this, the world has already moved on at its unrelenting and exponential pace. What you read here, or anywhere else, is history. Nevertheless, I am pleased and honored to add my perspective to the ever-growing mountain of knowledge about consciousness. I look forward to dialogue with like-minded seekers.

Section One

MAKING THE CASE

One

The Real Reason for Brains

Those who live in time,
descended from Adam,
and are made of earth and water.
I am not part of that.
I borrow nothing.
I do not want anything from anybody.
I flow through all human beings.
Love is my only companion.

~ "Neither This nor That," *A Year with Rumi*, 2006.

You Ordered the Dark Matter Soup

"Excuse me, Surge."

"Yes?"

"There's no soup in my bowl."

"You ordered the Dark Matter Soup."

"Yes, but where is the matter?"

"Right there, in the bowl. We make our famous Dark Matter soup with the very finest Dark Energy spices. The soup is so delicately flavored that it becomes tasteless, so to speak."

"I don't see anything."

"I just said that. Dark Matter Soup is invisible to our gross human senses."

"Well, I'm hungry and there is no soup in my soup!"

"Sure there is. Answer me this: are you made out of atoms?"

"Yes, of course, I'm made out of atoms."

"Well, that's wrong. Who told you that you were made out of atoms?"

"Miss Persephone, my 7th grade science teacher."

"Well, Miss Persephone lied to you. Atoms only make up five percent of the universe. The rest of the universe is made from Dark Matter Soup and Dark Energy Spices. You are mostly made of mysterious forces and insubstantial darkness."

"I'm not paying for non-existent soup."

"If 95 percent of existence is made of darkness, then we are 95 percent blind to all that there might be. Human beings can't perceive dark sentience; we are blind to our cousins, the Dark Sentient Ones. They are here now, by the way, all around us. Most of the Dark Ones can't perceive us either, but a few can. Hi, Cosmo."

"There's nothing in my bowl!"

"Did you think space was empty?"

"My bowl is empty."

"That's wrong. No wonder you are confused by existence."

"I am *not* confused by existence."

"You are confused because you don't know that space is sentient and has memory. I thought we discussed this. Space is alive. Space has emotions; be nice to space. Dark Spatial Intelligence far exceeds your puny light-based understanding."

"Do you have any other kind of soup? I need protein and flavor."

"You haven't tasted your Dark Matter Soup."

"There is no soup!"

"Have you tried it?"

"I am not going to dip my spoon into empty space and pretend that I am eating. That is stupid."

"Not really. Stupid is what visual arrogance gives us when we aren't aware of our pervasive blindness. Do you want croutons in your Dark Matter Soup?"

"Will I be able to see them?"

"Of course not. They are Dark Matter Croutons."

"Okay. Hold on a minute. Does any of this have to do with my book, with dual-cognitive processing?"

"Now we are getting somewhere."

"Okay, fine. Where are we getting, Surge? What does this discussion have to do with my book?"

"You say we have two minds. Very good, a nice beginning. You are on the right track. Keep going, Dutch; we have your back. One of the minds, the allocentric mind, creates the entity you earth creatures call space. What do you suppose is the raw material the allocentric mind uses to manufacture space?"

"Dark energy, I suppose?"

"There you go. And what are the consequences of being immersed in Dark Energy?"

"We are blind fish? Swimming around in total darkness?"

"Close enough. Humans are dimwitted tunas, unaware that the sea is full of information. The tuna-mind thinks that the water is made of water. But water lives in space. Water is alive. Be nice to water. Remember that space is wise and loving—or not—depending on how much love you have within you."

"This discussion is making me tired, Surge. Could we talk about food? I am pretty sure this is a restaurant."

"Space is a kind of *Dark Matter Animal*. Every space is a living creature. Here's how it works: if intelligent people occupy a space, then that space evolves intelligence. If the space is filled with compassion and wisdom, then the space evolves compassion and wisdom. If the space is a barroom on a Friday night, then your grandma's smelly underwear is more intelligent. If you don't believe me, just ask this space. The Third-Eye-Watching Vegan Restaurant contains the most intelligent and wise space in the Northeast Cosmos. You can talk to spaces, did you know that? They listen, and then they help you. Space knows the answer to any question you might ask. Go ahead, ask this space a question."

"No one will buy my book now."

"Relax. No one was going to buy your book anyway."

"Then why did I write it?"

"It's a Dark Matter Book, an empty bowl, made especially for you and the reader. Have a taste; your life will never be the same."

The Real Reason for Brains

I was happy enough to stay still
inside the pearl inside the shell,

Consciousness: A New Slant on an Old Conundrum

but the hurricane of experience
lashed me out of hiding
and made me a wave
moving into shore.

~ "Out of Stillness," *A Year with Rumi*, 2006.

In November, 2001, a TED talk by Cambridge University professor Dr. Daniel Wolpert caused a flood of associations to explode in my mind—as if a mental dam had burst.[1] I had spent over thirty years working with children who had what I came to call *navigational disabilities*. As an orientation and mobility specialist, I worked mostly with blind children, teaching them to find their way without the help of vision. I could see, after years of teaching, that there was something about blindness—removing vision from the navigational equation—that held important insights. I struggled to explain to others just what those insights were and why they were important for comprehending human behavior, especially consciousness.

Dr. Wolpert's TED talk was called "The Real Reason for Brains." He gets right to the point (italics mine):

I want to start with the easiest question and the question you really should have all asked yourselves at some point in your life, because it's a fundamental question if we want to understand brain function. And that is, why do we and other animals have brains? Not all species on our planet have brains, so if we want to know what the brain is for, let's think about why we evolved one. Now you may reason that we have one to perceive the world or to think, and that's completely wrong. If you think about this question for any length of time, *it's blindingly obvious why we have a brain. We have a brain for one reason and one reason only, and that's to produce adaptable and complex movements.* There is no other reason to have a brain. Think about it. Movement is the only way you have of affecting the world around you.

In the question and answer period, Dr. Wolpert made this observation:

> People have found out that studying vision in the absence of realizing why you have vision is a mistake. You have to study vision with the realization of how the movement system is going to use vision. And it uses it very differently once you think about it that way. ~ *"The Real Reason for Brains," TED Talk, Daniel Wolpert, 2001.*

Tying vision to movement is brilliant; it explains both the anatomical design of the vision system and its physiology. Indeed, there are two vision systems, one to isolate and explore features of the environment, and a second to establish a steady state background (a scene) that provides stability. The conclusion that follows from this logic is that one vision system is temporal and egocentric, and the other is spatial and allocentric. This is discussed in detail in Chapter Eight.

Essentially, Dr. Wolpert's conclusion is that animals evolved nervous systems so that they could *move with a purpose.* If there is no need to move through a domain, or manipulate tools, then there is no need for a brain, or for muscles, or for sensory systems.

After some reflection, I realized—with much gratitude—that Dr. Wolpert had redefined my entire profession, and put all my bewildering notes about human navigation in order. This will take some explaining, but the key idea is that purposeful movement is essentially the same as navigation. Therefore, from my perspective, brains and nervous systems *primarily* evolved so that animals could navigate from one location to another.

Dr. Wolpert gives the example of a remarkable creature called a Sea Squirt to drive home his point. The Sea Squirt begins life as a tiny fish, with a primitive brain and nervous system. It swims about the ocean as it grows. Then one day, perhaps as a teenager, it sprouts roots, flips upright, and attaches the new roots to a rock. Then *it absorbs its brain and spinal cord* as food! It was an animal, but then made the decision that living as a plant was a better idea for Sea Squirt survival. Why, Dr. Wolpert asks in his TED Talk, didn't the Sea Squirt need its brain and nervous system after it became a plant? Of course, it is because the Sea Squirt, as a plant, will stop navigating as a fish and live a stationary life—it will no longer need muscles, nerves, eyes, fins, and so on. Brains and nervous systems evolved so that animals could navigate; the Sea Squirt's life cycle provides dramatic evidence.

As I did research for this book, I found others in the neuroscience field for whom this revelation was common knowledge. Quoted below is Chilean biologist, philosopher, and neuroscientist Francisco Varela.[2] Dr. Varela is the co-founder—with the Dalai Lama—of the *Mind and Life Institute* that promotes dialogue between neuroscience and Buddhism:

> The entire history of the brain has to do with one simple fundamental thing . . . No motion, no nervous system. No motion, no behavior . . . It doesn't matter what kind of brain—hydra brain, cat brain, fly brain, human brain—basically they are the same thing. ~
> *Gentle Bridges; Conversations with the Dalai Lama on the Sciences of Mind, Francisco Varela and Jeremy Hayward, editors, 1992.*

This is common knowledge to many people in the biological sciences. However, scientists who study navigation know that two separate processes must be active for purposeful movement to occur: egocentric processing and allocentric processing. To my knowledge, the idea that two minds eventually evolved because of these two distinct processing systems is new; the further insight that these two minds developed into two kinds of consciousness, is also a new idea—as far as my research has discovered.

I will spend the next few pages discussing the evolution of the allocentric mind—comparing a tree, which is a plant, with a human being, a highly evolved animal. I will then briefly discuss the evolution of the egocentric mind. This entire book is about the distinction between the egocentric and the allocentric mind, so what follows is just an introduction.

Plants and animals share a common ancestry, a very ancient genetic heritage. Human beings are very much the same kind of creature as a tree, but we evolved a complex brain for navigation and that gave the illusion of separateness. When we look at all the ways animals and plants are similar, we see the derivation of the allocentric mind. When we look at all the ways plants and animals differ, we see the derivation of the egocentric mind. This will become clearer as you make your way through the book.

Trees—indeed, all plants—have allocentric minds. Therefore, they have a kind of allocentric consciousness. They are non-violent, peaceful creatures that sense the environment in complex ways and then respond to changes as they arise—in the moment and as needed. The whole tree responds to the environment, all at once. Animals also respond all-at-once to environmental changes, because animals, like plants, have allocentric minds.

How are Plants and Animals the Same?

The Derivation of the Allocentric Mind

Imagine a perfect and beautiful tree. Place that tree in a magnificent setting. I envision a Ceiba tree in my mind, at the edge of the jungle in Costa Rica. The Ceiba tree is shaped like a giant umbrella, five stories tall—a king among plants. You can use my tree if you want, or pick a favorite of your own. It is important to have emotional feelings for the tree. Now place your idealized self at the base of that magnificent tree. You and your tree friend stand side-by-side before eternity, the culmination of a miracle that is beyond all comprehension: the miracle of life, of existence.

Now have a friend, someone you love, take a photograph of you and that tree. Next have your friend take a motion picture of you next to the tree. You are waving, smiling; the tree's leaves are shimmering. There is a gentle breeze and you notice that there are other life forms in the tree, especially insects and birds. The Ceiba tree can be home to hundreds of species of plants, insects, and mammals—some that live only on this kind of tree. Perhaps individual human beings also have species of creatures that live only on their skin. Both you and the tree are hosts for other living entities. Save the award-winning photo and that vivid motion picture; we will refer to both later.

Consciousness: A New Slant on an Old Conundrum

How are you and the tree similar? Think it over for a moment before you go on to read my thoughts.

To begin, there was a time when you and that tree did not exist. Then, like magic both of you appeared. You are children of two life forces: yin and yang, male and female—a duality. The two forces swirled together and material substance ignited into a living speck that began immediately to grow. The tree grew at first within a seed pod, while you began inside a single fertilized cell. From single cells you and the tree quickly became embryos and you both began to evolve into a form.

Contained in each embryo was a molecular code, a genetic pattern that was a recipe for how to make a living creature. The seed held instructions for making a tree. Your embryo contained plans for creating a human being. Consequently, both you and the tree have a species-specific form. You look human and the tree looks like a tree. That code, a genetic instruction set, evolved for trees and humans, plants and animals, for billions of years.

Genes evolved long before the evolutionary split between plants and animals; therefore, humans and plants share a genetic ancestry. All complex living creatures contain DNA. It is just that the molecules can be arranged into recipes for creating different kinds of creatures. There still exist genes inside human beings that are the same as genes inside of trees. That tree you stand beside is your very distant cousin.

Furthermore, both trees and humans have a common developmental unfolding. The seed for the tree became a sprout just as you became an infant. The tree has a life cycle that stretches from the embryo to maturity just as you pass through childhood, adolescence, and reach maturity. Each lifeform emerges out of nature, grows, matures, and becomes an adult, after which the lifeform subsides back into nature. Nature, at each moment, is the culmination of all these small lifeforms moving through the same developmental cycle. Furthermore, you and the tree are alive at the same time in history. You share a universal time frame, a once-in-eternity brotherhood.

Both of you and everything else that lives at this moment are riding the crest of nature's wave.

You and the tree both breathe. The tree breathes in carbon dioxide and breathes out oxygen. Human beings breathe in oxygen and breathe out carbon dioxide. Therefore, animals and plants need each other. The breath of life is a pulsating sharing, a blending, and a necessary symbiosis. When Buddha sat beneath the sacred fig tree to attain enlightenment, it was a powerful marriage of two kinds of breath: what we breathe out, plants take in; what plants breathe out, we take in. Nature is constantly breathing. Furthermore, the air is recycled; you and the tree are breathing the same air that Jesus, Buddha, dinosaurs, Shakespeare, and your great Aunt Maude also breathed. You and your tree cousin are co-dependent; you need each other. We are all intimately linked, moment-by-moment.

The tree and you are both made of cells—humans have about 50 trillion cells, 15 billion of which are neurons in the brain. The cells become colonies that create tissues; the tissues join in super-colonies called organs. Organs form even larger colonies called bodies, or organisms. Each cell is a tiny universe, an amazingly complex entity that has evolved for billions of years. The cells inside a tree and the cells inside human beings have specialized, but they retain the same overall structure: a membrane holds the cell together and gives it a characteristic form; a "skeletal system" provides internal stability; organelles exist to carry out specific functions within a cell; internal transport systems move chemicals in and out of the cell; a gelatinous, saltwater-based interior fills the cell; machinery is in place for energy conversion; and a system exists for genetic repair and self-replication—to name just a few of the functional parts. You and your tree-cousin are built from the same cellular Lego kit.

The field of quantum biology is also discovering that cells are quantum factories.[3] All the strangeness of the quantum world, from quantum tunneling to entanglement, is showing up at the cellular level. There is increasing evidence that we are held together by a kind of frequency genetics that cells use to communicate and to carry out every molecular function.[4] This

12

quantum coherence is present not only in single cells, but it also operates in tissues, organs, and organisms. The evidence that we live in a quantum universe is coming in at a blistering rate. You and your tree-friend are quantum, and you live entangled in a quantum sea.

Without air, water, light, gravity, and nutrients in just the right balance, neither you nor the tree could survive. You and the tree are land creatures. You don't live in the ocean, nor do you fly about the atmosphere. You are both connected to the earth. As land-based creatures you are both highly successful. Furthermore, you and the tree have a genetic predisposition for clean air, clean water, and untainted nutrients that are not compromised by light pollution, air pollution, or water pollution. Plants grow strong and healthy in pristine environments, and so do we. Trees live in groves, humans live in families; there is some kind of primal social need, something comforting, that sentient creatures have for their own kind—there seems also to be a special bond between trees and human beings. Indeed, humans seem to have an affinity for all lifeforms; there is a bond among all sentient creatures sharing a life-space.

The atmosphere that surrounds trees and humans is filled with electromagnetic energies of various kinds. The tree and you are bathed in this electromagnetic sea. We have to be in harmony with these environmental frequencies or we perish. The space that surrounds trees and human beings contains information; consequently, no two spaces are identical. Both trees and humans have evolved ways to use spatial information to enable survival.[5]

From a universal perspective you and the tree are "carbon-based units" made from the same ingredients that compose everything else in the material universe. That we are stardust is not just poetry. Human beings and trees are composed of the primal quarks of the big bang.

Even at the beginning of evolution several billion years ago, cells could signal each other. All cells, ancient and modern, send and receive signals. Therefore, living creatures are made of cells that can harmonize, synchronize, and behave coherently. Plants, as well as animals, are made of these "cells that can communicate." The notion that only humans can communicate is not true. Trees have a heritage that is millions of years old. Consequently, they eventually developed complex social and communication capabilities, just like their animal cousins.

Trees may even have some kind of undefined mind that rivals our cognition. It could be that you and your tree friend are highly intelligent creatures, each using different varieties of intelligence. Indeed, the allocentric mind is a total-body perceptual system that responds in the moment to environment changes—to experiences. Human beings have experiences and trees have experiences—life is defined by what happens moment-to moment. Trees have characteristic bodies and they respond to atmospheric and geological disruptions. They have sensory systems that respond to frequency vibrations, and they respond to chemicals in the air and soil. They sense through the surfaces of their leaves, flowers, roots, and bark. Allocentrically, it is easy to postulate that trees have plant-like experiences—therefore, they have a kind of allocentric mind, just as human beings do.

I will make the case later that proprioception is the key to consciousness, both allocentric and egocentric. Trees, plants, animals, and insects have proprioception—they can differentiate their own form from the environment. Trees know if they are being invaded and need to take defensive measures. They also group in groves; they communicate with their kind through "root nervous systems" and through chemical exchanges through the airwaves. They are social in a way only plants can understand. We cannot assume that plants do not have a form of empathy, wisdom, and intelligence. We cannot assume that there are not levels of plant consciousness.

The evolution of the human allocentric mind, and the evolution of allocentric plant minds, goes back to the beginning of life—when single cells responded to the environment holistically. These single cells responded all-at-once to immediate changes in their surrounding domain. Plants and human beings also respond all-at-once, as if they were still single cells. This "all-at-once response system" *is* the allocentric mind. It gave birth to various kinds and degrees of animal and plant consciousness.

The tree will age, get sick, and die. That too is your fate. The dead body of a tree and the dead body of a human being quickly deteriorate and disappear from the scene of life as if they had never been born. When we are alive we have a characteristic temperature, a form, a chemical composition, and a set of frequencies that set us apart from our environment. But when we die we become the environment—we fall back into the canvas we call Nature. Your tree friend is as magnificent and tragic a creation as you, and when you die, you merge and become one.

How are Plants and Animals Different?

THE DERIVATION OF THE EGOCENTRIC MIND

Neither of you is dead yet (thankfully), so let us return to our vital idealized selves standing together, tree and human side-by-side in a lovely spot on planet earth—look at the photo and review the video. Now you can see what it is that differentiates animals from plants:

- Animals are bilateral, while plants are symmetrical.
- Animals have brains, nervous systems, sensory systems, and muscles because they navigate. Plants have none of that kind of architecture because they never leave home.

We know that trees turn their leaves toward the sun; they send their roots out in search of nutrients and water. Vines climb trees to get sunlight. The Venus flytrap opens and closes its pedals to catch flies. Plants, like animals, purposefully move. Therefore, purposeful movement alone does not differentiate plants from animals. However, there is a kind of purposeful movement, navigation, which does differentiate the animal kingdom from the plant kingdom. Navigation, the ability to move about without a root system, is what defines the animal world. Creatures who navigate need elaborate brains, networks of nerves, sensory systems, and muscles. Trees need none of these systems; they do not navigate.

We also know what happens when we don't move often enough. If we cultivate a couch-potato existence, if we limit our existence to a few routine spaces and rigid temporal habits, then muscles and neurons deteriorate. "Use it or lose it" is an apt folk description. Enriched or novel environments cause the mind to grow. Diminished and restrictive environments cause the mind and body to deteriorate. Muscles, sensory systems, and cognitive processing depend on purposeful movement for optimal health and efficiency.

Bilaterality is a strategy that evolved to allow for purposeful movement in *a forward direction*—toward things that hold a meaning for organisms. Bilaterality enabled animals to move one set of body parts forward while using other body parts for stability. This oscillation, the alternate swinging forward of a whole side of the body, caught on in evolution and has been part of the genetic endowment of animals for at least a half-billion years.

There are advantages and disadvantages that arise from bilaterality. Dividing the organism into two mirror halves allowed for faster, more efficient, straight-ahead locomotion, but it also caused the evolution of two sets of everything: two eyes, two ears, two vestibular systems, two nostrils, and so on. Because of this dual architecture, animals became binaural and binocular. They developed stereo-perception. This caused the two brain hemispheres to also differentiate. An entire neuro-mechanism had to develop to control the left side of the body, and another neuro-mechanism had to evolve to control the right side of the body. To ensure balanced and efficient movement, another neural system had to evolve to synchronize the activities of the two sides of the body.

Consequently, we evolved with an inherent anatomical and physiological duality. Interneurons had to develop solely for the purpose of coherently stitching these two sets of everything together. There had to be incredible synchronicity built into the neural architecture of the brain and body. Otherwise, without coherency, if we survived at all, we would be spastic, dysfunctional creatures. The neuro-blending of our mirror anatomical selves had to evolve an extremely fast circuitry for combining these separate worlds into a sensation of oneness.

Not only do we have mirror image halves, we also have front and back sides. This sets up a perceptual system that is future-oriented; the mind is always looking ahead to where it might go in the future. That is why the brain is sometimes said to have evolved for the purpose of prediction. We have a long-evolved system for projecting into the future to various spatial and temporal depths. We can also generate scenarios of what might happen if we were to make various choices. Consequently, our future-obsessed mental orientation is a product of the evolution of our front-and-back body architecture, and of our dual-navigational minds.

A major invariant that shapes every living creature is gravity. Plants and animals have a top and a bottom. This is obvious for trees that have a canopy above and a root system below. Likewise, human beings have heads above and feet below. Every cell in a body is affected equally by the downward tug of gravity. The first simple cells had to develop systems for adjusting to this force. Some bacteria, for example, use magnetic crystals that enable them— as a colony—to align with the earth's magnetic field. Likewise, the human vestibular system is an elaborate mechanism for adjusting to the gravitational pull of the earth. This ancient sensory system is crucial for smooth locomotion, for balance, and for posture.

Gravity may also have led to our spiritual understanding of ascending upward to Heaven, toward realms that are above our surface-of-the-earth reality. Gravity weakens as we move away from the earth; above us only creatures that can fly inhabit this rarified domain. Angels have wings because they fly in the highest, most ethereal realms. When we are "down to earth," we are practical, pragmatic; we don't dwell in those suspicious realms where the air is thin and the gravity weak. When we have "both feet firmly planted on the ground," we are less likely to float off into the nether-land of dreams and flaky speculation—one of my favorite realms.

Bilaterality, along with the development of the senses and muscles, defines the animal kingdom, differentiating animals from plants. But there is something else to be noticed here that is *the* essential differentiation. To move with a purpose, to navigate, requires two systems—one to manage movement, the other to manage stabilization or no-movement. Each of these systems eventually gave rise to a separate mind. One mind, the allocentric, we share with plants and animals, but the other mind, the egocentric, because of language, is not shared with plants and animals—it is unique to humans.

Whereas allocentric processing is a whole-body system, egocentricity is dependent on localized sensory systems—like eyes and ears—that isolate and focus on objects in the environment. The egocentric mind perceives the trees, but not the forest. The allocentric mind sees the forest, but not the individual trees. Egocentric processing identifies and remembers landmarks, specific invariants in the surround. Egocentricity gave rise to the ego, the personality, the sense of being separate from the world. There is much more to say about our two minds, but for now I offer the following summary:

- Navigation, the ability to go from place to place without a root system, is what primarily differentiates animals from plants.
- Navigation evolved a system for forward movement called *bilaterality* in animals. This caused an overall anatomical duality; the body was divided into two matching halves, two mirror images.
- Just like the rest of the body, the brain also evolved two halves. The left side of the brain controls the right side of the body; the right side of the brain controls the left side of the body. The left brain is responsible for perceiving space on the right. The right brain is responsible for perceiving space on the left.
- Perception was also affected by navigation. Vision became binocular, and hearing became binaural. Indeed, all the sensory systems contribute to stereo-perception.

- Creatures that navigate also evolved front and back sides, as well as a top and a bottom. These divisions also affected the evolution of perception. Human beings project into the future and reflect about the past because of their front/back architecture.
- The most startling development that arose from bilaterality and animal navigation was the gradual creation of dual cognition. Two minds eventually gave rise to dual consciousness.
- One of the senses, proprioception, may well be the reason that consciousness, in both minds, eventually evolved. This will be discussed in the next chapter and also in Chapter Eight.

The Duality Inherent in Navigation

Our evolution as sentient creatures took a giant leap forward during a time in evolutionary history called the Cambrian Explosion, over a half-billion years ago. During this era, and after several million years of rapid evolution, eyes and ears took a relatively modern form. It was also at this time that bilaterality, front-side/back-side, and top/bottom appeared as successful evolutionary designs. Sophisticated nervous systems evolved during the Cambrian age; the head—with bilateral eyes and ears—became especially important. Thus, the Cambrian Era was the great age when navigation became rapid and accurate. It was the age when creatures called *animals* emerged on the evolutionary stage. Because animals developed navigational capabilities that depended on neurological dualities, they slowly evolved two minds—that is the theory put forth here: not only do humans have dual minds, but so do all creatures capable of navigation.

One of the first revelations that came to me, now that I had this fresh perspective from Dr. Wolpert, was that the duality of the human mind paralleled a duality that was inherent in navigation. *Two attention systems are necessary for navigation. As we flow straight ahead we have two choices*: we can ignore solid objects and flow around them, or we can approach and consider objects-of-regard to determine their relevance.

Neural structures evolved to decide between these two alternatives: should I stop and investigate, or should I continue on my way? To investigate requires sustained attention and long-term memory. This "stopping to investigate" forged the egocentric mind, which gave birth to meaning and relevance. When we do not stop to investigate, when we flow from place-to-place, we are using our ancient allocentric mind. The allocentric mind does not stop to consider meaning and relevance as it moves from one location to another; its role is not to gather knowledge or to use landmarks. Its purpose is to efficiently navigate through a very specific environmental domain.

As I emphasized earlier, the historical landscape within all professions, from philosophy to education, from nuclear physics to neuroscience, is littered with the perplexity of duality. Understandably, each discipline tries to explain this duality in the language of their profession. The result of this multi-disciplinary specialization is that we now have a linguistic mess. Each discipline defines our inherent duality using different words for the very same concepts.

This linguistic mess came into better focus for me when I realized that the ability to navigate required two fundamental neural architectures; two separate processing systems had to slowly co-evolve. Essentially, *two ways to pay attention* had to be co-created within the animal kingdom. My new perspective concerns *how animals gather the information needed for navigation*. This insight—that we are bi-dimensional beings—helps us see why every field of study would eventually discover and puzzle over our fundamental duality.

Two Ways to Gather Information

What apparently happened in evolution, the proposition set forth here, is that the two attention systems which evolved to enable navigation did so independently of each other. Two neural systems *evolving side-by-side* became so elaborate and sophisticated over time that they can be said to have evolved into two minds. These two neural networks, over time, gave birth to two different kinds of mind. Each mind evolved its own brand of what we call consciousness. Therefore, I contend that there are two fundamental ways to perceive, two minds, and *two kinds* of consciousness.

One mind, the egocentric mind, evolved the ability to manifest objects from a formless background. Think of the figure-ground dichotomy. The egocentric mind is all about the *figure-side* of that equation. The allocentric mind, however, is all about the *background-side* of the equation. Two separate processing systems had to evolve to enable two ways to pay attention: either we can attend to figures, or we can defocus and attend to a surround—*but not both at once.*

Cognitive scientists understand egocentric processing and allocentric processing as two kinds of reference points for understanding space:

> When we consider how we find our way—what psychologists refer to as spatial cognition—we need to make clear what our reference point is. If we are talking about the location of things relative to ourselves, we are considering *egocentric space*. For example, as I write these words, my coffee mug is to my left. If I were to walk around to the other side of my desk and not move my coffee mug, it would be on my right. Egocentric space is the 'left—right' and 'front—back' space we use when specifying locations relative to our bodies. If our reference point were absolute space—where we are located relative to other objects in our surroundings—we would be talking about allocentric space. If I had a Global Positioning System on my desk, I could obtain the coordinates for my absolute, or allocentric, position on this planet. ~ *Why People Get Lost; The Psychology and Neuroscience of Spatial Cognition, Paul A. Dudchenko, 2010.*

Paul Dudchenko, in the quote above, is using a navigational perspective to differentiate *spotlight* from *floodlight* perception. When we use spotlight attention we are an ego, an entity around which the universe revolves. The spotlight is mostly focused straight ahead, but it can—with a little effort—swing left, right, above, below, and behind. In floodlight mode, however, the ego dissolves and sinks into the surround. "We" become part of a flat map in a relative position to all other solid forms. This is allocentric perception. The metaphor of spotlight perception versus floodlight perception will be discussed later—it is an image that I first encountered in the teachings of Buddhist philosopher and author Alan Watts.

The Twin Minds: Egocentric and Allocentric

Egocentric attention is always focused on something or someone. Whenever you lock onto an object-of-regard, then you are experiencing egocentric attention. Egocentric attention occurs as if you are the center of the world—everything is seen from the perspective of this ego. When you are trying to understand something, the egocentric mind is active and the allocentric mind is relatively inhibited.

To the contrary, allocentric awareness dissolves the ego, and you become part of a total gestalt—no longer the center of the universe. When you walk from place to place you are in allocentric mode. Moving is the best way to understand the allocentric mind because, as you move, you must relatively inhibit egocentric processing. Accurate navigation depends on both of these mechanisms—both minds—working together. A layman's summary of these two navigational abstractions is that *egocentric processing takes place one-thing-at-a-time*, while *for allocentric processing, everything-is-perceived-at-once*.

You can watch the two minds switch using optical illusions. My favorite is the *flower of life*. Get a copy of this image off the internet and stare at it. Notice that it is not possible to maintain the same image pattern for even a few seconds. This is because the brain is struggling to figure out what is background and what is foreground. You are experiencing a battle between your two attention systems: the egocentric mind fights to hold a 3-dimensional form, but the allocentric mind keeps trying to dissolve forms into a flat gestalt.

American psychologist William James said that we had two basic kinds of behavior that resulted in change. Either we followed the path of incremental learning, taking *one thing at a time*, accumulating knowledge and making judgments, or we could gain insights *all at once*, suddenly perceiving the big picture, without judgement.

The purpose of egocentric processing is to study the world of objects. The egocentric mind searches for *object relevance*. The allocentric mind, however, has a different kind of relationship to meaning. As we move about we must flow around objects that would impede our forward progress. The solid objects in the surround are part of a scene in which objects have a dependable location. The meaning of an object for the allocentric mind is related to that object's position relative to the body and to other objects. Allocentric meaning is about *relationships* and about navigating around, toward, or away from solid forms. We are able to navigate because we trust that *location* is a usable invariant.

However, egocentric processing is different. It freezes flow (inhibits allocentric processing)—the egocentric mind stops to consider and to study. Egocentric processing turns objects into landmarks, which are "forms" that hold meaning for the traveler. The word "meaning" is used differently by the two minds. The allocentric mind wants to know the *relevance of a scene* so that it can navigate through an environment without collisions, and so it can get where it wants to go. The egocentric mind seeks to understand the essence of things, asking, for example, "What is this?" "Who is this?" "How can I use this?" "Should I remember this?"

Notice that when you pay attention using your egocentric processing system, when you are using landmarks to navigate, for example, you must relatively suppress the allocentric mind. However, when you are using the allocentric mind to navigate, the egocentric mind must be relatively suppressed. In other words, *it is impossible to totally attend to the whole surround the same time as you attend to a specific object within the surround.* Consequently, these two processing systems *take turns* attending to the world. You are either in one state of attention or the other. The egocentric mind needs to examine objects for meaning, for purpose; it is best served if the whole background doesn't suddenly show up and melt all objects into a flat gestalt. Likewise, if the organism is monitoring the whole of the environment, primarily watching for movement, for change, then the sudden obsession with a figure, will close down allocentric awareness. Think of the figure and the ground again. You must have both. There has to be a background out of which figures can manifest. But you cannot perceive the whole background the same time as you closely examine some part of the whole.

The more focal our attention, the less we are aware of a surround. Contrary to this, the more we take in broader regions of a scene, the less detail we extract from the surround. However, the attention system that is being ignored is still operating subconsciously. In other words, when we focus intently on something—what we are reading, for example—allocentric processing drops into a subconscious monitoring mode. Likewise, when we *experience our surroundings* allocentrically, we must inhibit egocentric processing. The egocentric mind does not disappear, it is just momentarily subconscious. So the two systems maintain their connection, but they take turns on center stage—they ebb and flow in a linked manner.

Here is another analogy to clarify the distinction between allocentric and egocentric processing: think of the words you are reading on this page. The egocentric mind is decoding the patterns buried in the language of the letters, words, and sentences, extracting meaning from these symbols. But the words could not exist were it not for the page they are part of. The page (or screen) is the background. You cannot have a foreground without a background. You must have both the background page and the foreground words—they depend on each other. In the same way, a painting may contain a lovely landscape but it could not exist without a canvas. If you are listening to these words, the same analogy works: out of a background of silence comes a language—distinct packets of vibrations encoded in a symbolic structure called language. Auditory information is part of a background field of silence. You must have both, silence plus sound patterns, because they are inseparable.

Here is another image that will help you comprehend the connection between the two attention systems: imagine yourself in a totally white world without any features. Your white body is floating in a sea of whiteness. Nothing has manifested out of the whiteness so there are no figures to use as landmarks, nothing to go toward or to navigate around. Take away the whiteness and leave only a void; now the starkness of this featureless world contains nothing to study, nothing that has meaning for you. This is an allocentric world, all background with no foreground. Now imagine yourself in a world of objects floating randomly all around you—above, below, to the sides of you. The objects are moving at various and unpredictable speeds. There is no horizon, no earth to stand upon, and no sky above. Nothing is simply vertical or horizontal, and objects move randomly. This is an egocentric world of objects without a stable background.

Embedded in a Common Background

For you to be grounded, for you to navigate, it is necessary to have invariant solid forms in the environment. Navigation for land-based creatures requires an earth to walk upon and a sky above. Vertical objects arise from the ground and "grow" toward the sky. Now you have a background. If the objects in the background were, let's say, all various sizes of red rectangles, then there would be no need for an egocentric mind. Only when the world is populated with complex and variable "objects" is there a need for a system that studies these forms to extract relevance. The egocentric mind asks: What relevance does this form have for me? Can this form be eaten? Will it eat me? Can I mate with this form? How is this form constructed? How does it work? Can I use it as a tool?

Eastern religions hold that *we are all one*. They mean that the background and foreground are interdependent. Figures cannot exist without a background, and backgrounds without figures hold no information. *We are all one* because we are embedded in a common background that we cannot escape. I believe that much of human thought—religious, philosophical, psychological, and so on—has been fundamentally shaped by our two sophisticated minds, by the background/foreground dichotomy.

The key point here is that *attention systems* determine background and foreground. Egocentric attention evolved to highlight figures, to pull 3-D images out of a flat world. Allocentric awareness flattens the world and melts forms back into the surround so that we can move through scenes.

Figure and ground are relative terms, dependent on scale or perspective. A figure can become a background if we move our perception closer, if we magnify the image. Likewise, a background can become a figure in a larger scene if we minimize the image.

Although, from a quantum perspective, egocentricity is as old as allocentricity, it can also be understood as the newer mind, as the latest, cutting edge invention of nature. This is because language development and our complex social relationships—alongside the evolution of vision and hearing—has resulted in a recent ascendancy of egocentric processing. As such, we have a capacity for focal concentration that is extremely powerful. We constantly exercise our egocentric attention system. We are always reading, or watching screens, listening, or looking at a person as we communicate—we are always concentrating *on something* so that we can extract relevance.

However, allocentric awareness requires that egocentric concentration be relatively suspended. This is very hard to pull off for modern humans since the ego keeps interjecting its focus, its obsessions. So powerful is this egocentric attention system that we have come to believe that we are *only* egocentric creatures. The allocentric mind has been suppressed for so long that it has sunk deeply into the subconscious; it has become the hidden mind. The ego will hardly entertain the possibility that a twin mind might exist. Ironically, the harder it searches for its twin, the more hidden she becomes because the ego must dissolve itself before the allocentric mind can show up!

The allocentric mind monitors the environment for change, especially for activity that might be dangerous. Most of the time there is no danger, life is okay, and so there is no need for hyper-awareness. Therefore, in our normal daily activity, allocentric consciousness is relatively hidden—overshadowed by egocentricity. However, when we are in *shock*, a focused egocentric state is dangerous and not conducive to survival. Animals fascinated by the gleaming teeth of the Sabretooth Tiger, for example, left no trace in evolution. When the body is in danger, or perceived danger, the whole biochemistry of the organism alters—it goes into hyper-allocentric mode. Shock can be physical, emotional, cultural, religious, drug-induced, and so on. It doesn't matter what caused the shock because the reaction is the same: the egocentric mind shuts down and the allocentric mind goes into high alert. Mystical traditions try to push us into a hyper-allocentric state wherein we can feel extreme calm, sharp sensory acuity, and the powerful feeling of being alive in the moment. That is the purpose of rites-of-passage: to awaken the allocentric mind.

The human nervous system uses two mechanisms for controlling movement: a feed-back system, and a feed-forward system. In other words, the brain uses bottom up (feed-forward) processing, as well as it uses top down (feed-back) processing. This is a dual-control system found in machines—like robots—as well as in all animals. These two redundant signaling systems are a way to maintain stability and balance in an always changing environment; feed-forward processing must be in balance with feed-back processing.

The allocentric mind uses pure-perception in the moment; it then feeds-forward whatever it is experiencing. It doesn't wait to process information; it has an inherent knowledge of invariants and can anticipate what its actions will produce within its domain. The allocentric system also has a set of invariant motor behaviors built-in that it uses for automatic responses. The allocentric mind uses the sum total of all hereditary knowledge to enable an organism to do the right thing moment-to-moment. Contrary to this, the egocentric mind uses a feed-back system; it is reactive, and it seeks to adjust behavior to fit the needs and circumstances of an organism. The allocentric mind is a bottom-up processing system, while the egocentric mind is a top-down processing system. The two minds need to be in balance.

It is very clear from the discussion above, that human beings use two evolutionary designs to enable navigation. Therefore, it follows logically—especially given billions of years of evolution—that these two designs eventually became sophisticated entities we call *minds*. Each mind evolved its own brand of consciousness. The duality that enabled navigation created our dual mentality.

Balancing Act: Giving Each Mind Equal Time

Each of us drives a chariot led by two horses.

~ PLATO

Plato has an elaborate discussion of our mental makeup in which he uses the analogy of a charioteer who is trying to get two dissimilar winged-horses to pull in unison. Plato's analogy is complex and I won't go into it here. However, I will steal his image because it fits nicely with dual-process theory.

The two horses in our story represent two minds, the egocentric mind on the left, and the allocentric mind on the right. The egocentric and allocentric horses have blinders on so they cannot comprehend that they have a partner who is also pulling the chariot.

For a smooth, swift, and enjoyable ride through life it is necessary for the two minds to pull in unison. Many disorders jerk human beings off the easy and smooth pathways when the horses are not balanced. Below, I discuss four kinds of balance that need to be in place for the ride to be smooth.

Movement Balance

As I gave Daniel Wolpert's ideas more consideration, I realized that there was a fundamental oscillation happening. To set the stage for navigation, the brain and nervous system had to *balance* movement with no movement, purpose with no purpose, and excitation with inhibition. In a way, these three processes are the same; they are different aspects of the same balancing act.

If the modern brain and nervous system evolved to enable purposeful movement, it is ironic that in order *to accomplish movement* nature had to develop an elaborate and equally powerful neurological system *for inhibiting movement* that alternates with the act of moving. This is very clear when we look at the activity of bilateral movement where part of the body is relatively stabilized while other body parts are moved.

To study an object-of-regard, to perceive *one-thing-at-a-time* using egocentric attention, we need to stabilize the body—to be relatively still. However, to actually move through the world, we need to perceive *everything-at-once* as we flow from one location to another.

Balancing Mind and Nature

What is the mind, and where is it located? The most obvious answer is that the mind is created by the brain. Brain damage alters the mind in very specific ways. There is not much disagreement, the brain is central to our understanding of the mind. However, the reason we have a brain and a nervous system is so we can move with a purpose, so that we can navigate. If the brain had no body to control, then it would be useless—its reason for existing would cease. The brain is an organ of the body, it cannot be dissected out. Since the nervous system is found throughout the body, the mind is also found throughout the body. That is why researchers speak of a gut brain or a heart brain.[6] Neural processing is always a result of total-body synchronization—the brain never acts alone. The mind is embodied.

Furthermore, as the nervous system evolved, it did so in harmony—and co-evolution—with nature. The brain and body could not exist without a very specific environment, and would be useless without a domain to navigate through. Purposeful movement presupposes an environment within which to move. Therefore, what we call *mind* can never be thought of as just in the head, or even just in the body. *Mind* is part of nature and cannot be dissected out. Therefore, *what we call the mind extends into—is intimately connected to—the surrounding milieu.*[7]

This is a rather profound realization. What we call mind is dependent on environmental circumstances, and it is also affected by the physical state of the body and brain. As the environment fluctuates from moment-to-moment, so too does the state and nature of mind fluctuate. Mind is a flowing *process* that is never the same moment-to-moment; it is never an *entity* that can be pinned down. This means that there can be no such thing as a permanent ego. I like this explanation from P. D. Ouspensky's book *In Search of the Miraculous* (Ouspensky is quoting the spiritual teacher G. I. Gurdjieff):

> "*Man has no permanent and unchangeable I.* Every thought, every mood, every desire, every sensation, says "I." And in each case it seems to be taken for granted that this 'I' belongs to the Whole, to the whole man, and that a thought, a desire, or an aversion is expressed by this Whole. In actual fact there is no foundation whatever for this assumption. Man's every thought and desire appears and lives quite separately and independently of the Whole. And the Whole never expresses itself, for the simple reason that it exists, as such, only physically as a thing, and in the abstract as a concept. Man has no individual I. But there are, instead, hundreds and thousands of separate small I's, very often entirely unknown to one another, never coming into contact, or, on the contrary, hostile

to each other, mutually exclusive and incompatible. Each minute, each moment, man is saying or thinking "I." And each time his I is different." ~ *In Search of the Miraculous, P. D. Ouspensky's, 1949.*

Ouspensky and Gurdjieff are referring to the egocentric mind at work. However, they are both aware of the allocentric mind, and they refer to it, in other contexts as the self, or soul. If there is a distinction between self and soul it is only spatial: the self is contained within the body, but the soul includes the environment *and* the body. The soul is that part of the mind that stretches through all time and space. The self is that part of the mind contained within the body—the self cannot be dissected out from the soul.

The egocentric mind is dominant in modern human beings, while the soul is inhibited or denied. Modern humans think of themselves as a permanent ego, a personality, an unchanging cognitive entity. Ouspensky and Gurdjieff say this is entirely wrong. The ego is composed of thousands of little manifestations, and these constantly contradict each other. There is no permanent ego steering the ship. In Eastern practices, it is said that *no one is in charge of the house*. There are many rooms in the house (the mind) but there is no master who looks after the whole.

Here is another quote from *In Search of the Miraculous* that underscores Gurdjieff's and Ouspensky's understanding of our two minds:

"There are," Gurdjieff said, "two lines along which man's development proceeds, the line of *knowledge* and the line of *being*. In right evolution the line of knowledge and the line of being develop simultaneously, parallel to, and helping one another. But if the line of knowledge gets too far ahead of the line of being, or if the line of being gets too far ahead of the line of knowledge, man's development goes wrong, and sooner or later it must come to a standstill." ~ *In Search of the Miraculous, P. D. Ouspensky's, 1949.*

The answer to the puzzle of multiple minds is to realize that we have two kinds of consciousness that must be held in balance. The *self* is one entity, one essence, but it has no voice, no personality, and no drive to control and compete. The *self* is about being, the *ego* about doing. The self is a rock to hold onto as environmental storms come and go. However, we need both our minds because they are symbiotic and co-dependent.

I find it amazing to imagine that *the self validates the existence of the ego, and the ego validates the existence of the self.* It is as if two illusory perceptual systems evolved that are able to look in a biological mirror at a reflection. I find this to be a humorous, but horrific, conundrum.

In the same way that ego and self validate each other, so too do we validate each other. Communication with others validates our reality and our very existence.

Balancing Frequencies

Researchers have found a resting state for the brain called the default mode network (DMN). From a dual-process theory perspective, the default mode network is anatomical evidence supporting the concept of two minds operating in balance. When the egocentric mind is active, the default mode network is inhibited. However, contrary to this, when the allocentric mind is active, the DMN is stimulated. What is significant about this research is that the DMN is a whole-brain state which contributes to a whole-body state. In other words, this research exposes a massive neural network that manages an oscillation between two overall frequency systems, which dual-process theory calls the allocentric mind and the egocentric mind.

Every organism has what is called a resonant frequency. This is a summation of all frequencies impacting the body at any one moment. Human beings have a certain overall frequency that holds the body together, and allows for coherent operation of all cells, tissues, and organs. In other words, our two minds are composed of whole-body frequency states in balance—this balance is our resonant frequency. Evidently, the default mode network creates and modulates our resonant frequency. Therefore, the entire body has a *summary frequency state* that is a balance between egocentric frequencies and allocentric frequencies.

This summation of frequencies is complex because the brain and body contain frequencies embedded within frequencies. In other words, frequencies overlap and cooperate in such a way that a variety of whole-body harmonic oscillations can occur. Speculation is that these harmonic levels of the brain and body correspond to states of consciousness.[8]

In Daniel Siegel's book *Mind* (2017), he concludes that the mind is the result of energy and information flow. In the chapter "What is the Mind?" he summaries his proposition:

> In this entry we have been exploring the notion of mind as emerging from energy and information flow. We've seen that neither skull nor skin is a limiting boundary of that flow, so that mind is both fully embodied and relational. At a minimum, the self-organizing aspect of mind would have this emergent embodied and relational property. As we've seen, information processing may be fundamental to that flow, attention being the process that detects and directs its movement within and between us. ~ *Mind, Daniel Siegel, 2017.*

I like Dr. Siegel's theory for three reasons. First, he can see that the mind is embodied and is not isolated in the brain—the mind must include the whole body. Second, he sees that the mind is relational, and as such it cannot be dissected out from the environment, especially from the other life forms in the surround. Third, as I discovered as well, attention is a key process to understanding what the mind is and how it works. Attention to invariants, and to the patterns that define them, is the information flow that Siegel identifies as significant to the embodied mind.

Dr. Siegel's research also led him to conclude that the mind is a self-organizing (self-assembling) system. The concept of self-organization has been observed in physics, chemistry, mathematics, and biology. It seems logical to conclude, as Dr. Siegel suggests, that the mind also follows the same laws. Out of seemingly random frequencies comes a coherent and stable state. A spontaneous order arises out of chaotic disorder. This coherent order is decentralized—it does not need an outside agent, or overseer, to direct the process. Self-organizing systems have been shown to be remarkably stable, to self-repair, and to endure waves of disruption.

Dr. Siegel is mostly concerned with the health of his patients. He has observed that when they are under severe duress, his patients either become rigid or chaotic. In other words, the breakdown of the self-organizing mind either enters a state of "freezing-up" or "sinking to chaos"—or sometimes, as in bipolar patients, both adaptations occur—they alternate.

What is fascinating to me is that what Dr. Siegel is mostly exploring is the allocentric mind—not the egocentric mind. The allocentric mind is, indeed, embodied, relational, and self-organizing. However, the egocentric mind is primarily in the brain. It is not as embodied, but, like the allocentric mind, egocentricity is also self-organizing.

Balancing Causation with Acausation

In a chapter of the book *Buddhism and Science* (2003) called "Imaging: Embodiment, Phenomenology, and Transformation," authors Francisco Varela and Natalie Depraz talk about forward and backward causation. They discuss these two terms in a way that appears to fit with my proposition that we have two kinds of attention mechanisms. The allocentric mind is a background out of which figures emerge. We call these manifestations by various names: figures, features, forms, objects, ideas, images, emotions, and sensations. It appears to us as if something *causes* the background mind to extrude these features.

If we think of our mind as a background, as a *potential* for thoughts, ideas, and emotions, we can then speak of thoughts as *arising* from the mind. We can also notice emotions arising in the mind and body. We can say that a mental substrate manifests—causes—thoughts and emotions.

This is forward causation: thoughts manifest—*appear to be caused by*—a global cognitive background. These ideas, emotions, images, and thoughts, can also melt back into the mind and disappear. This is backward causation as experienced from the perspective of these manifestations. Ideas, emotions, and sensations pop out of the mind—they manifest out of the silent background, but just as quickly as they come they melt back into the undifferentiated steady-state mind. Manifesting specific forms from a background is an egocentric process called forward causation. The allocentric process of dissolving forms back into the ground is called backward causation.

Consider that we can witness ourselves as we go about our behaviors moment-to-moment. This witness or watcher-system manifests in highly-evolved minds. The watcher is an executive function. If you witness your mind at work, you can watch as it manifests mental artifacts like thoughts and emotions. The Buddhists say to watch your thoughts come and go, like storm clouds moving across the clear sky, which is your allocentric mind. The watcher is a form of proprioception which I will discuss in later chapters.

The allocentric mind is a background canvas wherein cause and effect have no meaning. This background mind is pure potential, pure possibility; nothing is being caused in the pure background. From this state of pure potential, something mysterious happens (*cause* arises) and manifestation of form occurs. Forms then dissolve, after a time, and they disappear to make way for the next in the series of manifestations. There is a fundamental rhythmic alternation between backward and forward causation. A causal universe is egocentric. An acausal universe is allocentric.

Waiter, There's a Fly in the Soup

(Plop)

"Excuse me, Surge, there's a fly in my soup."

"Sorry to hear that. Although I think he just landed—look how the waves ripple out around the fly."

"I shouldn't have to pay if there is a fly in my soup."

"It's the fly's fault. When I brought your soup it was fly-less."

"What do you mean it's the fly's fault? That's outrageous!"

"I know! Because *that fly chose your soup*. Isn't that amazing! The fly had to decide to land in your soup; which means the fly had intent, a kind of will-power. It had to make a rapid pre-motor plan and construct a flight trajectory. Then, it flew directly at your soup and landed exactly where it intended to land. Life is endlessly fascinating, don't you agree?"

"The jury is still out, Surge."

"But it's baffling, isn't it? I mean that something the size of a housefly can navigate with such precision. Unbelievable, but then again, consider a smaller fly, the fairy fly, one of the smallest insects ever discovered. It is smaller than an amoeba! Yet the fairy fly has wings, eyes, guts, and genital organs. Invisible to the naked eye, it flies about its tiny domain, looking around all the while, and then it has sex, and later knows exactly the safest spot to lay eggs. There could be a million fairy flies laying eggs in your soup at this very moment and you wouldn't even know it.

"Thanks for sharing."

"My pleasure. Do you know about Augustus De Morgan, a mathematician who lived in the 1800s? He wrote a poem about fleas and such. It goes like this:

"Great fleas have little fleas upon their backs to bite 'em,
And little fleas have lesser fleas, and so ad infinitum.
And the great fleas themselves, in turn, have greater fleas to go on,
While these again have greater still, and greater still, and so on." ~
A Budget of Paradoxes, Augustus De Morgan, 1872.

"Yeah, wow, that *is* a wonderful poem. I love it!"

"Morgan makes a rather universal observation, don't you agree. It's kind of a repeat of the ancient wisdom of Hermes Trismegistus: *As above, so below*. Patterns repeat; fractals and all that. It's fascinating that nature uses the same recipe over and over again to make all kinds of creatures. All living things eat and then plop in your soup, so to speak. They get too hot or too cold and have to find shelter. They all have an urge to copulate and make little fairy flies. They sleep and wake up. And all their behaviors require movement, so they grow wings if their native terrain is the sky, or fins if their milieu is water, or they grow feet if gravity is going to stick them to the dirt. Funny thing about flies, though. They not only fly but they have legs as well—flies like to walk. I guess they don't know that you can't walk on soup. One of evolutions little screw ups, I guess."

"Are you going to charge me for this fly-infested soup?"

"I'm not sure yet, but you are a regular, so maybe not. I am concerned, however, with your lack of empathy for the fly. An innocent fly has drowned in your soup, and all you care about is getting a fresh bowl and a refund. It took nature billions of years to come up with that complex fly-creature, and you don't notice the sadness and irony of this moment. A fly has passed and no one gives a rat's ass about its tragic death. I mean, if you would have reacted quickly enough, you might have rescued the fly."

"Can I keep the fly as a souvenir, Surge? No one is going to believe this conversation. I want to take my precious fly home, have it bronzed and mounted on a small altar next to a tiny Buddha, Lord of the Flies."

"That's very clever, ha-ha, I detect some sarcasm. However, Buddha loves the image—if you don't mind my speaking for the Enlightened One. No problem though, the fly is free—take its dead body home and treat it with proper reverence."

"Could I ask a favor? Could you get me a fresh bowl of soup?"

"Of course, that is no problem. Do you want croutons?"

"No, thank you."

The Real Reason for Brains

"Do you want your fresh soup with or without paramecia?"

"There you go again."

"Of course, I am talking about the invisible one-cell organisms that live in warm soups, microscopic creatures that swim about and engulf bacteria and such."

"That's disgusting."

"Not really. It's mostly miraculous. I mean, think about it. How can you swim though warm soup, have sex with yourself, give birth, eat, defecate, move straight ahead, make turns, back away from danger, and write short stories for *Paramecium Magazine*, all without a brain or nervous system? I mean, come on, this one-cell-creature phenomenon has been going on for over three billion years."

"Paramecia are having sex in my soup?"

"Don't lose track of my point. These single-cell creatures move about with a purpose, but they have no brain and no nervous system. Doesn't that strike you as impossible? I mean, suppose you were writing a book about consciousness and made a statement that brains and nervous systems evolved so that animals could navigate. Then you discovered that single-cell organisms navigate just fine without brains, nervous systems, or big muscles. I am just saying."

"I might have made a similar statement myself once. It seemed reasonable at the time."

"I am sure you know what Thomas Huxley said about great theories like yours."

"I give up, what wisdom did Thomas Huxley share with us shallow types?"

"Huxley said that the great tragedy of Science was the slaying of a beautiful hypothesis by a single ugly fact. Your brilliant theory has been stabbed in the back, or so it seems."

"I am feeling some empathy for the fly now."

"There you go. Huxley also said that we shouldn't be ashamed about having an ape for a great grandfather, but I suppose that discussion should await a later meal."

"You think my theory about two kinds of consciousness is a dead insect, killed by an ugly fact or two?"

"Do you know what we call the soup you are eating?"

"The soup has a name?"

"This is our Meaning-of-Life soup; one of our more famous recipes."

"I am sure it's delicious. However, I only had one sip before the fly committed suicide—in the exact center of the Meaning of Life."

"Don't worry about your theory; you aren't wrong—there are, indeed, two kinds of consciousness. You just need to take a closer look at the smallest bugs, back near the beginning of life. Take a careful look at a paramecium, for example. Maybe I can help. I did my minor in Paramecia."

"Of course you did. I am not surprised. My coffee is also cold, by the way. Perhaps I could get a refresh when you bring me my fresh bowl of soup. You know what they say in culinary school, right? Be the waiter."

"Movement came first, before sensation; did you know that? The early cells just bumped into stuff for millions of years before nature built a sense organ or two. That's kind of profound, don't you think? I mean that bumping around blindly preceded eyeballs and such."

"Do you really care what I think!? But go ahead; I don't want to stop your flow."

"Okay, so in the beginning, there was just movement, probably random or maybe like a frequency, a pulse. Movement is all there was and all there still is. By the way, that's what the Buddhists say—that frequency, movement, is the primal mother. But we can talk about that at a later meal. Anyway, quarks, atoms, and molecules all vibrate, but they aren't alive. When molecules got together and life evolved, they transferred this frequency dance to the first cellular creatures like the paramecia. Therefore, these early cells used a simple frequency algorithm: move about until something impedes your flow, and then change direction. Move and bump. So, anyway, bugs are miraculous and important to understand. By the way, you need a license to kill bugs. I'm not going to report you this time, of course."

"I hope this doesn't get any more complicated or weird, Surge. No wonder this restaurant is empty. Is this discussion going anywhere? I'm curious."

"Let me confess: there aren't any paramecia in your soup, we boiled them to death. Do you know what paramecia like in *their* soup, by the way?"

"I give up."

"Bacteria. They eat bacteria. And bacteria are everywhere. As fast as you kill them, more just plop into your soup. We could say, and we would be accurate, that human beings are made out of bacteria. Indeed, there are ten times more bacterial cells in and on your body, than cells that are native to you. Indeed, it is also true, that in every single cell of your body there are mitochondrial organelles that are the children of ancient bacteria—you couldn't exist without the ancestors of bacteria. And just so we are clear on the importance of bugs, 99 percent of all genetic material in your body comes from the microbes that live in you and on you—well, that *are* you, but that is another story."

"All I wanted was some soup, Surge. So, okay, paramecia eat bacteria. Good for them."

"So, here's the kicker. Those early single cells really did have a brain and a nervous system; it just didn't look familiar to the scientists. There weren't any eyes or ears hooked up to a giant central processor encased in a head. But the single-cell nervous system was there, just in a different and rudimentary form; it was actually the beginning of whole-body processing, which as you know depends on the so-called hidden senses: proprioception, kinesthesia, photo-perception, and the vestibular system."

"I get that. I know it's a big deal. Aristotle threw us off track 2500 years ago with his insistence that we had only five senses. Teachers have dutifully passed on this false information. That's why we have hidden senses—because nobody looked for more than five. The god Aristotle had spoken. I understand that, Surge."

"Aristotle is a highly complex thinker. It's best not to quote too liberally about his views unless you are a scholar of the great man's works."

"Fine. Aristotle is a great man, but maybe he wasn't always entirely correct. Just saying. Anyway, what has this to do with whole-body brains?"

"Paramecia have a membrane, a boundary, which defines a rudimentary "self." The membrane is a key part of their proprioceptive system. They also move in a purposeful way, stopping, starting, going in a forward direction, turning, and flowing alongside and around obstructions in their path. So they have a primitive kinesthetic system. The whole cell adjusts to diurnal and incidental fluctuations in the amount and quality of light. That is their photo-perception system. In addition, single cells respond and move within a gravitational field. Paramecia spiral as they propel forward, and they move in a consistent sine-wave configuration. Therefore, they have a primitive vestibular system. Furthermore, their cilia, hair-like "oars" on their cell membrane, move coherently with a synchronized frequency. Paramecia also inherited—from bacteria actually—the ability to signal other cells; the paramecia have, in effect, a whole-body brain that can communicate. They could do all this three billion years ago. If you don't find this amazing, I am revoking your right to eat in slow-food vegan restaurants."

"You are hurting my mind, Surge. Okay, I am blown away that invisible bugs can communicate and eat lunch together and give birth to buglets. And this is important why?"

"It's important because of how this played out in evolution. One-cell organisms have a primitive form of navigation—we call it wayfinding. The system they used to move about became, after billions of years of evolution, one of your two minds: the silent and hidden twin, the allocentric background mind. Interesting that the hidden mind evolved from the hidden senses, don't you think?"

"Oh yeah, I'm overwhelmed with emotion. So, navigation has roots that go back to the dawn of life; that's what you're saying?"

"Well, it goes back before that, actually. Do you know what bacteria like to eat for supper?"

"I don't have a clue, but I bet you will tell me."

"Yeah, they eat viruses. So, do viruses navigate? Do they purposefully move? What do you think?"

"They get in my body somehow. Flu, common cold, and all that."

"Yup, but they hitchhike. They don't have self-mobility. They ride the frequencies of the domain, so to speak. They are morsels of food that just sit there waiting to be eaten—they travel by being ingested. Bacteria, however, *can* move with a purpose. So, the first self-navigators were bacteria, and our question begins with them: how did they find their way 3.5 billion years ago, and how do they still find their way? What is their brain and what is their nervous system?

"I am afraid I must talk with your supervisor, Surge. What with my cold coffee and tainted soup."

"Of course, that is to be expected. I will certainly get you fresh coffee and more soup, but we are at a very exciting place here in our discussion. I mean, are you as excited as I am?"

"You're kidding, right?"

"The answer is so important, so obvious, and yet we look right at it and don't see it. I am blown away by that. Aren't you amazed that the answer to Life, The Universe, and Everything is so obvious and yet unseen?"

"I wonder if there is a Tim Hortons in the Hotel California."

"Every bacterial cell is a quantum processor. And the quantum world is a world of duality: electro and magnetic. Eternity flickers on and off. That's your answer. Your two kinds of consciousness have their origin in the two sides of quantum oscillation. You have a magnetic mind and you have an electric mind. Well, not exactly, but that's good enough for now. The same oscillation is going on within your plant cousins and within insects. Actually, this quantum oscillation is the dual-bedrock of all sentient creatures."

"Go get my fresh coffee, right now, Surge, before I do a quantum adjustment on your forehead."

"I'll be right back with hot coffee, fresh soup, and my supervisor."

"Make that to go please."

———

"I am so sorry, sir. Surge gets carried away. I'm Bruno Salvatore. Nice to meet you."

"That's fine. Thank you, Bruno. Surge is a good guy. I like him. He is just not so good at being a waiter."

"He knows all about living creatures, but his passion is not customer service."

"That's fine, no harm done."

"But he does have a point about wayfinding starting with the evolution of movement. He did discuss quantum theory, right?"

"Oh, no! Not more quantum theory!"

"I'll just write this down; you can take it along. There is no charge for today's meal, by the way."

"That's very nice, thanks so much. Not that I ever got a meal. What are you writing down?"

"I'll read it to you. This is what we know so far; later it will be wrong—given the speed of change and the knowledge explosion—but go with this for now:

> "Primal wayfinding is based on the quantum wave mechanism, the two part phenomenon that is electro and also magnetic—the electromagnetic wave, a coherent duality. From this oscillating primal frequency came quarks, then atoms, then molecules. Everything is made of repetitions of this primal on/off frequency. As above so below. On earth as it is in heaven. Lesser fleas on littler fleas, and so ad infinitum. The end game so far is the *on and off dual minds* of human beings. It's all, every bit of it, made of the same frequency stuff. The purpose of this game is to make more of its "self," a kind of benevolent cancer on the landscape of whatever and forever. Everything that ever existed is trying to make more of its self."

> "That's nice, but I figured that out in the sixth grade when I was riding my bicycle, talking to my buddies Ron Waxell and Tom Reese. It's one of the first things junior philosophers figure out."

"Yes, of course, but now the young philosophers, as grownups, understand that they have two minds and so they can go through life accordingly. Human beings have two minds not because navigation requires two attention systems to work, as you propose, but because the source of all things is a duality that oscillates.

> "So, my whole thesis is wrong? Animals didn't evolve brains and nervous systems to enable purposeful movement?"

"You aren't exactly wrong; you just need to reframe the proposition. We are talking here about *movement and wayfinding*; however, in your book, you are talking about *navigation* which is *wayfinding with a cortical brain and nervous system*. All I am saying is that purposeful movement came first, as a creator—so to speak. Human beings, with their elaborate navigational capability, are Creation's most recent earthly design. Purposeful movement eventually created brains and nervous systems as it continued to clone its rhythmic self. There are many forms of evolved purposeful movement, and we, as animals, are one of them. Here is a sutra from the Mandukya Upanishad that might help you understand:

> "AUM is the imperishable sound [frequency], the seed of everything that exists, the past, the present, the future—all are but the unfolding of AUM [the fundamental frequency]." ~ *Mandukya Upanishad.*

> This is one of the most significant statements ever made anywhere on the earth at any time. It contains the whole secret of the mystic approach towards life. This small sutra contains the essence of the Upanishadic vision. Neither before nor afterwards has the vision been transcended; it still remains the Everest of human consciousness. And there seems to be no possibility of going beyond it. ~*Philosophia Ultima, Bhagwan Shree Rajneesh,*[10] *1983.*

"As above, so below." The ancient Eastern philosophers figured that out centuries ago. Existence is made out of the same repeating design, and so existence is one solid entity. It seems fair game to propose that a fundamental duality inherent in quantum fields evolved, like a fractal pattern, and that this unfolding eventually gave rise, after billions of years, to our two minds. Eastern religions have known about the allocentric mind for thousands of years. The rituals they design for their students are meant to reveal the hidden mind, the twin that is not the ego, not the personality, but something akin to our ancestral single-cell brethren, and akin to our plant and insect cousins. Wayfinding—proto-navigation—was a reality from the dawn of bacteria. It has been getting ever more sophisticated and complicated as the eons have rolled along."

"So, I can keep writing as long as I differentiate wayfinding from navigation?"

"Write away. Except you did get the part about purposeful movement as a Creator, right? That is where you locate enlightenment—at the edge of the flow—and where *love* is defined."

"I have a headache and indigestion. Perhaps I could come back later."

"There is no *later*.

"God have mercy on us, especially me."

"Of course, just let me explain this one little thing and then you can take a Tums. On the spiritual path, love is a much broader concept and a much deeper emotion than your feelings for Mary Jane and her perfect lips. Creation is happening as we speak. This constant flow that invents the universe moment-by-moment includes you. *You are that*, the mystics say. This process is called love. When you truly feel this creation unfolding, moment-by-moment, you become enlightened, you become one with the flow, and you notice that everything is connected on this cutting edge of creating."

"I have to go, Surge, even though I would love to hear more."

"Blindness is relative and can only be overcome through intentional attention. I hope you enjoyed your lunch."

"It was the most unusual lunch of my life."

"Glad you got a taste of the gourmet soup. Please come again."

Wayfinding versus Navigation

If we are to believe that neurons are the only things that control the sophisticated actions of animals, then the humble paramecium presents us with a profound problem. For the paramecium swims about the pond using numerous tiny hair-like oars—the cilia—darting in the direction of bacterial food or retreating at the prospect of danger. A paramecium can also negotiate obstructions by swimming around them. Moreover, these single cell organisms can apparently even learn from their past experiences—though this most remarkable of faculties has been disputed by some. How is this achieved by an animal without a single neuron or synapse? Indeed, being but a single cell, and not being a neuron herself, she has no place to accommodate such accessories. ~ *Shadows of the Mind, Roger Penrose, 1994.*

Paramecia are highly evolved creatures compared to the earliest primitive cells. Paramecia move in a consistent direction—swimming by spiraling through the water—using rhythmic oscillations of cilia that cover the organisms like fur. This means paramecia have an engine, an energy source that can move the cilia. It also means that these hair cells of the paramecium can synchronize; they can fire in rhythmic patterns that allow the creature to move with a purpose—paramecia swim in a sinusoidal pattern, as if surfing a wave. Some paramecia can mate, but most prefer to just split in two. They hunt down and ingest bacteria. They excrete what they don't digest. They have a cytoskeleton and organelles, including both a large and a small nucleus. When they bump into something solid, they backup or glide alongside and around the obstacle. They can signal other paramecia, and they can form symbiotic relationships. They have a rudimentary memory system for experiences. They also have a lifecycle and life-phases; they adapt to changes in their environment. Paramecia first appeared about 2.5 billion years ago, and they still exist on earth today as invisible creatures living in watery domains—obviously, like bacteria, their design was pleasing to nature and very successful. On the other hand, they are an evolutionary dead-end—they function the same today as they did three billion years ago.

Mathematician and philosopher of science Roger Penrose, who wrote the above quote, refers to paramecia as one-celled *animals*. Other writers also refer to paramecia as *animal-like* or as *original animals*. For my purpose, I will refer to complex, single-cell organisms like the paramecia as proto-animals. I contend that animals—unlike proto-animals—are creatures with brains and nervous systems. Animals have heads and bilaterality, appendages, and sensory organs like eyes and ears. Animals evolved these traits roughly half-a-billion years ago. This distinction between animals and proto-animals is important because many of the terms used in my discussion only have meaning for animals. For example, the word "navigation" refers to the use of a brain and a nervous system for negotiating a domain; navigation is what animals do to find their way.

Prior to the evolution of brains and nervous systems, creatures like paramecia could find their way without brains and nervous systems. I call this ability *wayfinding*, and the billions of years when wayfinding was predominate I call the *Wayfinding Age*. Concepts like "meaning" and "intention" make more sense when we speak about animals. For proto-animals, navigational skills were present in some proto-form, so we can speak of proto-meaning or even proto-consciousness. When animals emerged in the Cambrian era, the *Navigational Age* began.

There was a slow evolution over billions of years during which ever more complex creatures were built by nature. Our two minds have a parallel and a very long history of development. Allocentricity, for example, is about *immediate awareness* in a moment, followed by an immediate response. Allocentrically, we read environments with our whole body and respond according to our needs and desires *in the moment*. It should be possible to piece together, step-by-step, what the breakthroughs were in evolution that led to our advanced allocentric sensory-motor system—which has given us the ability to have a capacity we call *experience*.

Nature makes discoveries and then incorporates what works into future designs. This causes an evolutionary pressure toward greater complexity with each new design. For example, human beings were constructed—after billions of years of evolution—using all the successful designs that came before. Therefore, the study of evolution is an exploration of the building blocks that constructed life. Somewhere in evolution, very early, a software design was created that enabled self-propulsion (movement with a purpose). This design became a fractal which was used and modified within ever more complex creatures.

The terms "wayfinding" and "navigation" are often used interchangeably. For clarification in this context, I will use wayfinding to refer to pre-neural processing networks that evolved prior to the Cambrian Explosion

(650 million years ago). Navigation will be reserved for a process that came after the Cambrian age, after the cortex—the modern brain—developed and after eyes and ears evolved as unique sensory systems. Therefore, wayfinding is really proto-navigation. Likewise, purpose is pre-cortical, while meaning is a cortical activity. Thus, *purpose* is a pre-cursor to *meaning*. In other words, there was a series of evolutionary breakthroughs (steps in a process) that slowly defined what we mean by purpose, meaning, consciousness, and navigation.

Early organisms survived by staying within environments that supported survival. They had, for example, internal systems for monitoring temperature changes, fluctuations in the chemistry of the surround, pressure variations, and some even had eyespots for primitive ego-like perceiving. They also had cilia and flagella for locomotion and proto-touch. We can say, as I did previously, that these primitive internal sensory and motor systems constituted the equivalent of proprioception, vestibular sensing, and kinesthesia—but with the caveat that they were proto-systems. Fast, intentional, meaningful, and future-directed *navigation* did not evolve with sophistication until after the Cambrian era. There were many incremental steps (many designs tried) through time, out of which our human capabilities eventually emerged.

Single cell organisms can find their way in their watery domain; they move with a proto-purpose. Remember that these organisms have no eyes, no ears, no neurons, and no nerve fibers. They are using something holistic, a total-body sensory system. I suggest that the pre-neural network of simple cellular organisms has four internal components:

- A rudimentary vestibular system that "knows" upside-down from right-side-up and that can tell if the organism is steady and stable— a primitive gyroscope that became our "modern" *proprioceptive-vestibular system* for maintaining balance and posture.
- A simple *kinesthetic system* that "knows" if the whole organism is moving—usually in a consistent direction—or is still. Combined with vestibular input, this eventually allowed for a sense of head-tail,

up-down, center-side, inside-outside, and top-bottom. Eventually this led to bilaterality.

- A *proprioceptive system* that maps body structure and, therefore, "knows" where the cilia and other organelles are located. This is a very primitive sense of self that is mostly tied to a membrane that defines the organism as a specific form. The organism becomes a visible (discernable) object within an environment.

- A membrane that combined with internal "sensors" to monitor the quantity and quality of light. This *photo-responsive system* allowed organisms to move where light energy was available for fuel and to find a suitable temperature for survival. These early creatures were whole-body light sensors (seekers). They were solar-powered. There is a real possibility that these single cell organisms communicated using photons, not just chemical signals.[11]

The evolution of these four *internal* sensory systems—vestibular, kinesthetic, proprioceptive, and photo-responsive—eventually resulted in the allocentric perceptual system which, after billions of years of evolution, became the allocentric mind. Scientists are slowly unraveling the mysteries associated with this early wayfinding machinery.[12] Much later, during the Cambrian Age, egocentricity began a rapid evolution, mostly due to the creation of sophisticated eyes and ears. This advanced egocentric system for navigation was built on the foundation of the older and tougher internal, whole-body allocentric wayfinding apparatus. Indeed, our external five senses (vision, hearing, touch, smell, and taste) formed on the membrane of organisms long after the internal allocentric sensory system had evolved to a highly sophisticated degree.

Therefore, from the very beginning of cellular life, we witness a primitive "nervous system" that is allocentric. It is part of an invariant environment and has a *self* (but not an *ego*—which is a much later invention). Consequently, human beings have a remarkable ability to navigate because we have a three-billion-year-old allocentric processing system combined with a half-billion-year-old sophisticated egocentric processing system. Allocentric processing, because it is so old, runs subconsciously on autopilot; it is very fast and very effective. Sophisticated egocentric processing developed later in evolution to create the figures that populate the allocentric background.

Dual consciousness came about because of the evolution of what we call *attention*, (when referring to egocentric systems), and *awareness* (when the allocentric mind is operational). This differentiation between awareness and attention is important. When we are searching for meaning in a specific location of space, we are *paying attention* egocentrically. When we take in a gestalt, everything-at-once, especially when we are *moving*, we are in various *states of awareness*.

Redefining the Profession of Orientation and Mobility

One of the main reasons I wrote this book was to provide the profession of orientation and mobility (O&M) with a new foundation. That's a big task, and I am not so sure the profession is keen on changing just because I think it is a good idea. Many people dedicated whole careers to molding and nurturing the profession, so it is a bit presumptuous of me to suggest that my plan needs to be adopted. It is also hard to turn a rocket around while it is speeding through space.

O&M specialists are taught at a graduate school level to be experts in *blind* navigation. They leave graduate school knowing how to teach a person with no sight to use echolocation, landmarks, and the invariants within the environment to enable them to become proficient navigators. They do a job, in my opinion, that is far more complex than they are given credit for; most people have never even heard of the discipline.

The field of orientation and mobility grew up around the invention of the long cane which was specifically designed for blind navigation. A recipe was written down for how a long cane is supposed to be used. That was the foundation for an entire profession. As the years passed, however, the discipline got more sophisticated, but the cane and the recipe for learning to use it remained the heart of the curriculum. For me, it seemed that something was needed to give the discipline deeper credibility and nobility. In the eyes of other professionals, we were just cane trainers because we had no philosophical or scientifically sound reason for existing as a profession. Anybody with a cookbook (a strict set of unchanging rules) could teach the blind to use a cane to travel, or so it seemed to others outside the profession.

The science of navigation could have become the foundation for orientation and mobility many years ago, but it did not unfold that way. The profession was too focused on cane skills and on egocentric teaching practices; it missed the greater significance of the allocentric mind as it related to the evolution of navigation and consciousness. The profession "dipped its toes" into complex waters, but then went back to the cookbook curriculum. A revolution in thinking and practice is waiting for the right collection of leaders to bring about change. The profession needs to embrace a more sophisticated foundation. What would this new approach look like?

To begin with, the profession would come to understand the distinction between the allocentric mind and the egocentric mind. They would understand that animals evolved brains and nervous systems primarily for navigation. They would then realize that their training put them in a unique position to become experts in navigation generally—not just as it relates to blind individuals. From this new knowledge base, practitioners would see, as they looked back over the history of the field, that the past practices of the profession have been mostly egocentric. A different set of strategies and priorities would unfold as an allocentric curriculum was designed and implemented.

They would also be able to explain to non-experts why the O&M profession has such high expectations for blind individuals who are learning to move independently through the environment. Additionally, they would be able to explain, using research from the neurosciences, why certain blind individuals are amazing navigators, while others seem hopelessly confused. Indeed, it doesn't make sense to non-experts why someone who is blind could still navigate *at all* without vision. It seems logical that people navigate through space because they have eyes, and can see the world around them. If you can't see, so the commonsense logic goes, you should not be able to navigate. Understanding that the whole body was designed by evolution to navigate, helps the non-expert see that vision loss only affects one part of the whole-body navigational system. Having this new foundation of knowledge enables us to understand why it is that a human being could navigate with grace and precision even without vision.

A revolution is possible within the O&M profession if it embraces this new perspective, and especially as the field embraces an allocentric agenda. This would seem a daunting task except that most of the insights and practices have already been created by Daniel Kish,[13] a remarkable O&M specialist who has been blind since before the age of two. The profession could begin the journey of change by studying the philosophy of World Access for the Blind, the non-profit agency that Daniel set up as a vehicle for explaining his allocentric curriculum.

I wrote extensively about Daniel in my first book *Bugs, Blindness, and the Pursuit of Happiness*, published in 2016. In that book, I explained key concepts that Daniel helped define, like "the Freedom Formula," "Proxy Perceiving," and "Environmental Literacy."[14] These are all allocentric strategies for helping blind kids develop perceptual skills. Daniel has considerable credibility, not only because he is blind and has a graduate degree in orientation and mobility, but because he is also a developmental psychologist—he knows a great deal about human perception and child development. His theories and practices have a sound philosophical base. Daniel is called Bat Man by the world's media organizations because he is *the* global authority on human echolocation—an allocentric method for non-visual perception.

Daniel didn't need to research the neuroscience literature to support his own allocentric-based practices because the occupational therapy (OT) profession had already laid the groundwork. Decades ago, occupational therapists built an entire profession around allocentric processing, although they didn't use that terminology. As a young O&M specialist working at the Blind Children's Center in California, Daniel became familiar with OT practices and philosophy. Consequently, he built an allocentrically-focused curriculum for his O&M practice using the OT literature as a philosophical foundation.

What the occupational therapists *do not* have, of course, is an understanding of the navigational implications of Dr. Wolpert's research—they do not use the terms egocentric and allocentric to explain their perception-based approach. They are also not experts in blindness. Therefore, pulling from the substantial occupational therapy research-base is only half-satisfying from an O&M perspective. Additional perspectives need to be added to the OT philosophy for it to be completely acceptable to O&M practitioners. Daniel Kish has been busy building the O&M component to the OT philosophy.

Starting in the 1980s, Daniel began crafting a perception-based approach to teaching blind navigation. Therefore, Daniel has *already* redefined the field of orientation and mobility, and highlighted a path forward. However, the O&M field has not generally followed his lead.

Meanwhile, the profession of orientation and mobility is still drifting through academic space without a foundational anchor, without a sophisticated physiological underpinning. That can change now that Daniel Wolpert and Daniel Kish have provided the seeds for change, and frankly, this change is exciting and revolutionary. It is time to toss the cane trainer label into the trash bin of history and venture in a more reasonable and honorable direction. We are not, of course, tossing cane training out with this transition, only placing it within a subcategory of the new discipline.

If the O&M profession adopts the perspectives of Daniel Kish and Daniel Wolpert, then the profession will change in the following fundamental ways:

- Most significantly, the main focus will shift from the teaching of cane skills to the science of navigation. The research scope will suddenly explode. Universities will reduce spending on the study of cane design and cane technique. This research, valuable as it is, benefits only a few of the many navigationally disabled people who need help. The cane will become a proprioceptive extension of the body and will be seen, as Daniel Kish has been saying for decades, a tool of perception. It will remain important, but not the primary identity of the profession.

- The population served by mobility specialists will expand considerably. When I started teaching in a special education setting, I found myself surrounded by children who had diverse disability labels, not just blindness. Some kids had difficulty with navigation. The more severe their impairment, the more severe was their loss of navigational capability. Indeed, so many kids in special education had problems finding their way that I eventually declared that there was an unidentified disability in the special education pantheon: *navigational disability*. O&M specialists need to become disability specialists.

- O&M specialists will become experts in the new field of navigational disability. They will define and refine the concept. They will become experts in the navigational problems of different categories of children in special education.

- O&M specialists will also develop an understanding of allocentric and egocentric disabilities. Blind individuals have primarily an allocentric navigational disability, as do autistic kids. Deaf children, on the other hand, have primarily an egocentric navigational disability. Kids with physical and cognitive impairments have a mixed bag of allocentric and egocentric disabilities. This is an overgeneralization, but a useful and new perspective for understanding disabilities.

- The number of people interested in the O&M field will expand considerably. There will even be a reason to develop PhD programs since the scope of the profession will balloon to include all categories of special education.

- Research on the complexities of navigation, as well as research that focuses on navigational disabilities, will inform the study of human consciousness. The O&M profession will find itself face-to-face with the most profound and important questions of our time. Daniel Wolpert and Daniel Kish have thrown open a portal that takes the O&M profession from the backroom of the kitchen to the main floor of the ballroom. This can be an exciting time if we have the courage to redefine what it is that we do.

Notes

(1) Dr. Daniel Wolpert's TED Talk, "The Real Reason we have Brains," presented in 2011, can be found on-line at the *TED website*, or on *YouTube*. Dr. Wolpert is a neuroscientist, engineer, and a medical doctor. The following quote comes from the website of the Engineering Department, Computational and Biological Learning, at Cambridge University:

> Movement is the only way we have of interacting with the world, whether foraging for food or attracting a waiter's attention. Indeed, all communication, including speech, sign language, gestures and writing, is mediated via the motor system. Taking this viewpoint, the purpose of the human brain is to use sensory signals to determine future actions. The goal of our lab is to understand the computational principles underlying human sensorimotor control.

(2) Chilean biologist, philosopher, and neuroscientist Francisco Varela. As I did research for this book, the insights of Francisco Varela kept reappearing. What I loved about Dr. Varela was his understanding that the wisdom of Buddhism and the wisdom of neuroscience overlapped. He eventually became friends with the 14th Dalai Lama; together they helped create the Mind and Life Institute that promotes dialogue between science and Buddhism.

(3) The field of quantum biology is discovering that cells are quantum factories. The idea that cells employ quantum operations has not historically been taken seriously by biologists, even though cells are made up of atoms, which obey the rules of quantum mechanics. Quantum processes are extremely fragile, so it was assumed that the harsh biological world would be too destructive for quantum effects to be possible. This supposition is eroding as researchers and theorists find evidence of quantum operations within the cellular world. For example, quantum effects have been found in the retina of the European Robin, in the mutational processes of DNA, in photosynthesis, and in the energy conversion processes within the mitochondria of animal cells. To review the research studies that support the above observations, see the book *Life on the Edge: The Coming of Age of Quantum Biology* (2016), by Johnjoe McFadden and Jim Al-Khalili.

(4) There is evidence that we are held together by a kind of frequency genetics. The evidence I am referring to here begins with the common understanding that we are bathed in a sea of frequencies. From the electromagnetic waves of light to the pressure waves that create sound, ours is a world that is dictated by vibrational energy. Every one of our senses is specialized for a certain kind of vibration. There are patterns associated with these vibrations that are not random. These patterns scale up and down so that what we see on a gross scale also must be true for smaller scales. Using this "as above, so below" logic, we can postulate that there is a world of frequency genetics just as there is a world of biological genetics. However, giving this speculation, scientific validity is tricky because so many overlapping fields of expertise must be synchronized—we might have to wait for quantum computers before these correspondences can be demonstrated.

(5) Both trees and humans evolved ways to use spatial information. Whatever vibrational universe humans evolved within, it is the same universe that trees inhabit. Therefore, biological adaptations were required that are common to both plants and animals. Both must get water, nutrients, and light energy to survive and reproduce. In other words, plants and animals share a common domain and they must adapt to the conditions of that common domain—the environment matters. But this is not what I mean when I say that plants and animals share a common spatial universe.

Space is the "cup" that holds the environment. The domain must have a location within which to manifest. Unfortunately, space itself has been overlooked as an entity. There is a background that manifests the forms of nature. There is reason to suspect that this background has a quantum nature and that frequency genetics are operating. Whatever spatial domain underpins the environment affects both trees and humans equally.

(6) The gut brain and the heart brain. Neurons are found throughout the body, but in some cases neurons are clustered in large numbers in regions outside the brain. There are so many neural networks in the gut and in the heart, for example, that researchers speak of a heart brain and a gut brain. This suggests, for example, that our "gut feelings" have physiological roots. Likewise, all the poetic renderings associated with the heart also have a physiological foundation. The following quote about the so-called second brain, or *gut brain*, is taken from the John Hopkins Medical Center website:

> If you've ever "gone with your gut" to make a decision or felt "butterflies in your stomach" when nervous, you're likely getting signals from an unexpected source: your second brain. Hidden in the walls of the digestive system, this "brain in your gut" is revolutionizing medicine's understanding of the links between digestion, mood, health and even the way you think.
>
> Scientists call this little brain the enteric nervous system (ENS). And it's not so little. The ENS is two thin layers of more than 100 million nerve cells lining your gastrointestinal tract from esophagus to rectum.

The HeartMath Institute in California is famous for demonstrating that the heart is also a secondary brain, since it contains millions of neurons and has a direct connection to the limbic region of the brain. The following quote is from the HeartMath Institute's website:

> More intriguing are the dramatic positive changes that occur when techniques are applied that increase coherence in rhythmic patterns of heart rate variability. These include shifts in perception and the ability to reduce stress and deal more effectively with difficult situations. We observed that the heart was acting as though it had a mind of its own and was profoundly influencing the way we perceive and

respond to the world. In essence, it appeared that the heart was affecting intelligence and awareness.

(7) **Peri-personal space:** Like an aura, there is a peri-personal region of space immediately surrounding our bodies; it is the zone where objects are manipulated. Beyond peri-personal space everything is referred to as extra-personal space. One interesting aspect of peri-personal space is that whatever we grasp—a tool, for example—is immediately incorporated into the proprioceptive system as if the tool was now part of the body. For example, the cane used by a blind individual becomes an actual extension of the hand and arm, from the perspective of the proprioceptive system. As far as the brain is concerned, the cane is part of the body.

When teaching blind kids to understand space, an orientation and mobility specialist will teach the child about zones of perception. We often call the peri-personal space simply *personal space*. This space belongs to a child; no one may enter personal space unless it is okay with the child. Beyond personal space is *communication space* where sentient creatures dialogue. Beyond communication space is *landmark space* where objects can be perceived and used for orientation while navigating. Beyond landmark space is *beacon space* where distant landmarks are located. Landmarks can be approached and explored, but beacons are too far away to approach—the sun, intermittent winds, or a train rattling by, for example.

(8) **Harmonic levels of the brain correspond in some way to levels of consciousness**. This is a huge subject area and I don't have the time to give it a proper summary. I also don't have the knowledge base to do a proper job. Many people, with interests scientific and/or spiritual, have sensed the correspondences between musical scales, vibrational states, levels of consciousness, and even personality types (the Enneagram, for example). In his book *The Hermetic Code in DNA*, author Michael Hayes points out correspondences between the DNA code, the I Ching, the mathematical constant pi (π), the musical scale, and the importance of individual numbers in the musical scales—seven, nine, and three especially.

(9) The cells of our body sing in harmony. Every cell in our body is in a continuous state of harmonious vibration. Everything in the cell, the membrane, the cytoplasm, all the organelles and fluids, everything must be in the same harmonious state. Individual cells must also be in harmony with neighboring cells and this harmony must be continuous and synchronized. Disharmony is disease, which can eventually lead to cell and tissue death.

(10) Bhagwan Shree Rajneesh (Osho): Bhagwan was an important influence on me when I was a young man. I had friends who were sannyasins (followers of Bhagwan), and I traveled to India in the late 70s with his teachings fresh in my mind. I didn't end up in Poona, where his ashram was located, mostly because my wife Katherine was not comfortable with this life-direction. I have no regrets about that decision. I took many of Bhagwan's ideas to heart and even introduced my father to his philosophy—my dad read more of Bhagwan's works than I ever did. Bhagwan's many books (over 200) are clearly written and sensible. However, his followers got him into trouble in Oregon—where he had gone to establish a community—and his influence declined after that. He changed his name to Osho after the disaster in Oregon, and ended up back in India where he died in 1990.

I still find Bhagwan to be a powerful spiritual force. He taught the importance of meditation long before that became accepted in Western cultures. His followers were taught how to be loving, awake, and creative. He said that we should celebrate our moments with courage and humor. The international Rajneesh movement continued after his death and is active today; his ashram in Poona, India, is now the Osho International Meditation Resort.

(11) Single cell organisms communicate using photons (not just chemical signals). As an introduction, see the online article titled "Cell-to-cell signaling through light: just a ghost of chance?" published November 12, 2013 in the journal *Cell Communication and Signaling* (2013; 11: 87). The authors are Ondřej Kučera and Michal Cifra.

(12) Scientists are slowly unraveling the mysteries associated with early wayfinding machinery. As scientists study how single-cell organisms find food, mates, or shelter, they indirectly are discovering how these creatures move with a purpose. For example, bacteria monitor the number of other bacteria in the surround based on chemical signaling. When the signaling reaches a threshold level, all the bacteria in the population simultaneously change their position. Bacteria organize themselves (move with a purpose) to create complex structures.

(13) Daniel Kish: Daniel is a colleague and personal friend. Many of my best ideas come from dialogues with Daniel. He is a perceptual psychologist, an orientation and mobility specialist, and the CEO of World Access for the Blind, an organization that he founded and has directed since its inception. Daniel is a world authority on the use of echolocation as it is applied to human navigation. He travels the world to work directly with blind children and their parents. Daniel is totally blind yet he travels the world alone, without sighted assistance, and without a guide dog. He uses a long cane and his cell phone to supplement his echolocation skills. He is, in my opinion, perhaps the most remarkable blind traveler to ever live. As such, he is a national (global) treasure as important to the history of blindness as Louis Braille and Helen Keller.

(14) The "Freedom Formula," "Proxy Perceiving," and "Environmental Literacy." These concepts are available for study in my book *Bugs, Blindness, and the Pursuit of Happiness.* I also discuss these concepts in the last chapter of this book. The freedom formula is a set of guidelines for helping blind individuals develop perceptual skills. Daniel feels that self-exploration of the environment develops allocentric perceptual skills better than the guided instruction that is the norm of standard mobility training. Using the Freedom Formula frees blind individuals from the restraints established by well-meaning sighted "helpers." Proxy Perceiving refers to the many ways that blind individuals are assisted by the sighted. Daniel does not look upon this sighted help as useful. Real freedom comes through self-perception, not guided perception. Environmental literacy is a term that I coined, although I can no longer remember when; I do know that the concept came from discussions with Daniel. A blind person who can "read" an environment, and then use the information to independently navigate, is said to be environmentally literate. Those who are unable to read the environment are said to be environmentally illiterate.

Two

Dual-Process Theory

*One of the marvels of the world
is the sight of a soul sitting in prison
with the key to freedom in its hand.*

~ "The Sight of a Soul," *A Year with Rumi*, 2006.

Rumi-Approved Cuisine

"You know what Rumi says, right? About our mental diet? About our cognitive eating habits?"

"Not really."

"He says "Why have cooking pans if there is no food in the house?"

"I don't get it."

"We have fully stocked mental kitchens but nothing nutritious to eat. Our cognitive cupboards are mostly bare. We are starving to death in the Garden of Eden."

"Okay, Surge. Whatever."

"Here at the Third-Eye-Watching Vegan Restaurant we serve only Rumi-approved cuisine. Why would we do that?"

"Rumi has a cookbook you like?"

"I am talking about what is good for you and what isn't. See the sign on the wall? That's our motto:"

"You make our souls tasty like rose marmalade." ~ *"The Verge of Tears," A Year with Rumi, 2006*

"That's not a motto, Surge. That's a weird statement about jam. So, what's for lunch?"

"I'm not done explaining why Rumi is essential eating."

"I get it, Surge. It's very wise of you to quote guys who are clever. For example, e.e. cummings was also a cool poet who made quirky sentences and refused to use capitalization. cummings said 'One's not half of two; two are halves of one.' And Robert Frost said 'Two roads diverged in a wood, and I–I took the road less traveled by, and that has made all the difference.' Why is Rumi a better choice?"

"Poets sing in a common voice, Dutch. Each poet gives us 'gifts from the moment.' Accept the gifts. Poetry is not about decision-making or judgment. Poetry is about experience. I could quote cummings, or Frost, or Rainer Maria Rilke, but Rumi is different, a cut above the clever turn of phrase. Rumi lived his poetry. He *was* poetry. He *is* poetry. Rumi was enlightened; most Western Poets chain-smoked and died young. They never got beyond the words."

"I'm trying to remember the last time I ate here and left filled with food rather than words."

"Rumi saw that humanity could not distinguish between good cognitive nutrition and mental garbage. Can you imagine living with that kind of jaw-dropping brilliance and be surrounded by the crude mentality that walked the villages of the planet in the year 1250? I mean, look around at the present world. There are few people walking the earth right now who know they exist inside a soul. Most everyone lives with beliefs and ulcers. The whole world is peopled with hungry ghosts in search of things the ego cannot comprehend, like love, peace, joy, wisdom, thankfulness, and mindfulness. Very few people know what a soul is. Neither do you, by the way."

"How do you know?"

"Because your eyes are vacant, and your aura is micro-thin and pale. Waiters at the Third-Eye-Watching Vegan Restaurant can see your soul—it is as bright as the sun. But you keep this soul inside a dark, locked prison cell. Rumi says: stop being a slave owner."

"Okay. So, what is it you want me to say, Surge? I am stuck where I am. I can't find the prison cell where my soul is incarcerated. I can't see in the dark. I don't know where the key to the cell is. I am not even sure what I am looking for."

"Stop starving yourself. Your whole mental planet is ego-bound. The ego doesn't believe in souls; it won't look for the prison cell where the soul is alone, suffering, and dying from lack of nutrition. Egocentric living is intelligent, but it is not wise. The ego believes it is on firm logical ground, but this is not supported by the very science it uses to affirm its reality. The ego's outrageous behavior is not validated by religion, philosophy, poetry, psychology, or by common sense and common decency. The ego can be violent, suffocating, and mundane beyond comprehension. It is tasteless food. Even the dog won't eat it."

"Okay, Surge. Thanks. I will eat only nouns and verbs soaked in Rumi oil. But I have to go now. I made an appointment to leave the galaxy."

"Hold Rumi in your heart."

"I hear you. I'm going to buy his cookbook and make Rumi-biscuits on holy days."

"May your journey be safe from sarcasm, and may you have many soulful adventures. Namaste, my friend."

"Later, Surge. And I am still hungry, by the way. You forgot to feed me again."

"The ego is never full, Dutch. Eat less and be awake more often."

Ambiguity and the Human Mind

Locked in a cell, you grow bitter,
but out walking
in the morning sunlight with friends,
how does that taste?

~ "THE FACE," *A YEAR WITH RUMI*, 2006.

Many people have explored consciousness and unearthed fascinating clues to our common mystery. One such cognitive explorer was British philosopher Owen Barfield. In his book *Saving the Appearances; A Study in Idolatry* (1957), Barfield begins his introduction with the understanding that what he has unearthed so far is just a different slant on a common concern:

> There may be times when what is most needed is not so much a new discovery or a new idea, as a different 'slant;' I mean a comparatively slight readjustment in our *way* of looking at the things and ideas on which attention is already fixed. ~ *Saving the Appearances; A Study in Idolatry, Owen Barfield, 1957.*

Like Owen Barfield, all I am doing here is presenting a different slant on an old conundrum. I am looking at the evolution of consciousness from a different angle. Many factors came together—after 33 years of teaching—which enabled me to consider consciousness in a fresh way. My role in special education involved teaching children who had navigational disabilities. Therefore, I spent my entire career thinking about navigation. I eventually became fascinated with how all creatures move. I saw that even bugs with brains the size of a grain of sand scurry with confidence through their limited domains. How could it be that creatures so small, with no cortex and hardly a brain, could be such expert navigators? The answer to that question seemed important, so I began to dig into the knowledge base of various fields to find answers.

Along the way, I discovered just how ubiquitous the concept of duality was—I found a variation of duality in every discipline I studied. Eventually, I realized that navigation *required* a duality, and from that biological necessity came two perceptual systems that became two minds. After language and communication reached a threshold in evolution, two kinds of consciousness emerged from the two minds.

What slowly became clear, as I sorted through the research, was that I wasn't the only person struggling to comprehend duality. There was an ambiguity inherent in our deepest nature that was bewildering to humankind. Many others had found clues, in the research literature of their own professions, which suggested that human beings had two minds, but no one had definitive answers why this might be so. Most writers, however, did not shy away from offering a theory or two. Whatever the theory, it inevitably contained ambiguity.

I like this quote from Sarah Bakewell's book *At the Existential Café* (2016):

> Our condition is to be ambiguous to the core, and our task is to learn to manage the movement and uncertainty in our existence, not to banish it. - *At the Existential Café, Sarah Bakewell, 2016.*

"Our condition is to be ambiguous to the core,"—that sums it up nicely. The quote above by Sarah Bakewell refers to the philosophy of Simone De Beauvoir, the French existentialist, feminist author, and friend to many of the most inquiring minds in post-war Europe. De Beauvoir is acknowledging our fundamental ambiguity. Most significantly, she is asserting that we have to embrace our twin minds, and the inevitable conflict and confusion that arises, rather than deny one mind or the other. De Beauvoir was reacting to the materialistic world that had evolved since the Age of Enlightenment. She wrote most of her books soon after the Second World War ended. All around her, in Paris, and throughout a devastated Europe, she saw the consequences of giving the egocentric mind full reign.

Because science (Scientism[2]) considered spirituality to be nothing more than superstition and ignorance, the allocentric mind was ignored in Western history—the egocentric mind was thought to be all there was to humanity. Because of the belief structure of scientism and materialism, one entire half of our makeup—the allocentric mind—was denied validity. This egocentric perspective became a blueprint for confusion and despair that is still playing out in our modern world. Simone De Beauvoir could feel this alienation keenly. However, she also knew that it was equally insane to deny the importance of the egocentric mind, even though it could be destructive and unfeeling. To exist only in the allocentric mind is another kind of misleading blueprint that also leads into a jungle of ignorance and inertia. We need both our minds to be in balance.

In our era, science still rules and technology dominates our lives. However, this is slowly changing as the empathetic (allocentric) mind is recognized and given reign. The battle to preserve and save the spiritual mind has been ongoing since the Age of Enlightenment, since the rise of the scientific revolution. But our story, in this book, is not about the battle to deny or overturn the scientific revolution, which would be misguided. It is rather about the effort needed to balance our two minds for the common good. We have to save and celebrate our duality. Our condition is ambiguous to the core, but we have no choice except to embrace our inherent mental conundrum.

As I looked at the draft of this book, it dawned on me that I had written a very long literature review, page after page of documentation, chapter after chapter filled with quotes. This was necessary because the subject matter is too important, and evidence is scattered across numerous domains of expertise. I had to *show* that duality is pervasive in our literature. I had to show that many others had concluded that we must have two minds. I had to document how experts in their respective fields mused over our inherent ambiguity. I had to show how scientific disciplines contained hard evidence for our duality. The pages ahead are heavy with quotations and references. I hope you find my discussion helpful in your own domain of expertise.

Duality and Philosophy

I'll begin this first literature review with the insights of University of California, Berkeley professor, Alva Noë. Dr. Noë, who is a philosopher and author, understands that the mind is not in the brain, and he knows that the allocentric mind is a reality, even though he didn't use my terminology. Dr. Noë, therefore, is kindred spirit. Here is a favorite quote from his book *Out Of Our Heads*:

> My central claim . . . is that to understand consciousness—the fact that we think and feel and that a world shows up for us—we need to look at a larger system of which the brain is only one element. Consciousness requires the joint operation of brain, body, and world. Indeed, consciousness is an achievement of the whole animal in its environmental context. I deny, in short, that you are your brain. But I don't deny that you have a brain. And I certainly don't deny that you have a mind. To have a mind, though, requires more than a brain. Brains don't have minds; people (and other animals) do. ~ *Out Of Our Heads, Alva Noë, 2009.*

I agree with Professor Noë: minds are not confined to brains. Because our major external senses are located in the head, we get the illusion that our consciousness is arising from the head. However, we get a better understanding of consciousness, of the mind, using a rational process of elimination. For example, if we remove the brain, then we lose consciousness and there is no mind. If we remove the body, the brain has nothing to move, nothing to control, no way to have experiences, nothing to remember. Therefore, the body is part of what we call mind. If we remove the environment, the brain and body have nowhere to go, nowhere to have experiences, nothing to experience and remember. Therefore, the environment is also necessary for the concept we call mind.

What Professor Noë is referring to is the allocentric mind—he sees that we cannot ignore this aspect of cognition. He is rejecting the notion that the egocentric mind is all there is to human awareness. This discovery, that we have a second whole-body mind, is shared by many others. One of the most complex philosophers, one of my favorite authors, is Owen Barfield (1898-1997).

Owen Barfied spent his long career trying to explain the importance of the allocentric mind—a term not used when he was authoring his books. He saw our duality, and he perceived the existence of a whole-body mind that operated alongside an egocentric mind. It is appropriate that Owen Barfield's last book *Saving the Appearances* is a summary of his life-long study of the evolution of human consciousness. Barfield asserts that consciousness is due to a co-evolution of nature and mind. Mankind, according to Barfield, cannot be dissected out from the environment. People do not exist in isolation—apart from a natural domain. Animals are crafted by the environment, and in turn animals are altered by the environment. Barfield calls this two-way street "participation." We participate *within* the environment. The environment creates our form, and then that form alters the environment. It is a dance, an unending process.

The important point is that when we *participate* we are *experiencing* the world without analyzing or judging. The word "experience" is the key to comprehending much of what I have to say in this work. Owen Barfield felt that the mind that has experiences, the mind *in the act of participating* with the world around, is a different entity than the mind that identifies, judges, and analyzes. In other words, for Owen Barfield, and for many others including myself, *we have two minds.* One of the minds, the spiritual, empathetic, allocentric mind is hidden from the ego.

Furthermore, Owen Barfield is trying to save and explain this participatory mind since it is in danger of being forgotten, overlooked, leaving us with the belief that we are just rational egos. Like computer software, hard science would reduce us to neural wetware, to electromagnetic personalities, nothing more, nothing less. This is valid when we speak of the biological substrate for egocentric cognition, but it is misleading when we speak only of egos and personalities. According to Owen Barfield, science needs to turn its penetrating gaze toward that part of our essence that is not egocentric. As an example of science researching the allocentric mind, Barfield uses the work of the philosopher and social reformer, Rudolf Steiner.

Barfield was an authority on the philosophy of Rudolf Steiner. Steiner, rather than speak of experience, or phenomenon, or participation, refers to the *spiritual mind.* Steiner is very clear that this spiritual mind is in danger of being ignored and denigrated by scientific materialism. Indeed, science as dogma (scientism) refuses even to glance in that forbidden direction, toward anything spiritual. Both Barfield and Steiner fought for the preservation of the spiritual mind. Both men had complex intellects and are not easy to translate unless this concept of two minds is grasped.

The idea that we have two kinds of consciousness is, of course, nothing new. In *Saving the Appearances*, Barfield tells us that Aristotle and Plato held this perspective 2500 years ago:

> [We can] illustrate the difference between a "knowledge" which does depend on alpha-thinking [rational thought] and a different kind of knowledge altogether which does not. Plato and Aristotle, and others . . . taught that there was such knowledge and that it was accessible only to participation [experience]. ~ *Saving the Appearances, Owen Barfield, 1957.*

According to Aristotle and Plato, we can gather knowledge using a cognitive system that depends on rational thought, or we can gather a different kind of knowledge when we interact with our environment—when we participate with nature, when we have experiences. The ancients knew the secret long before the philosophers of our era rediscovered our fundamental duality.

As I did research for this book, the philosophical discipline called phenomenology became increasingly relevant and fascinating. Phenomenology was the life-work of Edmund Husserl (1859-1938), a distinguished philosopher and mathematician. Husserl became obsessed with the *science of consciousness*. Indeed, thanks to Husserl's contributions, phenomenology is now called the "first-person science of consciousness."

Husserl came to believe that our brains were not just registering the outside world. Our reality, he taught, was an interaction between out there—the environment—and in here, our body and brain. In other words, like Alva Noë, Owen Barfield, and Rudolf Steiner, Husserl concluded that *experience* was the key to understanding consciousness.

Furthermore, Husserl used the word "empathy" as if it was *a way of knowing*, as if there was a part of the mind that didn't use logic to understand existence, a separate brain process that could *know* in an experiential way, but was not rational.[3] After years of philosophical exploration, Husserl essentially suggested that human beings have two minds.

Husserl is also important to this discussion for another reason. He said that *navigation and consciousness both have movement as their primal root*. He appears to have made the same connections that I am making in this book: movement is the key to understanding how our minds work.

If, as Husserl and others suggest, there are two minds inside of us, two anatomically and physiologically distinct processing systems, then logically there must be two fundamental kinds of consciousness. It also follows logically that there must be two ways to pay attention, two perceptual systems, two kinds of memory, and two relatively separate anatomical and physiological neural networks inherent in our biology. I contend that this is, indeed, the case. That case is laid out in the chapters that follow.

The phenomenologists, especially Edmund Husserl and Martin Heidegger (1889-1976), saw our duality clearly. Heidegger stated that we had lost contact with the part of us that was experiential. We had lost sight *of being*, he said. In effect, these phenomenologists implied that we had lost contact with one of our two minds; the faster, intuitive mind had been relegated to automatic processing, below the level of ego-awareness. Ironically, but very importantly, in my opinion, our two minds became so differenti-ated through eons of evolution that they can no longer understand that they are twins. The consequence of this "internal blindness" is a landscape littered with our puzzling duality. Today we split issues into science versus art, science versus spirituality, logic versus creativity, exoteric thought versus esoteric thought, and so on. It is as if one mind is locked in constant debate with the other mind.

After Edmund Husserl, French philosopher Maurice Merleau-Ponty (1908-1961) wrote a book called the *Phenomenology of Perception* (1945). In this important work, Merleau-Ponty again defined two kinds of con-sciousness: representational consciousness, and intentional consciousness. Representational consciousness builds a spatial representation of the world. It floodlights whole scenes and is the mind that experiences. Intentional consciousness is a spotlight mechanism focused so that we might derive meaning and relevance from the solid things in our environment. Merleau-Ponty, therefore, also suggests that human beings have two minds; he simply generated different terminology for the same concept.

We are also blessed with 2500 years of intense first-person study of hu-man consciousness through the meditations and philosophy of Eastern reli-gions. The Buddhists and Hindus brought introspection to a highly refined state. They laid out strategies for exploring and using the human mind. A duality emerged as they did their introspecting. Consequently, we have 2500 years of evidence in support of dual-process theory. This is discussed in detail in Chapter Six.

Philosopher and author Alan Watts[4] (1915-1973) has a classic lecture series archived on *You Tube*, called the Tao of Philosophy, in which he speaks of "spotlight" attention and "floodlight" attention, which I alluded to ear-lier. I find his analogy to be especially effective when trying to comprehend our two minds since both have their roots in the evolution of attention. We think we are spotlight creatures, Watts says. Whatever we set our spot-light upon is embraced by our egos, recognized and remembered. But we are mostly floodlight creatures according to Alan Watts. This floodlight perception is our true self, but we don't know we have a tool to perceive as a floodlight. We don't know that we have a *true self*.

The floodlight mind illuminates the vastness of our *being*. We have no idea of our potential and power because we have defined ourselves narrowly as mere *spotlight organisms*. Watts agrees, then, with Barfield and Husserl. All three men are using different words for the same idea. Spotlight attention is the egocentric mind at work, while floodlight attention is the allocentric mind at work.

The German philosopher Friedrich Hegel (1770-1831) pointed out a symbiotic relationship between what he called particulars and universality. He stated that a lifetime of experiences (the particulars) fed the rational (universal) mind. These two cognitive processes need each other, but they are not the same thing. There is a back-and-forth relationship between experience and conceptual (egocentric) universals. These two ways to understand existence need each other—one cannot exist without the other. Hegel is another great thinker who discovered the contrast between the allocentric mind and the egocentric mind. He saw a universal background mind out of which figures emerge and became an egocentric mind.

Tyler Burge wrote a chapter called "Two Kinds of Consciousness" in the 1997 book *The Nature of Consciousness: Philosophical Debates*. Dr. Burge discusses two kinds of consciousness called "access consciousness" and "phenomenal consciousness," terms coined by Ned Block, one of the book's co-editors. Again, a different language is being used but the concept is the same as that discussed by Barfield, Husserl, Merleau-Ponty, and Watts: access consciousness is an egocentric rational system, while phenomenal conscious is about experience.

In the introduction to *The Nature of Consciousness: Philosophical Debates*, co-editor Guven Guzeldere presents a summary of debates on consciousness. He also identifies two general types that he calls *"consciousness is as consciousness feels"* and *"consciousness is as consciousness does."* One mind simply *experiences (feels)*, while the other mind *does*. In religious terms, one mind is concerned with *being* (called the spiritual, empathetic mind), while the other mind (called the rational, problem-solving mind) is concerned with *doing*. This is an often repeated observation but what is different here, in my discussion, is that I see an evolutionary, anatomical, and physiological derivation for these two minds that is based on the evolution of navigation.

Professor Gerald Edelman (1929-2014), in his book *wider than the sky* (sic), refers to "primary consciousness" and "higher-order consciousness" to explain our duality. Dr. Edelman was a Nobel Prize winner and a highly respected expert in neuroscience. He shows in the quote below that neuroscience takes our duality as a fundamental starting point for the study of mind:

> . . . we need to make a distinction between primary consciousness and higher-order consciousness. Primary consciousness is the state of being mentally aware of things in the world, of having mental images in the present. It is possessed not only by humans but also by animals lacking semantic or linguistic capabilities whose brain organization is nevertheless similar to ours. Primary consciousness is not accompanied by any sense of a socially defined self with a concept of a past or a future. It exists primarily in the remembered present. In contrast, higher-order consciousness involves the ability to be conscious of being conscious, and it allows the recognition by a thinking subject of his or her own acts and affections. It is accompanied by the ability in the waking state explicitly to recreate past episodes and to form future intentions. - *wider than the sky, Gerald Edelman, 2004.*

Notice that Dr. Edelman speaks of two kinds of consciousness. This implies that we have two minds going on within the same body. He also uses the word "aware" to speak of the primary consciousness. I, too, make this distinction: the allocentric mind is "aware;" the egocentric mind "pays attention." In other words, attention and awareness are two forms of perception used by the two different minds. Dr. Edelman also sees that primary consciousness (the allocentric) exists in the present; it has no ability to probe the future or consider the past. Furthermore, the allocentric mind "does not know that it knows." However, the egocentric mind does "know that it knows."

To give another example found in a cognitive psychology textbook published in 2010, Roy Baumeister and E. J. Masicampo propose a division between a lower phenomenal consciousness and a higher rational consciousness:

> First, there is phenomenal consciousness, which "describes feelings, sensations, and orienting to the present moment" (Baumeister and Masicampo, 2010, p 945). It is a basic form of consciousness. Second, there is a higher form of consciousness probably not available to other species . . . It "involves the ability to reason, to reflect on one's experience, and have a sense of self, especially one that extends beyond the current moment." ~ *Cognitive Psychology, a Student Handbook, Roy Baumeister and E. J. Masicampo, 2010.*

Baumeister and Masicampo state above that phenomenal consciousness involves *orienting to the present moment.* This is a key understanding because it links navigation with allocentric consciousness. Our sense of self is tied to our orientation within an environment—*what* we are cannot be separated out from *where* we are.

In his first book *The Spectrum of Consciousness* (1977), psychologist Ken Wilber[5] has an entire chapter on "Two Modes of Knowing." Below are a few examples from his book that demonstrate how different people at different times in history have all arrived at the same conclusion: we have a duality, two minds, within us [I inserted the terms "allocentric" and "egocentric" where it seems appropriate]:

> Western philosophy is, by and large, Greek philosophy, and Greek philosophy is the philosophy of dualisms.

> We have, then . . . two basic modes of knowing . . . one that has been variously termed symbolic . . . or dualistic knowledge [the

egocentric mind]; while the other has been called intimate, or direct, or non-dual knowledge [the allocentric mind].

The way of liberation called Taoism recognizes these two general forms of knowing as conventional knowledge [egocentric] and natural knowledge [allocentric].

. . . as it states in the [Hindu] *Mundaka Upanishad* "There are two modes of knowing to be attained—as the knowers of Brahman say: a higher and a lower." The lower mode, termed aparavidya, corresponds to what we have called symbolic-knowledge [egocentric] . . . The higher mode, called paravidya, is reached . . . all at once . . . intuitively, immediately [allocentric].

Insights similar to these abound in Christian theology—Meister Eckhart, for example, called symbolic-map knowledge "twilight knowledge" [egocentric]. The non-dual [allocentric] mode he called "daybreak knowledge."

In Mahayana Buddhism, the symbolic mode and the non-dual mode . . . are termed vijnana and prajna, respectively.

Perhaps no modern philosopher has so stressed the fundamental importance of distinguishing these two modes of knowing as has Alfred North Whitehead. Whitehead pointed out most forcefully that the core characteristics of the symbolic form of knowing are abstraction and bifurcation [egocentric] . . . Opposed to this mode of knowing is what Whitehead called Prehension, which is an intimate, direct, non-abstract, and non-dual "feel" of reality [allocentric].

William James said: "There are two ways of knowing things, knowing them immediately/intuitively [allocentric], and knowing them conceptually or representationally [egocentric].

The recognition of the symbolic mode and the non-dual mode of knowing also figures prominently in the work of Henri Bergson (intellect versus intuition), Abraham Maslow (intellectual versus fusion knowledge), Trigant Burrow (ditention versus contention), Norman O. Brown (dualistic versus carnal knowledge) . . . Andrew Weil (straight versus stoned), Krishnamurti (thought versus awareness), Wie Wu Wei (outseeing versus inseeing), Spinoza (intellect versus intuition), not to mention the seminal work of Dewey on transactionalism—and these to name but a very, very few. ~ *The Spectrum of Consciousness, Ken Wilber, 1997.*

Creation myths, from all over the planet, similarly tell the same story about our inherent duality. From a time when there was only a unity, there arose a crisis, and out of the chaos came the duality that characterizes human existence. This archetypal mythology is found in the Hindu myth of Indra and Vritra (in the *Rigveda*), in the Babylonian creation myth (*Enuma Elish*), in the Canaanite myth (Baal and Yam), and in Egyptian mythology (Seth and Apophis).

The differentiation of our two minds is often attributed to hemispheric differences in the brain, which clearly has two halves. Author Leonard Shlain (1937-2009), for example, has written eloquently in his books about our two minds using hemispheric differences as the anatomical and physiological explanation for duality. I don't agree that our mental twins primarily arise from hemispheric differences (current research doesn't support it either) but I love Dr. Shlain's discussion in his books as he draws from literature, evolution, and philosophy to illustrate the landscape of duality.

My journey is similar to Dr. Shlain's; he had a medical background and was fascinated, as I am, with the mind and the brain from a scientific perspective. His research into our duality parallels my own journey. In his last book, *Leonardo's Brain* (2014), he wrote:

The right-and-left-brain functions are commonly associated with the dualities of masculine/feminine, active/passive, particular/general, focused/holistic, and rational/intuitive. The arrangement of a masculine side of the brain and a feminine side promotes a psychic hermaphroditism in both men and women, making the human sexes unlike any other species. Every man has an animus and an anima just as every woman has an anima and an animus. ~ *Leonardo's Brain, Leonard Shlain, 2014.*

Animus and anima are concepts used in the psychotherapeutic theories of Carl Jung. They are an obvious duality, a balance between creative energy and rational energy. Jung, like so many others, has unearthed our fundamental duality and has given the two cognitive worlds names that suit his fascinating perspective.

Another way to view the operation of our two minds is to say that the egocentric mind *differentiates* while the allocentric mind *integrates.* The egocentric mind is busy dividing whole networks into sub-networks while, to the contrary, the allocentric mind is busy assembling sub-networks into larger networks. Ego is tearing down, while allo is building up. The egocentric mind uses the cognitive tool called reductionism, reducing wholes into constituent parts, while the allocentric mind uses holistic thinking, putting all the parts together to make a whole.

Psychologists also use the terms "observer memory," and "field memory," to label our dual ways to store information. Here is how Evan Thompson, professor of philosophy at the University of British Columbia, describes the distinction in his book *Waking, Dreaming, Being* (2015):

> These two perspectives—first person and third person—also appear in memory. Take a moment to remember your last birthday party or some other memorable event. Do you see yourself from an outside point of view (the way someone would see you), or do you see things from the perspective of your own eyes (as you do now reading these words)?
>
> Psychologists use the terms "observer memory" and "field memory" to describe these different ways of remembering. In the observer

mode you're an outside spectator on yourself; in the field mode you recall what happened from within. ~ *Waking, Dreaming, Being, Evan Thompson, 2015.*

Here again is the recognition of our duality. The egocentric mind uses a self-centered, first-person viewpoint and remembers from that perspective. The allocentric mind uses an overhead, third-person view and remembers from that perspective. Dr. Thompson has a fascinating history and was introduced as a child to a steady stream of brilliant minds, many of them Buddhist practitioners. Thompson was home-schooled at the Lindisfarne Association, a think tank and retreat established by his father, William Thompson. Evan Thompson's writing is a blending of Eastern and Western perspectives—many of the visitors to Lindisfarne were practicing Buddhists.

Before we leave this discussion, I want to make an important summary observation. Yes, there is ample evidence to support duality coming from diverse fields of study. Yes, there are two kinds of consciousness. However, what hasn't risen to the top of the agenda is the understanding that the allocentric mind is *the equal to* the egocentric mind. The allocentric mind has been hidden, ignored, belittled, and rarely is it taken seriously (by the ego). It is a non-rational mind, so why would rationality take it seriously? The allocentric mind is also the source for empathy and love. The ego is not capable of either. Consequently, the ego is hostile and defiant when faced with the proposition that *love is the equal to reason.* Our lopsided world does not have love and empathy on an equal footing with reason—and that is a disaster that defines our era. The cognitive ship is leaning seriously to starboard and we are about to capsize unless we accept and use balanced cognition—love and reason must coexist as equals. We have to bring the allocentric mind onboard to right the ship.

This brief look at duality from different professional perspectives is just a small sampling. In the book that follows this one, *The Confusion Caused by Being Your Own Twin*, I look with greater depth at each discipline.

Dual-Process Theory, the Science of Consciousness

We are all just prisoners here
of our own device.

~ SONG LYRIC FROM THE EAGLE'S "HOTEL CALIFORNIA"

The idea that we have two minds is being explored and validated by a scientific discipline called dual-process theory. What follows is a simple literature review that highlights some of the history and terminology of this relatively new field. This is also a quick look at duality from a research-based perspective. I continue to use grounded theory to select research from my various stacks of notes, holding them to your face saying "Look! Science also knows about our two minds!" Therefore, this is not a chronological history, nor is it a logical flow. It is more a history of what I came across during my investigation of the scientific literature, and it contains evidence that I found compelling. I will start with the textbook that introduced me to dual-process theory.

In 2009, a textbook was printed by Oxford University Press, a compendium of articles on the duality of mind. It was called *In Two Minds: Dual Processes and Beyond*. The co-editors, Jonathan Evans and Keith Frankish, state that dual-process theory has a long history, and disciplines are often not aware of the terminology used by other fields:

> . . . philosophers and other academics have proposed versions of the two-minds hypothesis over many centuries. The idea that there are fast, automatic and unconscious thought processes that combine and compete with those that are slow, deliberate, and conscious, has been proposed many times by many authors, often in ignorance of each other's writings. ~ *In Two Minds: Dual Processes and Beyond, Jonathan Evans and Keith Frankish, 2009.*

Researchers can see that there are two overarching processes available to human beings, one that is fast, the other slower in comparison with the first. One process is under conscious control; the other process is not conscious. In a way, science is simply waking up to what philosophers have known for centuries. What science will rediscover, as dual-process theory develops, is simply the age old division between *being* and *becoming*. Modern science is only now providing the hard evidence for our fundamental duality.

In the Western world, long before Socrates, the Greek thinker Parmenides of Elea (born 502 BC) made a clear distinction between being and becoming. He describes these two views of reality in a fragmentary poem—all that survives of his writing. Parmenides says that one view of reality understands the world to be universally connected. In my terminology, Parmenides identifies a background reality of pure being where existence is timeless, uniform, and unchanging (the allocentric mind). His second view of reality speaks of a world of appearances; our senses define this world and lead us into false concepts and false belief structures—this is a product of the egocentric mind. The ideas of Parmenides were probably influenced by contact with Hinduism and Buddhism. At any rate, he articulated the same basic truth that monks and gurus discovered over 5,000 years ago. The idea of mankind's fundamental duality stretches beyond recorded history. Dual-process theory is the entrance of science into this ancient dialogue. *Becoming* is egocentric and slow. *Being* is instantaneous and allocentric.

As I stated previously, awareness of our cognitive duality is found throughout our scientific, cultural, and philosophical literature. Dual-process theorists have made the leap from science to mind. They now have no problem speaking about the existence of two minds inside each of us:

Dual-process theories hold that human thought processes are subserved by two distinct mechanisms, one fast, automatic, and unconscious, the other slow, controlled, and conscious, which operate largely independently and compete for behavioral control. In their broadest form, they claim that humans have, in effect, two separate minds. - *In Two Minds: Dual Processes and Beyond, Jonathan Evans and Keith Frankish, 2009.*

Dual-Process Theory

Initially, dual-process theory referred to *System One* and *System Two*, as first defined by cognitive scientists Keith Stanovich and Richard West. These are *cognitive* divisions. They refer to brain processes directed at thinking, problem solving, and social interaction. I call these two divisions the allocentric mind (System One) and the egocentric mind (System Two) because I trace the development of the two minds to the origin of navigation. Others see the significance of this duality from within the wisdom of their own disciplines; the popularity of the concept is expanding as ever more experts find application in their respective fields of study. I have to confess, whimsically, that my own mind—many others have also made this delightful association—keeps quoting Dr. Seuss who understood decades ago that we had two minds: Thing One and Thing Two (from *Cat in the Hat*, 1957). These two minds take turns getting us into trouble.

One of the earliest researchers to postulate that we had two minds was Arthur Reber, a cognitive psychologist who studied and compared implicit and explicit memory. As Reber puzzled over the differences between two memory-systems with their different learning styles and knowledge base, he eventually saw a clear distinction that suggested two brain processes at work. His research-based ideas showed that humans had two systems for processing and remembering:

- System One [type one, allocentric] was old in evolution; pre-conscious or unconscious; animal in nature (all animals were conscious in this way); used implicit memory; was automatic and very fast; was processed in parallel; had a high capacity; was intuitive, pragmatic, and associative; and was independent of general intelligence.
- System Two [type two, egocentric] was the opposite of System One in every way. It was new in evolution; it was conscious; wasn't shared with animals (was unique to humans); used explicit memory; was controlled and slow; processed serially; had low capacity; was reflective, abstract, logical, and rule-based; and was linked to general intelligence.

The above distinction was taken from a chart provided in "The Duality of Mind: An Historical Perspective," by Keith Frankish and Jonathan St. B. T. Evans (available online).

It is noteworthy that dual-process theory might have started with the study of attention. That view was quickly overshadowed as the field became dominated by the social sciences with their emphasis on memory systems. My contention that our dual minds evolved because of attention brings the focus back to where it began (italics mine):

> . . . the origin of modern dual-process theories is sometimes cited as stemming from the distinction between controlled and automatic processes *in attention* made by Schneider and Shiffrin (1977; also, Shiffrin and Schneider 1977). It is true that this work provided a major stimulus for the development of dual-processing accounts in social cognition from the 1980s onwards, but it actually played no part at all in the development of dual-process accounts of learning and reasoning which predated this publication. ~ *"The duality of mind: An historical perspective," by Keith Frankish and Jonathan St B T Evans, 2009.*

The research conclusions of Richard Shiffrin and Walter Schneider, mentioned in the above quote, are almost exactly what I am saying today, 40 years after their research results were published. The researchers find two attention systems at work feeding two different cognitive processing streams (the two minds of humanity):

> We reported (Schneider & Shiffrin, 1977) the results of several experiments on search and attention that led us to formulate a theory of information processing based on two fundamental processing modes: controlled and automatic. In the context of search studies, these modes took the form of controlled search and automatic detection. Controlled search is highly demanding of attentional capacity, is usually serial in nature with a limited comparison rate, is easily established, altered, and even reversed by the subject, and is strongly dependent on load. Automatic detection is relatively well learned in long-term memory, is demanding of attention only when a target is presented, is parallel in nature, is difficult to alter, to ignore, or to suppress once learned, and is virtually unaffected by

load. ~ *"Controlled and Automatic Human Information Processing: Perceptual Learning, Automatic Attending, and a General Theory." Psychological Review, Vol 84, No. 2, March 1977.*

Richard Shiffrin and Walter Schneider were spot on, half a century ago. Allocentric awareness is global and automatic, a kind of floodlight awareness. Egocentric attention is controlled and focal, a kind of spotlight perception.

Researchers in the dual-process field generally differentiate the two minds in the following way:

Since the 1970s, *dual-process* theories have been developed by researchers on various aspects of human psychology, including deductive reasoning, decision making, and social judgement. These theories come in different forms, but all agree in positing two distinct processing mechanisms for a given task, which employ different procedures and may yield different, and sometimes conflicting, results. Typically, one of the processes is characterized as fast, effortless, automatic, nonconscious, inflexible, heavily contextualized, and undemanding of working memory, and the other as slow, effortful, controlled, conscious, flexible, and demanding of working memory. Dual-process theories of learning and memory have also been developed, typically positing a nonconscious *implicit* system, which is slow learning but fast access, and a conscious *explicit* one, which is fast learning, but slow access. ~ *In Two Minds: Dual Processes and Beyond, Jonathan Evans and Keith Frankish, 2009.*

My own perspective on dual-process theory differs somewhat from the standard explanation defined in the above quote. For example, I find that both our minds have their origin before the dawn of evolution (they are quantum-based), and they are necessary twin systems that had to evolve side-by-side to enable purposeful movement. They also had to work together to enable the evolution of functional creatures. In other words, from my perspective, current studies of dual-process theory seem too narrowly focused.

Consciousness: A New Slant on an Old Conundrum

Physiologically and anatomically, the egocentric mind is built to examine the features of an environment and to extract meaning and relevance. This process takes time, and so the mechanism is described as "slow, deliberate, and conscious." From an evolutionary perspective, egocentricity is the new kid on the block because of language and the evolution of the external senses, eyes and ears. However, its origin is as old as allocentric processing. Egocentric processing is slow only when we compare it to the lightning fast processing of the allocentric mind. Allocentric processing is on autopilot; it is habitual, immediate, rote, and mostly innate. Totally dependable, the algorithms used to create allocentric minds are deeply nested. The modern egocentric cognitive system, however, especially since the evolution of human language, is slower and more deliberate.

In 2011, Nobel Prize winner Daniel Kahneman wrote a *New York Times* bestseller called *Thinking, Fast and Slow*. In the paragraph below he explains how he uses dual-process theory as the basis for his book:

> The distinction between fast and slow thinking has been explored by many psychologists over the last twenty-five years . . . I describe mental life by the metaphor of two agents, called System 1 and System 2, which respectfully produce fast and slow thinking. I speak of the features of intuitive and deliberate thought as if they were traits and dispositions of two characters in your mind. In the picture that emerges from recent research, the intuitive System 1 is more influential than your experience tells you, and it is the secret author of many of the choices and judgements you make. ~ *Thinking, Fast and Slow*

Kahneman is differentiating intuitive thought, which is unconscious and experiential, with the slower, more deliberate processing of the rational mind. To use Alan Watt's terminology, Kahneman differentiates spotlight attention, System Two, with floodlight attention, System One.

The original authors who coined the terms System One and System Two, Keith Stanovich and Richard West, now use the terms *Type One* and *Type Two*. They made the change after they observed how the debate about their theory was received in their professional communities. They found it hard to justify the existence of two anatomical and physiological *brain systems*. It was easier to show evidence for *two types of human behaviors,* and ignore speculation about the biological substrate for the two processes.

It might be harder to prove that there is a biological substrate that supports duality, as they observe, but in my opinion, disagreeing somewhat, there is plenty of anatomical and physiological evidence for our two minds. I will explore the hard evidence later. On the other hand, agreeing with Stanovich and West, what I clearly see when I look at human navigation is two very distinct ways we pay attention as we navigate using egocentric and allocentric processing.

Ohio State University Professor Richard Samuels wrote a chapter in the book *In Two Minds: Dual Processes and Beyond* called "The Magic Number Two, Plus or Minus: Dual-process Theory as a Theory of Cognitive Minds," (2009). Dr. Samuels warned that use of the word "systems" was problematic, especially if we insisted that only two existed. As an example, he says that the vision system reduces to ever smaller sub-systems for color, movement, etc. Therefore, multiple systems come into the discussion for any cognitive process, not just two. So perhaps it is better, he suggests, if we talk about two *types of behaviors* rather than use the word "system." It would also be better, according to Dr. Samuels, if we restrict dual-process theory to refer selectively to reasoning.

Although I can see the logic used by Dr. Samuels, this shrinking of dual-process theory to be exclusively about reasoning seems to be where dual-process theory has settled and, in my opinion, where it has walled itself in. I also see that biological systems are nested, as Dr. Samuels points out. However, I know from my own professional studies that there are two overarching biological and quantum processing systems that serve navigation, and which eventually enabled our dual consciousness.

Consciousness: A New Slant on an Old Conundrum

In 2008, neuroscientist Guilio Tononi introduced a theory suggesting that consciousness has two different jobs, integration and differentiation. Dr. Tononi called his perspective Integrated Information Theory (IIT). Using a movie analogy, he explains that we can see highly distinct frames of a movie (differentiating the parts) or we can watch the movie (integrate the frames into a whole). These two processes require two different cognitive mechanisms. This is about as close as you can get to the dual-process theory that I am explaining. However, my perspective adds evolutionary evidence that connects navigation with consciousness. In other words, I have a slightly new slant on the age old debate about consciousness, but it meshes nicely with the sophisticated research of Dr. Tononi.

In his book *Mind and Its Evolution; a Dual Coding Theoretical Approach* (2007), Dr. Allan Paivio (1925-2016), former professor of psychology at the University of Western Ontario, Canada, assembled a convincing case for the evolution of two minds, one mind which is nonverbal and a second mind which uses language:

> Nonverbal mind and verbal mind thus became interlocked in a synergistic relation that evolved into the nuclear power source of our intellect ~ *Mind and Its Evolution; a Dual Coding Theoretical Approach, Allan Paivio, 2007.*

Dr. Paivio's book is an update on his dual-coding theory of cognition (DCT); it is based on years of research. He postulates that there are two mental systems with two different kinds of memory which are stored and managed in separate locations within the brain. The human mind can form images, or it can form words of the same entity, but the image and the word are processed differently and remembered differently. The crux of Dr. Paivio's theory rests on the difference between language processing and nonverbal (non-language) processing:

> The theory is based on the assumption that thinking involves the activity of two distinct cognitive subsystems, a verbal system specialized for dealing directly with language and a nonverbal system

specialized for dealing with nonlinguistic objects and events. ~ *Mind and Its Evolution; a Dual Coding Theoretical Approach, Allan Paivio, 2007.*

According to Dr. Paivio, memory is the engine of cognitive evolution. He says that:

> In a general sense, all evolution is memory. ~ *Mind and Its Evolution; a Dual Coding Theoretical Approach, Allan Paivio, 2007.*

The following quote points out the importance of studies pioneered by memory researchers like Dr. Paivio. Dual-process theory developed largely because of studies about memory:

> It was, indeed, the study of memory that more or less launched the dual-process theory of mind. When cognitive psychology was just being formulated in the 1960s and 1970s, a shift occurred in research focus.

> With hindsight, we can see that . . . researchers mostly shifted from studying implicit to explicit forms of memory. We now know that there are multiple memory systems in the brain, some of which are implicit and others explicit, a fact established beyond doubt by numerous neuropsychological studies (Eichenbaum and Cohen 2001). There is an explicit learning system located in the hippocampus and quite separate implicit learning systems residing in regions of the brain associated with motor skills and emotional processing. These can be dissociated from each other by specific kinds of brain damage: for example, patients with hippocampal damage, known as amnesics, suffer impairment of explicit learning and memory, while retaining the ability to learn new skills and habits. ~ *"The Duality of Mind: An Historical Perspective," by Keith Frankish and Jonathan St. B. T. Evans, 2009*

Dr. Paivio feels that memory rather than attention is the key concept for understanding cognitive duality. Attention is just the "gathering mechanism," the front end of a process that ends with memory. If the two attention systems had no way to store and retrieve what had been collected through the senses, then two minds never could have evolved. I agree with Dr. Paivio about the importance of memory, although there is never a time when attention and memory are decoupled. We are embodied creatures, all our processes work together. The important idea is that there are two systems with two ways to attend, and two ways to remember. After studying memory for a while, dual-process theorists could see that implicit memory was part of the system called Type One (the allocentric mind), while explicit memory was part of the Type Two system (the egocentric mind).

Let me pause to point out something that is rather obvious, yet often overlooked. When Dr. Paivio says that we have an image-generating mind and a word-generating mind, he is actually exposing the brain mechanisms for vision (image-generation), and the brain mechanisms for hearing (word-generation). I also discovered this distinction when I looked at navigation and purposeful movement. Vision builds most of the background through which we navigate; hearing creates most of the egocentric concepts that give meaning to the objects in scenes. However, as I will make clear later, both vision and hearing have egocentric and allocentric components.

Dr. Paivio and I have gone down a similar pathway; our journey arrived at the inevitable conclusion that we must have two minds. His conclusion is based on a study of language and mine is based on a study of navigation. In both cases there is an oscillation between two processing streams wherein we can be in one state or the other, but not both at once.

In a recent bestselling book, *Mind, A Journey to the Heart of Being Human* (2016), Dr. Daniel Siegel defines the duality of the mind using the terms "embodied mind" and "relational mind." Using my terminology, Dr. Siegel has identified the embodied egocentric mind, and the relational mind, the allocentric processing system.

Dual-Process Theory

In cognitive science, under the umbrella of theories of mind, there is a long standing debate between representationalism and eliminativism. In my opinion, these mind-hurting abstractions are nothing more than a debate between our egocentric and allocentric minds. Both theoretical camps are correct. Those that support representationalism see the mind from an egocentric perspective. This group of thinkers has not embraced the allocentric mind as a separate entity; they see the mind as mostly, or completely, egocentric. Eliminativism is based on the American Naturalism movement that led to behaviorism and to J. J. Gibson's theory called Ecological Psychology; these are allocentric theories.

Confusing the issue further, both of these theoretical camps embrace the idea of embodiment. From my perspective, when we consider the whole-body as a single unit, we are dealing with the allocentric mind. When we consider cognition as primarily a head-space phenomenon (not whole-body), we are dealing with the egocentric mind. Both sides of this debate see the logic of the opposite argument, but they feel compelled to take sides since it appears that we must decide between these two mutually exclusive perspectives. I contend that we don't need to take sides; this is just another case where we have identified our conflicting duality.

Given the discussion above, it is clear that we have two minds. This is not a new idea; it is not revolutionary. Indeed, my differentiation between the egocentric mind and allocentric mind could be just another set of labels added to a long list. However, what no one could produce was the evolutionary reason why we ended up with these two minds. My new slant on this old conundrum is different and worth consideration. Evidence is mounting in anatomy, physiology, and evolutionary science that we have an inherent duality hardwired within us; the reason for the duality can be traced back to purposeful movement.

This observation, linking navigation to cognition, is important for two reasons. First, from a personal perspective, it provides a historical background for the field of orientation and mobility—it is part of the justification for a new way to define the O&M discipline. In order to teach navigation, we need two curricula, one for the egocentric perspective, and another for the allocentric perspective. Second, and more importantly, the link between navigation and consciousness gives us a new perspective on the mind; we cannot ignore the evidence, we are dual creatures with dual responsibilities because we navigate.

The participatory universe of the philosophers got substantial support as the implications of quantum theory began to be understood. Seeing the world solely from a biological perspective was challenged and weakened by quantum theory. The quantum world is composed of potential and probability. There is nothing substantial to grasp, no cellular matrix to hold or perceive at the quantum level. Objectivity took a serious hit when quantum theory moved into town. Here is how physicist John Wheeler put it:

> Useful as it is under everyday circumstances to say that the world exists "out there," independent of us, that view can no longer be upheld. There is a strange sense in which this is a "participatory universe." ~ *"Law without Law," Quantum Theory and Measurement, John Wheeler, 1983.*

So here we have a famous physicist saying that we cannot be separated from our domain. We live in a participatory universe. What we do, where we place our attention, affects everything. Observation disturbs reality at the most fundamental quantum level. We shall see later that quantum theory also supports the idea that we have two cognitive systems, two minds, two kinds of consciousness. The strangeness of the quantum world is probably the reason why our biological duality evolved.

Notes

(1) *At the Existential Café:* I got an understanding from this book about existentialism and phenomenology that was clear and fun. Sarah Bakewell

is an excellent writer who makes otherwise dry and abstract philosophical complexities interesting. I found myself becoming fascinated by the real people behind their historical masks. Simone de Beauvoir and John Paul Sartre, for example, come alive, as do phenomenologists like Merleau-Ponty, Edmund Husserl, and Martin Heidegger.

(2) **Scientism** asserts that the scientific method and empirical science are supreme methods for arriving at truth when compared to any other viewpoint. Scientism is a dogmatic endorsement of scientific methodology and reductionism, and it holds that only knowledge which is measurable can be admitted as evidence for truth. Therefore, science becomes the only reliable source of knowledge. Scientism is a belief structure that comes from the egocentric mind. It invalidates, or fails to recognize, the allocentric mind. This is a problem because abstractions like love and justice are not precisely measurable—our values, morals, aspirations, and emotions, for example, are relatively beyond serious scrutiny when precise measurement is a prerequisite for truth.

However, it is important not to overlook the importance of the scientific method in our zeal to counter scientism. We should embrace the scientific method as a well proven pathway for uncovering "facts" about the physical world, even as we reject scientism.

(3) It was actually **Husserl's student Edith Stein** who championed empathy as a subject worthy of attention. Stein's doctoral thesis (later a book) was called *On the Problem of Empathy*.

(4) **British philosopher Alan Watts** was a student of Eastern philosophy; he helped bring Buddhist and Hindu ideas to the west. He was highly articulate, and he was able to turn complex ideas into common sense. His TV show, sponsored by the Electronic University, aired in the 1970s. The TV show, and his many books, influenced a generation that was undergoing an evolution in consciousness.

Watts talked about the importance of the concept of figure-versus-ground more than 50 years ago. He says that the background is not just potential and probability (as quantum thought would propose), but it is actually "nothing." The figure that arises from the ground is "something." From emptiness (the background) comes manifestation.

(5) Ken Wilber is a prolific genius who has written extensively on individual and collective consciousness. His integral psychology uses an approach that synthesizes knowledge into four divisions. This four-part quadrant clears up many of the misunderstandings that arise when we debate complex issues. The quadrant is difficult to explain in a summary fashion; please see my detailed explanation in *The Confusion Caused by Being Your Own Twin*, or refer to Wilber's many books and internet posts. Wilber is a student of Buddhism and also Western psychology and he has synthesized vast amounts of knowledge from these two realms. Wilber's ideas have heavily influenced my own perspectives.

Three

We are the mirror,
as well as the face in the mirror.

~ "MIRROR AND FACE," *A YEAR WITH RUMI*, 2006.

Mirror Image Brains

"What's for breakfast, Surge?"

"Eggs over easy with double yolks, served with two mirror-image side dishes. I suggest for side dishes that you get the Cauliflower Brain Lobes with Differently Pickled Halves, or the roasted Janus-Head Pumpkins with matching Halloween faces."

"Got any porridge? I don't like the idea of double yolks—it's like eating twin baby chickens. I thought this was a vegan restaurant."

"We use vegan eggs from vegan chickens."

"That makes no sense at all, Surge."

"Yes. Thanks. No. We don't have porridge. You get two choices. You can get the double-yoked vegan eggs, or you can get the double-yoked vegan eggs. Which do you want?"

94

"Could I get the eggs with the double-yolks and the cauliflower halves?"

"Very good choice. What do you want to drink?"

"Just water, please."

"We don't have any water. How about two sides of orange juice?"

"How can orange juice have two sides?"

"That's what makes this restaurant special. Most places serve their OJ mixed, but we separate the juice into matching halves. If you sip from the left side of the glass, the juice has a sweet, compassionate taste, but if you sip from the right side of the glass, the taste is slightly bitter and has hints of Middle Eastern spices, like cardamom and smoked paprika."

"What if I stir the drink with a spoon?"

"It doesn't matter. The sides never blend. Put the juice in a food processor if you want, or drop a nuclear bomb on the glass, it won't matter. Nature is the mirror image of nature; that is just the way things are. You have two choices. You can accept this dual reality, or, if you want, you can accept this dual reality. The choice is yours—what with your wonderful free will."

"Could we talk about my book for a minute, Surge?"

"Yes, of course. What's on your mind, Dutch?"

"Why do you insist on calling me Dutch?"

"Dutch is your shadow self, a mental sidekick. Everybody has a Shadow Dutch-Man. Humans think they are egocentric *Lone Rangers*, but that's not true because humans have a shadow-self that perceives everything all-at-once. That's who you really are, Dutch—a soulful comedic sidekick to your Lone Ranger ego."

"Okay, sure. That's as clear as the rest of this dialogue, Surge. Let me get this straight. I am a shadow Dutchman who hangs out with my own ego, The Lone Ranger—who wears a mask and shoots people with silver bullets."

"Very nice summary, Dutch. I am pleased that you understand."

"So what about my book? Can we talk about that?"

"Which book?"

"This book."

"This is two books. Which book do you want to talk about?"

"The one that doesn't have Surge-the-waiter being obtuse and sarcastic."

"Oh. Very well, No problem. Here you go:"

Dr. Jekyll, and at Other Times, Mr. Hyde

> *He said: You are not mad enough.*
> *You don't belong in this house.*
> *I went wild and had to be tied up.*
> *He said: Still not wild enough to stay with us.*
> *I broke through another layer into joyfulness.*
> *He said: It is not enough. I died.*
> *He said: You are a clever little man.*
> *full of fantasy and doubting.*
> *I plucked out my feathers and became a fool.*
> *He said: Now you are the candle for this assembly.*
> *But I am no candle. Look. I am scattered smoke.*

~ "SUBLIME GENEROSITY," *A YEAR WITH RUMI*, 2006.

From about 1970 to the present, popular science authors attributed human duality to differences between brain hemispheres. If you examine a brain on a lab bench, you clearly see that it is composed of two mirror halves. The two sides of the brain are held together by a strong cord made of nerve fibers called the corpus callosum. To arrive at their dual-mind theories, popular authors used studies based on what is now called split-brain research. Essentially, surgeons cut the corpus callosum to control epilepsy. This made it difficult for the two brain hemispheres to communicate with each other. When the corpus callosum was severed, strange behaviors resulted. The epilepsy was controlled, but patients now acted as if they had two contrary minds in constant conflict. Watching these split-brain patients go about daily tasks resulted in a set of assumptions about how our minds must work.

A person with a brain that has been split into two halves by surgery behaves as if two minds exist in the same head. Each half-brain seems to exist without the other. The two split-minds often have conflicting ideas of what should be done. For example, if one brain directs the muscles of the arm and hand on the left to put a cigarette into the mouth, the other brain might direct the muscles in the right arm and hand to take the cigarette out of the mouth—a battle can ensue between the two hands. Examples like this confirm that hemispheric differentiation is real—although it is not as exact as popular notions hold.

The right half of the brain was seen as the location of a hidden mind—something the ego, the left brain, could not perceive. The left half of the brain, with its language capability and its egocentric personality was seen as a dominant brain. Strokes in the left temporal lobe leave many people mute, and in severe cases, the personality is altered. When this happens, the hidden mind shows up, speechless, but animated. Logically, these observations appear to be a physiological explanation for our fundamental duality. Here is how popular science author Dr. Leonard Shlain described the dual mind in his book *Leonardo's Brain*:

> In truth, we are Siamese twins conjoined at the corpus callosum—the broad band of fibers that connects the right and left halves of the brain in all vertebrates. Each of the halves of a human brain can generate opinions, perceptions, likes and dislikes different from those of its yoked twin across the way. There is a long history in literature and art of two personas existing within one body, epitomized by Robert Louis Stevenson's two characters operating within

the confines of one body that manifest as Dr. Jekyll, and other times as Mr. Hyde. ~ *Leonardo's Brain, Leonard Shlain, 2014.*

The confusion over the roles of the two hemispheres of the brain comes about because there is some truth to lateral-specialization. However, subsequent research showed this conclusion to be less clear-cut than it was first assumed. Here is how Bernhard Schroeder puts it in his book *Simply Brilliant: Powerful Techniques to Unlock Your Creativity and Spark New Ideas*:

> From self-help and business success books to job applications and smartphone apps, the theory that the different halves of the human brain govern different skills and personality traits is a popular one. No doubt at some point in your life you've been schooled on "left-brained" and "right-brained" thinking—that people who use the right side of their brains most are more creative, spontaneous and subjective, while those who tap the left side are more logical, detail-oriented and analytical.

> Too bad it's not true.

> In a new two-year study published in the journal *Plos One*, University of Utah neuroscientists scanned the brains of more than 1,000 people, ages 7 to 29, while they were lying quietly or reading, measuring their functional lateralization—the specific mental processes taking place on each side of the brain. They broke the brain into 7,000 regions, and while they did uncover patterns for why a brain connection might be strongly left or right-lateralized, they found no evidence that the study participants had a stronger left or right-sided brain network. "Despite what you've been told, you aren't 'left-brained' or 'right-brained.' ~ *Simply Brilliant: Powerful Techniques to Unlock Your Creativity and Spark New Ideas, Bernhard Schroeder, 2016.*

The biggest fallacy of the split-brain hypothesis is that the hemispheres work in isolation. This is not supported by research. The brain works as a whole;

the corpus callosum is not severed in normal brains. There is *never* a time when the normal brain is using just one isolated region in only one hemisphere. The brain always acts as one unit coherently firing in networks that connect both sides of the brain. According to co-authors Stephen Kosslyn and G. Wayne Miller in their book *Top Brain, Bottom Brain,* there was never enduring evidence that the brain hemispheres were the answer to our duality:

> Researchers have known for decades that none of the sweeping assertions about left brain/right brain differences are supported by solid evidence. Although they were not shouting from the mountain tops, these scientists had unimpeachable evidence that the popular culture versions of the left brain/right brain theory do not capture how the brain really works.
>
> For example, the left hemisphere is often described as verbal and the right as perceptual—but this distinction doesn't hold up as a generalization. In reality, both hemispheres typically contribute to both sorts of activities—but do so, often subtly, in different ways.
> *~ Top Brain, Bottom Brain by Stephen Kosslyn and G. Wayne Miller, 2013.*

In his book *The New Executive Brain: Frontal Lobes in a Complex World* (2001), neuroscientist Elkhonon Goldberg postulates that the right hemisphere gradually took on the identification and interpretation of novelty—patterns never before encountered became a specialty of the right hemisphere. As these new patterns became more practiced and embedded, the left hemisphere took over the task of dealing with routines. The left hemisphere became the arena for automatic processing and the right hemisphere became the arena for processing novelty:

> The novelty-routinization hypothesis of hemispheric interaction implies a leading role of the right hemisphere at early stages of dealing with a new task, and a leading role of the left hemisphere at later stages. Does this imply a literal right-to-left hemispheric

information transfer, a neural trans-callosal railroad of sorts? Not necessarily. More likely, mental representations develop interactively in both hemispheres but the rates of their formation differ. The right hemisphere is more effective at early stages of learning a cognitive skill, but this is reversed in favor of the left hemisphere at the later stages. ~ *The New Executive Brain: Frontal Lobes in a Complex World, Elkhonon Goldberg, 2001.*

Here, Dr. Goldberg has been cautious not to assert that there is absolute hemispheric specialization. The truth about our amazingly complex brains always seems to need caveats. The brain works as a single unit, always as part of a body in a specific environment. Every kind of sensory–motor activity relies on a pattern, a specific neuronal network. As one pattern becomes active, other sets of neurons, other patterns, are relatively inhibited. This is a total brain process.

One of the most prolific and insightful students of consciousness was English writer Colin Wilson (1931-2013). Wilson wrote over one hundred books, many of which examined mental processes, especially consciousness. In at least two of his works of non-fiction, *Access to Inner Worlds* and *Frankenstein's Castle*, he discussed the proposition that human beings evolved two brains which, consequently, created two minds. Wilson knew about split-brain research, and like others who got hold of this information, he made associations between contrary human behaviors and the roles of the two hemispheres. The quote below, taken from *The Secret History of Consciousness* (2003), by Gary Lachman and Colin Wilson, is an example of a correct understanding of our duality, a description of allocentric and egocentric perception, but with an overemphasis on hemispheric differences. Lachman was discussing Colin Wilson's understanding of dual minds when he wrote the following:

As we have seen, the left-brain "scientist" focuses on the isolated "fact," while the right-brain "artist" absorbs pattern and "meaning." Usually the two are in conflict, and in our own fast-paced, highly outer-world oriented culture, the left brain is dominant, with the result that we can "scan" the world, picking out the "facts" and absorbing little "meaning": we see a lot of trees but rarely the forest. Yet, in those infrequent moments when the two brains work together—either at times of deep relaxation or when the right brain, which moves slower than the left, is "speeded up" through interest or excitement—reality suddenly acquires an additional dimension . . . ~ *The Secret History of Consciousness, Gary Lachman, and Colin Wilson, 2003.*

Seeing the trees and not the forest *is, indeed, the job* of the egocentric mind; the egocentric mind cannot see the forest—it was not designed to perceive backgrounds (gestalts). However, the allocentric mind *can* observe the forest, but has no ability to extract meaning—it cannot perceive the trees as objects. So Lachman has the description partially right, but there is no "left brain scientist" or "right brain artist." Also, there are *no moments* when the two systems work together—they are forced by the laws of physics to alternate. The evolution of navigation dictated the need for dual processing, for an oscillation between perceptual functions.

To walk, to move forward, to navigate requires whole-body neurological alternation and synchronization. The brain is a major part of this whole-body oscillation—the two brain hemispheres also alternate and synchronize neural impulses in harmony with what is happening in the whole body. This alternating rhythm between the two sides of the body and brain fluctuates depending upon the speed of navigation—the speed of purposeful movement—and on limited executive control coming from the highest centers for cognitive processing.

One way to think about navigation and its impact on our cognitive development is to consider the whole body as a system that is either attracted or repulsed by environmental situations. If circumstances are positive, we navigate forward. If circumstances are not favorable, we move away. When we are not "pulled or pushed," we are in a state of balance or indifference. When we are not sure, we stand immobile in anticipation.

Consider also that we can hop on one foot, balance on either foot, and we can sit without moving any large muscles. In other words, we can cognitively modulate hemispheric alternation. We cannot completely shut down the reverberating synchronization because the body and brain need an overall tone, posture, and balance, but we can bring the large skeletal muscles under limited and partial control. This basic coherence of the human body and brain—brought about because we navigate—is the result of a balance between inhibition and excitation.

This basic concept of dual-process theory—that navigation dictates neural behavior—is typically not taken into account when researchers explore the roles of hemispheric brain regions. For example, neuroscientists consistently discover that the hemispheres have different roles. Often the entire hemispheres are differentiated—for example the right hemisphere is said to be where new behaviors are developed and where creativity arises, while the left hemisphere regulates automated behaviors and is rational rather than innovative. More specifically, neuroscientists look at one region of a hemisphere and compare it to a corresponding region in the other brain hemisphere. When they do this "split-brain" research, a brain region in the right hemisphere is often found to have a contrary, mutually exclusive role when compared to its sister region in the left hemisphere. To give one example, the right amygdala and the left amygdala have been found to have opposite functions:

> The right amygdala is specialized for fear and anxiety as well as pain and stress. The left amygdala is involved with positive emotions, like elation and happiness, and even the maternal happiness that comes with holding one's infant. It recognizes positive emotions in others. ~ *Sacred Pathways: The Brains Role in Religious and Mystic Experiences, Todd Murphy, 2015.*

This kind of conclusion is misleading when it ignores the constant rhythmic oscillation between all matching brain organs—one in the right hemisphere, the other in the left hemisphere. We can make three summary comments about this biological oscillation:

1. As we navigate, there is a powerful, all-the-time oscillation between the left and right halves of our bodies. This is also manifested in the two brain hemispheres as a powerful oscillating rhythm. When one side is stimulated, the other side is inhibited. This suggests that the separate organs of the brain—hippocampus, amygdala, prefrontal lobes, and so on—also contribute to and are in harmony with this alternating rhythm.

2. Researchers consistently find that the matching regions of the brain have opposite functions. There is always a left organ that corresponds to a right. For example, the right hippocampus has separate and opposite functions from the left hippocampus, the right frontal lobe has different functions from the left frontal lobe—this apparently holds true across all brain regions.

3. Brain organs oscillate in a mutually exclusive rhythm, like an on/off switch. When one area is on, the other corresponding area is off. This is a rapid oscillation that is not available to perception. When researchers isolate a brain region in one hemisphere, for example, the hippocampus, they are setting up unnatural conditions because, under ordinary circumstances, the left and right hemispheres of the brain never act in isolation. The allocentric/egocentric balance is artificially disrupted in split-brain research—especially when there is little understanding of dual-process theory.

Bilaterality and navigation are also inherently and intimately connected with our emotions. What we call positive emotions are simply triggers, signals, which get the body ready to approach. Therefore, positive emotions are intimately linked to forward navigation. Contrary to this, negative emotions are triggers signaling the body to navigate away from unpleasant environments and situations.

There is a balance between positive and negative that can be isolated (observed) in any two organs of the brain. This has led to a false labeling of brain organs, like the amygdala, as specializing in either positive or negative attributes. For example, in the limbic region of the brain there is a balance between what we call positive and negative emotion. In reality, what we are observing is a navigational response that is part of a whole-body neural-net.

Think about emotions from a navigational perspective. An emotion is linked to a motor response. For example, if there is danger, fear, or the possibility of bodily harm, emotional alarms cause an aversion—the whole body reacts by navigating away from the danger. On the other hand, if we are attracted to food, a mate, warmth, or safety, we have positive emotions. Behaviors, purposeful movements, are intimately tied to the emotions. This is a push-pull phenomenon: attraction versus repulsion; moving forward versus moving away; navigating toward or navigating away. In other words, emotions create navigational behaviors.

We know this alternation exists because of the consequences of strokes in the brain. A stroke on one side of the brain can paralyze parts (or all) of the opposite side of the body: a stroke on the right side of the brain affects the ability to move the left side of the body, while a stroke in the left brain affects the ability to move the right side of the body. This is the standard explanation of what happens as a consequence of a stroke. However, using dual-process theory, we can see that it is actually a disruption in the oscillation that causes paralysis. Each hemisphere alternates between activation and inhibition. A stroke interferes with the oscillation, so both sides of the brain and body are affected.

We often think of the two brain hemispheres as being connected by the corpus callosum. This is misleading if we leave out other hemispheric connections like the anterior commissure, which connects the two temporal lobes. The corpus callosum is over 10 times larger than the anterior commissure—we cannot downplay its major role—but the direct connection between the two limbic regions via the anterior commissure (deep within the temporal cortex) is very significant, especially for understanding emotional responses and their relationship with navigation. If just the corpus callosum is severed during surgery—and not the other commissures—we get a false understanding that the brain has been split totally into two isolated regions.

Besides the corpus callosum and anterior commissure, connecting the two hemispheres are the hippocampal commissure and, indirectly, subcortical connections. The anterior commissure works with the posterior commissure, subcortical pathways, and the hippocampal commissure to link the two cerebral hemispheres. The corpus callosum is the primary link between the brain hemispheres, but it is not the only important pathway.

As we move forward with this discussion, keep in mind this basic whole-body oscillation and synchronicity. This oscillation is reliable, consistent, and synchronized; it is a fundamental part of the background steady-state that enables the allocentric mind.

Top Brain and Bottom Brain

Authors Stephen Kosslyn and G. Wayne Miller, in their book *Top Brain, Bottom Brain,* strongly reject the theory that our dual nature can be explained adequately by hemispheric differences. They offer instead a theory that they call "Top-Brain and Bottom-Brain." It makes more sense, they tell us, to divide the brain into upper and lower sections because these *actually do* correlate with our dual nature.

The bottom brain contains mostly the occipital and temporal cortex plus the bottom part of the very large frontal cortex. This brain region—this combination of three brain lobes—is concerned with processing and organizing sensory input, both the perception of objects and the perception of scenes. The top brain contains the parietal lobe plus the top section of the frontal lobe. This brain region is involved with the processing and controlling of movement—including speech and attention—as well as reasoning, short-term memory, and executive functioning—for making plans and anticipating the consequences of actions:

> . . . the top brain is involved in setting up plans that involve moving
> in space, and hence information about location is crucial, whereas

the bottom brain is involved in classifying and interpreting objects and events. ~ *Top Brain, Bottom Brain, Stephen Kosslyn and G. Wayne Miller, 2013*

The top brain-bottom brain architecture is an overall design that supports dual-process theory. The bottom brain is often called the "what is it?" processing stream because it processes language and concepts; it could be the neural substrate for the egocentric mind. The top brain is often called the "where is it?" processing stream because it processes spatial relationships; it could be the neural substrate for the allocentric mind. Navigation requires both processing streams working symbiotically. The bottom brain provides us with landmarks that have location-based meaning that we use for personal navigation. The top brain processes movement and enables us to go around solid objects as we navigate—it is concerned with processing "flow" during the act of navigation. I will discuss this anatomical divide and the implications further in Chapter Eight.

Colin Wilson, Gary Lachman, and Faculty X

Gary Lachman and Colin Wilson wrote at a time when the anatomical and physiological differences between our two brain hemispheres were considered to be the source of dual cognition. They emphasized six fundamental ideas:

- Human beings have two minds. The circumstantial evidence is everywhere in our literature and in our sciences.
- There is anatomical and physiological evidence why two minds evolved. This evidence resides, so say these authors, within hemispheric differentiation.
- The two minds have different functions which are mirror images of each other. If one brain is rational, the other is intuitive. If one brain is verbal, the other brain is non-verbal, and so on.
- Profound and universal implications (responsibilities) arise because of our inherent duality.

- Spirituality, the evolution of religions, and psychic (paranormal) phenomenon arise from the right brain.
- The scientific method, materialism, and rational discourse arise out of the left brain.

I agree with Gary Lachman and Colin Wilson that human beings have two minds and that there is an evolutionary, anatomical, and physiological reason that we have this duality. I also agree that the two minds are oppositional twins—if one says "move," the other says "don't move," etc. However, the assumption that hemispheric differentiation is the major explanation for our dual minds is incorrect.

Colin Wilson and Gary Lachman were friends and colleagues; they influenced each other's ideas. In Lachman's 2014 book *Revolutionaries of the Soul,* he begins chapter one with Colin Wilson's ideas about duality and consciousness. Below is a summary.

Wilson defined a capability called Faculty X, which causes a sudden elation that transcends everyday boredom and angst. Wilson wondered where this good feeling came from. What might be a physiological explanation for elation which arrives and departs suddenly and without provocation?

Wilson's answer is that reality is not confined to the present moment. He feels that we are not the space-time creatures that we suppose ourselves to be. We are not imprisoned in the moment. Indeed, Wilson states that the brain and nervous system were created in evolution so that human beings could be in two places at the same time. In other words, we have two minds, one grounded in everyday reality, and another mind that is free of the constraints of the body. Faculty X is the ability to leave the mundane, boring routines of daily life and mentally flee to more exciting and fulfilling locations—places that are exhilarating, joyful, and life-affirming.

Wilson also holds that thoughts or daydreams *can actually* transport us to other locations. We totally forget the present moment—the time and the space our body occupies—and we disappear into another realm.

In other words, to call this time travel "daydreaming" or "imagination" is missing the reality: we *actually* depart to a distant place, to an alternate time. In fact, according to Wilson, we are capable (because of our twin hemispheres) of being in two places at once. Wilson suggests that evolution, using the physiology arising from hemispheric differentiation, enables this time travel. Only in retrospect do we realize that we have actually been elsewhere. This ability to will ourselves to be in two places at once, according to Wilson, has yet to fully evolve—some people can do it, most cannot.

There are many problems with Lachman's and Wilson's interpretation of our duality and with how the brain functions. For example, Lachman says that meaning resides in the right hemisphere and that the right cortex is specialized for pattern recognition. Both of these views are questionable. In fact, the whole cortex, left cortex as well as right cortex, deals with pattern recognition, and there are two kinds of pattern recognition, spatial and temporal. Also, there are two kinds of meaning, egocentric and allocentric. "Meaning making" (discovering meaning) is not the function of a single hemisphere—it is a whole brain process.

In dual-process theories, like the one I am proposing in this book, the argument is that two processing streams (two massively complex neuro-networks) evolved into two kinds of consciousness. Brain lateralization cannot be left out of this theory. We know that people are left-handed or right-handed and also have a preference for a dominant eye—some people are right-eyed and some people are left-eyed. Some people have cross-dominance wherein their hands or eyes are equally capable (alternately preferred). The brains of individuals who are right-handed, for example, have larger sensory-motor brain areas in the left hemisphere. The opposite is true for left-handed people. These facts show that we must factor in hemispheric differences as we slowly unravel the mysteries of the duality of consciousness. However, we have gone astray—implied too much—with our left-brain versus right-brain theories.

Dr. Jill Bolte Taylor's Stroke of Insight

Further evidence for two minds that exist anatomically and physiologically apart can be found in the fascinating book *My Stroke of Insight* by Dr. Jill Bolte Taylor, a Harvard trained neuroscientist. In her thirties, Dr. Taylor had a stroke in the language centers of her brain—in the left hemisphere. As a scientist, she watched part of her egocentric mind slowly disappear as the language neural-net in her brain faded away from lack of oxygen. She was left, using my terminology, with allocentric consciousness. In other words, her misfortune showed the world that one twin, the egocentric mind, could be selectively impaired but something was left that looked like what saints and seers had been trying to describe for centuries. Fortunately, Jill Taylor got better; her language returned, and she was able to articulate her experience. Her TED talk nicely summarizes her book.

In early evolution the two hemispheres had balanced functions, but over time they became asymmetric in function, almost like two continents that had split in half, after which they evolved different "flora and fauna." Hemispheric differences were, for Dr. Taylor, the neurological substrata for two separate minds inside the same head—that was the perspective of the scientific community at the time she had her stroke.

Dr. Taylor identified the *ego* with the function of the left brain, while the hidden, invisible mind, according to her reasoning at the time, resided in the right hemisphere. I agree with her identification and characterization of the two minds, and obviously hemispheric differentiation is part of the total picture—her descriptions are brilliant and articulate—but the origin of our two minds is clearly to be found in the evolution of wayfinding and the consequent need for allocentric (background, space) processing, and egocentric (figure, time) processing. Below are some quotes from Dr. Taylor's excellent book *My Stroke of Insight*. Her description, like Leonard Shlain's and Gary Lachman's (quoted earlier) is accurate as it explains egocentric and allocentric processing, but the hemispheric-differentiation perspective is not entirely accurate:

My right hemisphere is all about right here, right now. It bounces around with unbridled enthusiasm and does not have a care in the world. It smiles a lot and is extremely friendly. In contrast, my left hemisphere is preoccupied with details, and runs my life on a tight schedule. It is my more serious side. It clenches my jaw and makes decisions based upon what it learned in the past. It defines boundaries and judges everything as right/wrong or good/bad . . .

My right mind is all about the richness of this present moment. It is filled with gratitude for my life and everyone and everything in it. It is content, compassionate, nurturing, and eternally optimistic. To my right mind character, there is no judgment of good/bad or right/wrong, so everything exists on a continuum of relativity. It takes things as they are and acknowledges what is in the present . . .

My right brain character is adventurous, celebrative of abundance, and socially adept. It is sensitive to nonverbal communication, empathetic, and accurately decodes emotion. My right mind is open to the eternal flow whereby I exist at one with the universe. It is the seat of my divine mind, the knower, the wise woman, the observer. It is my intuition and higher consciousness, my right mind is ever present and gets lost in time.

My right mind is open to new possibilities and thinks out of the box. It is not limited by the rules and regulations established by my left mind that created the box. Consequently, my right mind is highly creative in its willingness to try something new. It appreciates that chaos is the first step in the creative process. It is kinesthetic, agile, and loves my body's ability to move fluidly into the world. It is tuned in to the subtle messages my cells communicate via gut feelings, and it learns through touch and experience.

. . . my left brain is equally amazing. My left mind is responsible for taking all of that energy, all of that information about the present moment, and all those magnificent possibilities perceived by my right mind, and shaping them into something manageable.

My left mind is the tool I use to communicate with the external world. Just as my right mind thinks in collages of images, my left mind thinks in language and speaks to me constantly.

Our left brain truly is one of the finest tools in the universe when it comes to organizing information. My left hemisphere personality takes pride in its ability to categorize, organize, describe, judge, and critically analyze absolutely everything." ~ *My Stroke of Insight, 2009.*

Dr. Taylor says it clearly: we have two minds, with two different mandates and two different skill sets. When she wrote those words, as a talented neuroscientist, the common understanding was that our two minds were the result of our split hemispheres. But further research did not support this conclusion. Our duality has deeper roots and is more complex than can be explained by separate brain hemispheres. Nevertheless, Dr. Taylor's description of allocentric and egocentric consciousness (using my terminology) is valuable and clear.

Zoltan Torey: The Connection between Consciousness and Proprioception

There are many examples of writers who "discovered" two minds and attributed this to hemispheric differences. Jill Bolt Taylor is one example; Colin Wilson and Gary Lachman are two others. To give another example, in his book *The Crucible of Consciousness*, psychologist Zoltan Torey gives his version of our duality based on hemispheric differentiation. Torey, like others, found two minds and tried to show what they were, and why they came to be separate islands in the same skull.

To greatly oversimplify, Torey points to research that isolates percepts (images) in the right hemisphere and language in the left hemisphere. This is similar to Allan Paivio's Dual Coding Theory (discussed in Chapter One) in which Dr. Paivio differentiates between a mind specialized for image processing, and a second mind specialized for language processing.

Torey speaks of *connotation* coming from the right hemisphere where percepts are silently defined by their attributes, by associated emotions, and by physical characteristics. While *denotation*, according to Torey, occurs in the left hemisphere where language evolves into concept formation—the left hemisphere provides names for objects and concepts, and then allows a dialogue about them. The brain processing that allows percept formation is automatic and unconscious, so Torey calls this a hidden mind. He says that the two minds cannot manifest at the same time, so the brain hemispheres rapidly oscillate (take turns). This is fascinating thinking, even though, as I have emphasized, the differentiation between hemispheres is no longer unquestionably supported. However, in my opinion, Torey's major contribution to the theory of mind[1] is what seems at first to be a side comment to his major work.

Torey arrives at the interesting and compelling conclusion that consciousness has arisen from the inner senses, from proprioception. This conclusion is rather profound. It seems to explain not only our monkey minds—the voice inside our heads that won't shut up—but consciousness itself—both of them. Therefore, according to Torey, consciousness is a product of proprioception, just as vision is a product of the visual system, and hearing a product of the auditory system. If vision manufactures space, and hearing manufactures time, then proprioception manufactures consciousness.

Proprioception can be understood from a global perspective or from a narrow perspective. Globally, it is a term that includes all the internal hidden senses, like kinesthesis and the vestibular system. For example, when we say that proprioception gives rise to consciousness, we are using the term in the global sense. However, in a specific sense, as *one* of the *internal* sensory systems, proprioception is the whole/body mechanism *that maps* the location of every part of our anatomy—it remembers relative spatial locations.

Proprioception works with kinesthesis to know which muscles are working and where—so it registers whether the body as a whole is moving or not. Proprioception also works with the vestibular sense to monitor the posture of the body relative to the pull of gravity. Hold your hand behind your head and proprioception knows where your hand is. Close your eyes in any scene and proprioception can still reach out and find things that are relative to the body. At first glance, proprioception is a monitoring and memory system for knowing the relative position of ego-to-other (egocentric perspective) and other-to-other (allocentric perspective).

To work efficiently, proprioception must have constant feedback from every part of the body. This feedback system can never stop if we are to do purposeful movements of any kind. Now consider what happens as we use language to communicate. When we form words and sentences, muscles move in an exact pattern (a motor pattern). Our jaw, lips, mouth, tongue, and head alignment must smoothly coordinate. As the motor pattern is executed, proprioception knows exactly what every muscle is doing (how the jaw, tongue, and mouth move in harmony with head position and with body posture). Proprioception stores these complex sets of coordinated movements in brain maps that enable muscle memory. Indeed, proprioception is a kind of muscle memory. The next time we repeat the same speech—when we use the same sentences or emotional tone, for example—words flow faster, with less stammer or inaccuracies because the motor pattern has been embedded in memory. Proprioceptive memory of motor patterns enables ever more articulate speech as we go through our developmental years.

Because sensory memory is also being recorded in the mind at the same time as proprioceptive (motor-pattern) memories, visual and auditory patterns are synchronized with the muscle patterns. Proprioception enables the mind to access muscle memory patterns and play them back inside the brain at will. Therefore, we can silently rehearse a speech before we give it—as many times as we want. An athlete can rehearse the muscle sequences to be used during the execution of a skill set without actually preforming motor skills. The result of inner-rehearsal—the recall and playback of sensory-motor patterns—is inner-dialogue, the voice in your head. All of our mental life, thoughts, sensations, and emotions, are the result of proprioception—the recall and playback of muscle memories. Our inner-dialogue, the voice in the head, is often assumed to be what we call consciousness (the egocentric mind). It is associated mostly with egocentric consciousness because it uses language. Therefore, in summary, *proprioception is the perceptual system that enables subjectivity.* The eyes see, the ears hear, the full-body proprioceptive system "knows."

Notice how difficult it is—impossible actually—to use the inner voice without subtly moving the muscles of the body—especially the small muscles of the tongue, lips, and face. Try to use the inner voice with no lip, tongue, jaw, or body-head quiver. Try to keep every muscle rock-still and then try to "think." When we do this muscle-inhibition exercise, the inner voice stops; thoughts and emotions cannot arise in the mind unless there is proprioceptive feedback (playback). This is one reason we sit so still while meditating, to stop the inner voice. When we quiet the proprioceptive system, we also quiet the egocentric system; we allow the hidden mind, allocentric consciousness, to be "seen." This is, in my opinion, Zoltan Torey's great contribution to the theory of mind: our inner voice, indeed, consciousness itself, is a proprioceptive phenomenon.

Buddhists compare the physical world with the mental world. In the mental world, objective input, or physical reality, is not necessary, except as memory. Therefore, the external senses are blind, or unused, during cognition. Indeed, the external senses, all of them, are inhibited when internal cognition is occurring.

To perceive the physical world, we use layers of perception. For example, eyes range over great distances (near-to-far), ears reach to lesser distances, smell to even closer-in spaces, and taste and touch are surface perceptions. Therefore, our external senses probe through layers of space depending on where we move our attention. In other words, objective reality is spatial and temporal depending on how we use our attention. However, internal reality is devoid of objective space-time. Inside the mind, temporal and experiential processing occurs, but objective space does not exist within the mental arena. *The internal senses continue to inform the internal mind even with no objective information.*

The Buddhists, of course, have known this for 2500 years. They just did not use the word proprioception. Instead, they stated that *the mind was a sixth sense.* What they mean is that a different sensory system, besides the external sensory systems, is operational during the process of thinking. Here is a quote by the Dalai Lama explaining this perspective:

> I was intrigued when I came to discover that in modern Western psychology there is no developed notion of a non-sensory mental faculty. I gather that for many people the expression "sixth sense" connotes some kind of paranormal psychic ability. But for Buddhists it refers to the mental realm, including thoughts, emotions, intentions, and conceptions. - *The Universe in an Atom, the Dalai Lama, 2006.*

In other words, the world of thought is created by an internal sensory system, by proprioception. "Thought" is a perceptual entity, an "internal imaging system." *Notice that once a specific sensory-motor memory is laid down in the brain, it becomes available as an object-of-regard itself.* We can now create memories of memories, layered proprioceptive patterns. For example, once the monkey mind is created, once the voice in the head is fully operational, proprioception can use the same patterning mechanism to build a higher mind that can observe itself or that can become self-conscious. Also, notice that what we call levels of consciousness, or the evolution of consciousness, might very well be the result (evolved ability) of particular internal minds creating ever more refined memories of memories—proprioception, a system that creates brain/body maps, can grow in sophistication and capability over time.

Notice that the voice in the head does not age. When we say that we don't feel any older even though the body is aging, it is because our inner life is a recording system that is always fresh. Also, the ability to mirror the behavior of others is proprioceptive. Children copy the sensory-motor activities of adults because children have a vigorous, youthful proprioceptive system that is rapidly learning. Many adults also can observe, remember, and playback the mannerisms and gestures of others. Comedians are especially good at reproducing the personalities of others (usually in caricature).

Here is an observation that requires volumes to explore: If there is such a thing as an intelligent universe, one that is conscious, then creatures from other realms would have a way to speak through us (or to us) using our proprioceptive system: a voice that is not ours could contact us and provide insights—as long as we do not inhibit the incoming messages. This insight is beyond the scope of this book, so I leave this line of thinking to those who study the occult, or those who perceive the universe as intelligent and filled with intelligent entities. I am going to lunch to talk with Surge. This whole line of thinking about *proprioception as consciousness* makes me anxious.

Musing is Forever, Dutch

"Hey, Surge. I hate noise pollution, did I mention that? Here at the Third-Eye-Watching Restaurant, vegan silence is precious. Thank you for the quiet."

"You are welcome, Dutch. Vegan Spaces do not suffocate your native heartbeat. There are no Big Brother screens in vegan space."

"Thank you, Surge. Outside these walls, Big Brother is advertising from giant digital screens. Fools are lined up around the block to spend their paychecks. I can't breathe in the noisy spaces of Tinsel Town."

"I see. Well, yes, the spaces of the so-called real world are disruptive to the harmony of the soul. What will you have today, Dutch?"

"I need something to resolve the paradoxical nature of my existence. What meal do you have to resolve paradoxical indigestion?"

"Ah, you want a life without paradox, is that what I hear you say? No problem. For a main dish, we have Vegan Briskets sautéed in Rattle Snake Venom. This is served with Cyanide Biscuits slathered in Arsenic Butter. For a side dish we offer marinated Death Angel Mushrooms. For dessert we have Actual Death by Chocolate Pudding.

"I feel kind of nervous about that meal, Surge."

"There is no life without paradox, Dutch. What you are asking for is a final meal."

"No, I'm not! I have contradictions to ponder. I want to ease my dilemmas, not end my life."

"Oh. Sorry, I misunderstood. You don't want the Death with Dignity Last Supper, then? What you want is a meal you can ponder, is that right?"

"Yeah. That's it; a meal to muse over."

"No problem. Today, for those who wish to go on living, we have delicately sautéed Pondering Parsnips saturated in our famous Musing-Fools Sauce. This is served with Think-it-Over Donut Holes and Run-That-by-Me-Again Asparagus Spears. On the side, we serve Middle-aged Carrots in a Rough Conundrum Sauce. I suggest, for your complimentary drink, that you have cognac blended with Perplexed Pomegranate Juice. How does that meal grab you?"

"That's perfect, Surge. I'll have that."

"Good choice. However, we are concerned about your book, Dutch. Perhaps we should chat before you eat."

"What's bothering the book-nymphs, Surge?"

"Your claim that proprioception is the sensory system that results in consciousness. That is troubling. Many experts in the field of consciousness are grumbling and breaking wind. We understand the analogy: eyes are the portals for vision, ears the portals for hearing, and likewise, proprioception—so you say—is the portal for consciousness. But we don't see that you have adequately explained this conclusion. The reader is perplexed."

"Me too, Surge. That's why I came here for dinner."

"You need our help, is that it?"

"Bingo."

"Read me what you figured out so far. Run through the logic again."

"Okay, sure. Well, brains and nervous systems came about so that animals could navigate."

"That's plausible. We accept the premise. What else?"

"Navigation requires both background and foreground processing; so the brain and nervous system had to evolve two simultaneous, yet mutually exclusive ways to attend. To navigate, we need two paradoxical, oppositional twin processing systems, one to control movement and the other to control no-movement. One system generates a stable background; the other system generates action figures."

"Okay. You might get away with that. You did spend considerable time explaining the proposition. The book-nymphs feel that your observation is plausible. What comes next?"

"Then, after eons of evolution, the two alternating, oppositional attention systems became two minds. In other words, we can say that evolution made these two neural systems so complex that, over a long stretch of time, they became separate universes, so to speak. We can now label these two complexities *minds*.

"The educated public might buy that. However, the suppositions are piling up, Dutch; I am not sure how many theoretical cards we can stack on top of each other before the whole thing collapses. Anyway, what comes next in your logical sequence?"

"Then pre-humans stood upright, evolved opposable thumbs, started using tools, and eventually they formed social groups and started grunting at each other. Add all that together and you get language. The ability to precisely communicate made human beings unique creatures among all the animals. Language is what pushed both minds beyond anything that had ever come before. Language accelerated socialization and cooperation; ever greater degrees of cultural evolution became possible and began to unfold."

"That's pretty much accepted knowledge, Dutch. Then what happened?"

"Then, proprioception—our ability to remember the muscle sequences that allow for repetitive behaviors—created brain maps that store, remember, and recall behaviors. Proprioceptive maps are memories of memories. Some of the neural maps turned out to be quite special; these evolved to store the muscle sequences that enable speech. This caused the inner voice to arise in evolution. We are the only species to evolve with the ability to go insane talking to an illusory self—it's kind of like two illusions validating each other's reality, which is tragically funny, in my opinion. The internal voice slowly evolved into self-awareness, self-consciousness, personality/ego, and the ability to witness the mind and body at work. Thus was born rocket science and mental illness."

"Be sure to credit Zoltan Torey for this insight: proprioception created the internal mind and that led to the ego and internal dialogue. You've already explained this before, so where's the dilemma?"

"Well, I can see that proprioception created the ego, the personality, and the need to eat donut holes over and over again. But the miracle of proprioception is that it is also the main contributor to the background mind, to allocentric consciousness. It is the main creator of internal space. But I am missing something, Surge. I have chronic angst. What am I missing?"

"Well, let's muse for a while and see what floats out of the background void. Here are a few thoughts. First of all, proprioception measures changes in muscles. There can be no movement, no stabilization of movement, no navigation, and no animal behaviors without muscles cooperating with muscles. Therefore, it should come as no surprise that the primary suspect for consciousness is a muscle-monitoring mechanism. Second, proprioception isn't an external sense, like the eyes and ears that tell us what is going on beyond our bodies. Proprioception is an internal sense; it monitors the whole body, and it keeps track of the location of body parts. Consciousness is not "out there." It is a subjective phenomenon which requires a powerful subjective sensory system. Again, this puts proprioception under the spotlight. Proprioception is your key witness, the system that has been hidden from scrutiny until now—no one has looked at the whole body and at the internal monitoring system as the source of consciousness."

"That makes sense. You are, of course, defining proprioception as the sum of all the internal monitoring systems, including the vestibular system, kinesthesis, and photo-perception. Is that it? Are we done musing, Surge? Because, I still have angst."

"Musing is forever, Dutch. Relax. Let's try another avenue. Think about the sense of smell for a moment. To smell something requires a conscious intake of breath. This intake-breath is the egocentric component of olfaction—it is a search for meaning, for an understanding of objective reality. However, contrary to this egocentric investigation, the out-breath is not taking in olfactory information. This suggests that the in-breath is egocentric, while the out-breath is allocentric. Two hundred years ago, the German philosopher Goethe famously used the phrase "expire and expand." He understood in the 1700s that the allocentric mind was related to the outbreath and to spatial expansion—pretty brilliant. Notice also that as you breathe in, as you use the egocentric mind, the inner voice cannot function. Only when the breath is still, or while exhaling, while using the allocentric mind, can the inner voice function."

"What's that have to do with proprioception and dual-process theory?"

"The inner voice dominates when you are not breathing at all, when the muscle systems are relatively quiet—when there is no noise pollution. It is a gap-phenomenon; it shines most brightly between the in-breath and the out-breath. The inner voice also functions as you are breathing out—although not optimally. It works best in the phase transition, in the silence, as movement slows down toward zero, as the pendulum reaches the turn-around-point, as one threshold fades and the next is breached. By the way, you are not getting enough oxygen to your cells when you sit and think too much. Egocentricity is a cousin of entropy."

"That's interesting, Surge. Thank you, even though you didn't answer my question. The reader is still bewildered. You kind of lost us when you started talking about breathing."

"Your angst, tell me where is that coming from?"

"I don't know, Surge. It's just a whole-body, low-intensity dread."

"I see. You feel that something is incomplete. Is that it? The circle is not closed. The logic is missing steps. The puzzle has holes in it. You don't feel full, satisfied, or satiated. Is that what your angst is about?"

"That's it, Surge. I am filled with fear and loathing; there is no joy in Mudville, Mighty Dutch has gone down swinging."

"So, where is the circle incomplete, Dutch; where are the gaps? Where is the logic not airtight? Where are the missing puzzle pieces?"

"That is where we started Surge, speaking of incomplete circles. That's why I came for dinner. To get your help."

"Why didn't you say so?"

"I did say so, Surge. Anyway, never mind the criticism. I get enough of that in everyday life."

"Define the puzzle. Show me the landscape. Draw me a picture."

"Okay. Look under the microscope. See that paramecium? It is a single cell. It has no neurons, no muscles, no eyeballs, no ears, and no brain. Yet, it finds its way."

"We already talked about the paramecia."

"Where is the paramecium's brain, Surge? That's what I want to know. Where are the muscles that move the cilia? Where is the nervous system that innervates the non-existent muscles?"

"Scientists figured that out a long time ago, Dutch. The whole paramecium is a brain; it is a swimming, sensing, reactive, whole-body, self-contained cognitive system. Indeed, the paramecium is a model for all cells, including neurons. Every cell in your body is a brain unto itself. These cellular brains can form colonies that become super brains. For example, body organs are super brains. Collections of intelligent body organs can become intelligent whole-body organisms. What we call "human being" is a collection of intelligent organs that cooperate to make a super being. That's what the allocentric mind is, a whole-body super brain.

"I think I get it now, Surge. All the cells in our body are self-contained brains. Add them together and you get a self-contained huge brain called a human being—or a squirrel, or a fairy fly. That whole-body brain *is* the "real self," the soul, the allocentric mind, the background. So the sensory system that is in charge of monitoring and regulating this super-brain is called proprioception. Is that it?"

"Yes, that's it. Proprioception began at the level of the single cell. Paramecia are very sophisticated single cells with a sophisticated proprioceptive system—they are cellular super-brains. How does that conclusion feel?"

"Much better, Surge. My angst is subsiding."

"That's good, glad to hear it; although we have only begun to think this through, Dutch. Don't forget that cells are quantum creatures. Therefore, whatever is created by adding cells together is also quantum. Cells are quantum, clusters of cells are quantum, and tissues are quantum. The organs of the body are quantum. Therefore, every organism is quantum. But here is what has been missing, Dutch: *cells communicate. Everything exists in relationships. Everything communicates.* Quarks, atoms, molecules, single cells, and groups of cells exist only in constant communication, constant relationship. Nothing exists in isolation. Summarize these new puzzle pieces, Dutch. How do you suppose these insights fit with dual-process theory?"

"Okay, here I go. Intelligent, quantum, single cells joined together, during eons of evolution, to create ever larger and more complex multicellular organisms—bigger and more sophisticated whole-body brains are created over time. These creations could only exist through communication and relationship. This binding together became proprioception, the allocentric mind. Eventually, animals evolved. Animals can navigate. They have highly evolved but separate dual-attention systems that became two minds which then became two kinds of consciousness. One of these minds, the allocentric, is a whole-organism proprioceptive system, a big brain that harmoniously orchestrates every cell in the body. Consequently, animals have a kind of whole-body wisdom, what we now call allocentric consciousness. In addition to this whole-body mind, a focal processing system—using external senses—developed—we call this the egocentric mind. The evolution of vision and hearing, along with the evolution of language and socialization, created this highly sophisticated egocentric mind. Therefore, we can conclude that proprioception gave rise to both of our minds. As the sum total of the actions and reactions of the whole-body brain, proprioception gave rise to the allocentric mind. Second, proprioception gave rise to the internal voice. That voice eventually gave us the ego, personality, and self-consciousness. Proprioception was also the source of the egocentric mind. How did I do, Surge?"

"Close enough for now, Dutch. Except you keep missing one fundamental understanding."

"What's that?"

"How are multicellular organisms built? How do bacteria form colonies? What is the frequency glue that keeps them cooperating? How do they build relationships, bond, make friends, and build communities of the like-minded?"

"I see where this is going. It reminds me of the Christian trinity."

"Exactly. Two cells can join together but only if a force keeps them bonded. Think of the figure and the ground again. The background is amorphous, but then forms emerge from this formlessness. After a while, forms fade away, back into the ground. There must be a force that can create and dissolve forms. This force is also responsible for cell division, for organism maturation, and it is the force that enables communication between cells. Proprioception doesn't work unless cells can communicate *and cooperate*."

"What is the force, Surge? What do we know?"

"The mystics call the force *love*. Physicists call the force *causality*. Christians call it *grace*. Catholics call it the *Holy Ghost*. The Christian Holy Father is the background from which manifestation happens. In Christianity, the manifest figure is Jesus. The force that creates, that turns patterns from potential to reality, is called love, or grace. Saints and philosophers could see that the background also pulls its creations back into itself. Some looked on this reverse of manifestation, this death of a form, and proclaimed that it was the process of returning to the Father—going back to Heaven. For the form itself, for that which is dissolving, it is death, a kind of entering the hell of non-existence."

"What do the physicists and the hard-nosed skeptics call this cellular communication, this life force?"

"They call it cellular communication and the life force."

"Very funny, Surge."

"What's so funny about the truth? By the way, the idea that a Holy Father is the background essence is patriarchal madness—only Holy *Mothers* give birth. If you have to anthropomorphize God, it is best that She be female, given her birthing role—just saying."

"Okay, Surge. Since we have wandered over and under and all around the topic, let me ask about esotericism. Can this communication-fractal create actual Holy Ghosts? If quantum effects manifest at a cellular level, if quantum tunneling and spooky influence at a distance are normal cellular processes, if the ability to be in multiple locations at once is part of cellular routines, if faster-than-light signaling is possible, and if fractals build ever more of themselves, well then, how has this manifested at our scale of existence? Can I receive signals from spooky invisible guys at a distance? Can they communicate through tunneling into my memory banks and deposit thought bombs and insights? Can my soul leave my body and visit the great Probability Mother? Can I send love to my wife who died in 2007? Can she send love back to me? Can my father, who died in 2005, still father me, and can I finally say *I love you dad; thanks for loving me when I was a kid*?"

"Times up, Dutch. Thanks for a thoughtful meal, and please come again."

"Where are you going, Surge?"

"We're closed. Did you think the Third-Eye-Watching Vegan Restaurant stayed open on the weekends? Wow, are you naive. Lights out, Dutch. Sweet dreams."

Dr. L. M. Archer's Heart of Consciousness

One of the most fascinating perspectives on consciousness, and the place where I found the most satisfying definition for consciousness, comes from an article in the *Journal of Medical and Dental Science Research*, a South African publication. In an article called "The Heart of Consciousness," Dr. L. M. Archer defines consciousness as follows (italics mine):

I would define consciousness in the biological context, as the individual lifeform's ability to create a mental picture of its self in relationship to its surroundings, both external and internal. *It is important to appreciate this is a perception of reality and not reality.* One immediately can understand the enormous evolutionary advantage, if these perceptions are close enough to reality, to enable the lifeform to determine the environmental situation it finds itself and behave appropriately to ensure survival. - *"The Heart of Consciousness," Journal of Medical and Dental Science Research, 2015.*

In other words, the ability to see your body as an entity existing within a specific domain has an enormous survival advantage. You *are* your space-time location. You are a mental construct of a "self" in a place. The brain's job is to represent the surrounding domain accurately enough, with you in the scene, to ensure survival. Using this definition, of course, confers consciousness on all creatures that move with a purpose.

Dr. Archer's goal is to provide a biological explanation for consciousness. He finds this explanation not only in our two brain hemispheres, but within the very nature of bilaterality. We are creatures with two mirror image halves. But evolution did a curious thing: it cross-wired the brain. The left side of the body is controlled by the right hemisphere, while the right side of the body is controlled by the left hemisphere. There are massively complex neural highways that connect the two hemispheres—the cross-talk is rich and constant.

Now Dr. Archer makes a rather profound, and a bit startling, observation. He says that consciousness can *only* occur in duality; *there is an illusion of singularity that arises based on having two mirror image sides.* Nature cross-wired itself into a conundrum: it had no choice, given the way the nervous system is designed, except to create two mirror image minds that became "an illusion of singularity:"

So consciousness can only occur in duality, as an illusion of singularity. The closest everyday experience to help comprehend this concept is standing in front of a mirror. A neurological mirror of two symmetrical beings fused together is required to create this illusion and creation of consciousness. This is what conscious beings are! ~ *"The Heart of Consciousness," Journal of Medical and Dental Science Research, 2015.*

Basically, what Dr. Archer is saying is that two entities *are required* to give us the notion of singularity. *The images validate each other* because "they can see each other" in a neurological mirror. To be conscious implies that something is watching, something is looking in the mirror at its twin, saying "ah, there you are! You are indeed real!" Except that both images are looking at each other from their own perspective, and each is validating the other. *Two unreal, brain-made images are each validating the reality of the other:*

The CNS [central nervous system] is a neurologically structured mirror of two systems facing each other and passing information back and forth, not only internally, but externally as well, through sound, visual light and touch. The understanding of humans as two rather than one entity is not difficult in modern biology. The CNS is two halves joined together, the right and left side of the peripheral nerves and cranial nerves entering each halve. If it was not for the crossover of nerve fibers connecting the two sides, we would be two separate beings. These two individual halves, I believe, would lack the ability to be conscious. I would go as far as saying, if you could prevent any information crossover of the CNS, there would be no consciousness ~ *"The Heart of Consciousness," Journal of Medical and Dental Science Research, 2015.*

Dr. Archer speculates that there is a Center of Perception of Consciousness (COPOC), a location in the brain where consciousness is regulated. He finds this in the brain region called the thalamus, which not only receives input from all the external senses, but also sends and receives nerve fibers to the cerebral cortex. Remember that there is a left-brain thalamus and a right-brain thalamus. Information from each of these centers is relayed to the cortex. The cortex is where the two mirror images are fused into the sense of unity.

Consciousness and Proprioception

Selecting the thalamus as the location for the center of perception is a logical choice. The right and left thalamic regions are located directly between the eyes, in the center of the brain. There is a very rich communication system between the thalamus and the cortex—signals go back and forth. Meanwhile, *all the senses,* internal and external, feed into the thalamus—it is a grand central station for receiving, reordering, and transmitting information from the whole body, that is then sent to the cortex for determination of relevance.

After a stroke, some people start to ignore half of their world. For example, a patient may neglect to eat half the food on a plate, despite still being hungry. Or the patient may draw a clock face showing only the numbers 12 to 6, leaving the other side of the clock blank. This is called hemi-spatial neglect; it happens when damage to one brain hemisphere causes a person to lose awareness of one side of the space around them. The person with hemispherical neglect can no longer process information received from that side of the body opposite the brain injury.

That our bilaterality might be the foundation for our dual consciousness is a fascinating idea that deserves further exploration. To my knowledge, there has been little attention paid to Dr. Archer's theory. However, his concepts fit with those of Zoltan Tory, Jill Taylor, Colin Wilson, Gary Lachman, and many others, who see anatomical bilaterality as a biological explanation for our cognitive duality.

Before I leave this discussion, I want to point out the importance of "validation." This looks like another fractal, a set of repeating patterns—as above so below. Could the egocentric mind validate the allocentric mind, and vice versa? Both minds may ultimately be illusions, but together, mirror-to mirror—they provide the sense we call "reality." This would make the concept of "validity" very powerful, especially when we communicate with each other. When we pay attention to another human being, or to another sentient creature, we validate their reality for them. They also, through their attention, validate our own sense of being real, of having validity. Isolation is terrifying because it erodes are very sense of being real, of having validity. We might be actually "killing" another person's soul by ignoring them.

Validation may be a cosmic principle. The background and the foreground would validate each other's existence. A God would validate his creation, just as creation validates God. Or perhaps, because our duality, our bilateral nature, is built into our body structure, what we perceive as real must always contain this validation principle. The reality that we create—be it foreground versus background, God versus creation, or allocentric versus egocentric—may be simply the result of the way we were constructed by nature through the process of evolution.

From a psychological point of view, the importance of validation, respecting and honoring the voice of others, cannot be overlooked. To "see" another person, is to breathe life into them. We validate each other's reality. We validate each other's worth and beauty. As they react positively to our gift of validation, they send waves of appreciation and love back toward us and thus validate and honor our sense of reality.

The Theory of Neuronal Group Selection

The idea that our cognitive duality comes from hemispheric differentiation was expressed by popular authors like Gary Lachman, Colin Wilson, and Leonard Shlain. Likewise, scientists Jill Bolt Taylor, Zoltan Tory, and L. M. Archer also used hemispheric differentiation to explain their perspectives. These authors are joined by many others who used hemispheric differentiation as a foundation for exploring human duality. However, there were major problems with this perspective. Indeed, research eventually invalidated this line of thinking, and scientists were forced to look elsewhere for an understanding of our duality.

Although we can divide the brain into left and right halves, or top and bottom halves, or into modules, scientists now understand the brain to be a totally interconnected meshwork that "fires" coherently to enable behaviors. A leading proponent of this perspective was neuroscientist Gerald Edelman, a distinguished author, and Nobel Prize winner. Dr. Edelman saw the mind as arising from the cooperation of groups of neurons that discharge together in patterns. These neuronal groups fire in variable, transient patterns, and they evolve—they strengthen through use, or weaken through disuse. This "new" way to think about the brain and mind does not rely on left-right or top-bottom anatomy, nor does it support the idea of fixed modular organs in the brain. Edelman advocated for a Darwinian-based philosophy called the Theory of Neuronal Group Selection (TNGS).

The brain is an organ that is *constantly crafted by movement.* This makes perfect sense for a brain designed by evolution to enable navigation. As the brain grows developmentally, it constantly creates new neural connections and prunes others. Indeed, that is what the brain does from birth to death: *it is always moving.* As Dr. Edelman repeatedly pointed out, this is not a description of a digital computer, where circuits are fixed and the overall motherboard is forever-the-same. The brain is alive, energetic, and on the move. Neurons are constantly responding to ever-changing stimulation.

According to Dr. Edelman, the brain is also said to be self-organizing and plastic—which just means that it can change or adjust to circumstances. Furthermore, the brain could not learn, nor could it add memories, if it was unable to "rearrange the furniture." Therefore, no two brains, no two minds, and no two kinds of consciousness can ever be exactly the same. Furthermore, human beings are made of moving parts; "we" are never exactly the same moment-to-moment.

Dr. Edelman calls his perspective Global Brain Theory. He postulates that assemblies of neurons fire in patterns that eventually correspond to levels of consciousness. He adds the additional fascinating speculation that the development and adaptation of these neuronal cell assemblies follow the Darwinian Theory of natural selection. It is as if cell assemblies are creatures themselves that undergo selective pressures and so evolve to enable the survival of the organism. He calls this perspective *neuronal group selection*, a kind of in-the-skull battle of the survival of the fittest.

Dr. Edelman's theory starts with an understanding that groups of neurons fire in patterns. These patterns can be mapped—as they are firing—using imaging technology, so they are called neural brain maps. For example, in the brain there is an area for processing color. There are millions of neuronal connections in this region of the brain which fire together when presented with input arising in the retina. This region will not respond to black and white images, nor will it fire in response to sounds. The color processing area in the occipital lobe has a specific function related only to color perception. However, nerve fibers connect the color "center" to other areas of vision—to at least 30 other visual brain maps, as well as numerous motor and sensory maps. These other areas are informed that the color center is firing and they feedback their status to the color region using massively parallel and reciprocal connections. Dr. Edelman has a name for "brain maps communicating with brain maps;" he calls such a system "reentrant." He means that brain *regions* interact; they cooperate and synchronize. Maps are connected to maps in such a way that the whole brain is informed of activity as it happens.

A helpful analogy is to think of brain regions, like the color center, as storm clouds of neurons. It is hard to find the edges of a storm cloud, and no two clouds are ever the same even though they are located in the same region of the sky. The activity inside the storm cloud is variable; lightning flashes seem almost random. Neuron groups (brain maps) are like that; they have no clear boundaries and every image of a neuron cloud firing is unique. In neuron assemblies, the "lightning" is analogous to discharges that connect the storm cloud with other storm clouds of neurons. These other clouds can be nearby or anywhere in the brain. The clouds that are "hit by lightning" send an answering lightning strike back to the original neuron cloud. The signaling that goes back and forth between neuronal clouds is called *reentrant* communication.

Notice that although the cloud is never the same twice, it is still a cloud, and it is still relatively regional. The neuron cloud is never the same twice when we snap a picture of it in action, and yet it gives the same functional result, for example, a color is perceived. The name for this "consistently reliable expression" is called "degeneracy," which I find to be an unfortunate term because there is nothing "degenerate" about the process.

Consciousness and Proprioception

Dr. Edelman postulates two kinds of perception which support dual-process theory. Primary consciousness, according to Edelman, uses a mechanism for perceiving in the moment, while higher-order processing deals with conceptual categorization. Both primary and high-level consciousness systems have their own specific maps. Using my terminology, allocentric maps are created for primary processing, and egocentric maps are created for higher-order processing.

Before the cortex evolved and before complex neuronal groups were organized at a cortical level, primitive brains were driven by survival needs. Organisms need food, companions, sex, and a supportive environment. Dr. Edelman refers to a human "value system" which attempts to satisfy basic needs. Human brains are not like computers, according to Edelman, because human beings are primarily ruled by these value systems, which are often in conflict with cortical processing. A duality is created between emotion and logic as our two minds try to maintain balance. This is another perspective on allocentric processing, which is non-verbal and closely linked to the limbic brain, and egocentric processing, which is verbal and closely linked to cortical processing.

Reentrant communication, degeneracy, value systems, and the Darwinian selection of specific, organism-like populations of neurons are key concepts in Dr. Edelman's theory.

Dr. Edelman studied Darwin's work and he understood and appreciated the great man's complexity. Here is a quote from Edelman's book *Bright Air, Brilliant Fire; On the Matter of the Mind* (1993), in which he quotes Darwin:

> "The possibility of two separate trains going on in the mind as in double consciousness may really explain what habit is." *~ Bright Air, Brilliant Fire; On the Matter of the Mind, Gerald Edelman, 1993.*

I found this quote to be delightful since it shows that Darwin was musing about the possibility that the mind had two "trains" running through it and that from these emerged two kinds of consciousness. I suspect that Darwin was aware that automatic habits came from a primary consciousness (the allocentric mind), while intentionality and novelty came from a higher consciousness (the egocentric mind).

The key question for dual-process theory is how neuronal cell assemblies relate to duality, especially to the proposition that only two general kinds of processing are happening in the brain.

One way that Dr. Edelman's ideas relate to dual-process theory is his conviction that philosophy, psychology, and cognitive science have all failed to find a reason or process for consciousness because they have failed to take into consideration evolution, especially Darwin's ideas. Dual-process theory, as I propose, is based on the evolution of navigation (purposeful movement), and so is based on natural selection.

An interesting idea here is that *super brain maps* can be created from smaller brain maps, or we might say that we can have memories of memories—*layers of cognition* can be created. Simple maps can be combined to make more sophisticated maps and these can be combined to make super-sophisticated brain maps, and so on. I would contend, in keeping with dual-process theory, that we have two kinds of super-maps, one map that is temporal and egocentric, and a second map that is spatial and allocentric.

Proprioception defined broadly as the combination of all the internal senses, is a key player in the idea that brain maps can be layered. Once the brain has an algorithm for a successful mapping strategy, it uses that same strategy to make more sophisticated internal systems. This is like internal Darwinian evolution, but on a much faster scale.

Dr. Edelman defines primary consciousness as an ability that evolved for the creation of scenes (gestalts). This is exactly what I have found; the allocentric system is primarily visual. It is a background against which objectivity can emerge. Here are Edelman's words:

It is the evolutionary development of the ability to create a scene that led to the emergence of primary consciousness. - *Bright Air, Brilliant Fire; On the Matter of the Mind, Gerald Edelman, 1993.*

For primary consciousness to work, it had to evolve its own kind of memory system (episodic memory)—a memory for experience. Episodic memory enabled humans to have what Edelman called a "remembered present." We are aware in the moment.

Before I move on, let me summarize where we are on this dual-process journey. Notice that there is very little debate about duality. That we have two minds is almost a starting point. However, authors differ on why this might be the case. Why do we have both a self and an ego? Why do we have an empathetic, allocentric mind, as well as a rational, egocentric mind? Most modern theories see this conundrum and start their speculations from that assumption.

Many authors found the results of split-brain research and concluded that there must be two separate minds, one in the left hemisphere, and one in the right hemisphere. When that perspective was discredited, neuroscientists like Gerald Edelman proposed that the brain was a dynamic pattern-generating system that operated in a primary (allocentric) mode and a higher-order (egocentric) mode. Dual-process theory is another way to consider how our two minds work. It incorporates elements from theories that came before, but at the same time rejects any approach that does not embrace duality.

In the next chapter I will more precisely define and differentiate the characteristics of our dual consciousness. However, before I do that I want to give one more example of applied dual-process theory. If we go through our existence as dual creatures, shouldn't we also die in two different ways? That is the question that author Peter Novak addressed in his book *The Division of Consciousness; The Secret Afterlife of the Human Psyche*, 1997.

Peter Novak's Death and Duality

Author Peter Novak lost his wife shortly after the birth of their daughter. This tragedy left him with an overwhelming need to understand death. I also lost my wife when our three kids were young adults. Like Peter, I came to know, deep in my essence, that the death of your mate is a shadow that cannot be emotionally escaped.

When someone you love dies, you want death not to be final. You dream of the person who is now nowhere to be found in this world. A part of you has also died, and you want to be whole again. You pray, even if you have never prayed before. You talk to God even if you don't believe in God. You become angry at materialism for being cold and logical, and for telling you (smugly) that this death is like a fly smashed with a flyswatter—death is meaningless, heartless, and final. If you are fortunate, friends rally around and make you food, do chores for you; they stop over just to be near you and to give you hugs.

Then one day when you are deep into your personal grief, you look into the eyes of your child and see the pain there. You then know that your own grief has to be set aside because the children are deeply hurt and confused—they need you to stop being absorbed in your own grief.

Peter Novak needed some way to cope that gave each day some purpose. He needed some reason (besides the all-important love for his child) to keep going. So Peter Novak went on a quest to understand death. He wrote a book to share his insights called *The Division of Consciousness; The Secret Afterlife of the Human Psyche.*

I took a similar path to Peter's; I picked a subject and went after it with passion. Instead of the mystery of death, I went after the conundrum of consciousness. I went on a quest to explore the mystery of living. Along the way, I also began to write books. I know, as does Peter, that writing is a catharsis, a diversion. I am grateful to Peter Novak for his adventure and his insights. Often, the heartfelt journey of another person helps us cope with our own struggles.

For ten years, Peter studied how human cultures explain death. What he eventually saw was a stark, perplexing duality. In Eastern cultures, it is believed that the dead come back after dying; they are reincarnated. However, they return with no memory of who they had been in previous lives. They come back with a different personality, a different body, sometimes as a different creature. Past-life regression, through hypnosis, seems to support the idea of reincarnation. Some very young children (in diverse cultures) report a previous life. Occasionally, monks and gurus also report past-life memories.

In contrast, Western cultures believe that the dead do not come back. They go instead to heaven or to hell. Once in heaven, they get to keep their personalities, and they are issued new bodies that don't age (but look and feel like the old bio-suit). Near-death experiences seem to support this view: when we die, we race down a tunnel of light to meet our "still-living" loved ones. They are whole and happy. They have been hanging out with old friends, and with God, in heaven (or if they don't show up in the tunnel, it's because they are in hell for eternity). Your dead relatives did not reincarnate—they are having tea every morning for eternity with the breakfast saints. Here is how Novak puts this dilemma:

> Half the world believes in the Judeo-Christian-Muslim heaven-or-hell scenario, an afterlife containing a judgement followed by an eternal reward or punishment, while the other half believes in the Hindu-Buddhist-Taoist reincarnation scenario, in which people are continually reborn into new bodies, forever forgetting their past lives and identities. Instead of one of these traditions ultimately proving more valid than the other, they have remained stubbornly locked in a stagnant debate for at least the last several thousand years . . . ~ The Division of Consciousness; The Secret Afterlife of the Human Psyche, Peter Novak, 1997.

These two views are diametrically opposite. Logically, either one belief is correct, or the other, but both cannot be correct. Peter Novak looked at this dilemma, mused about it, studied his dreams, looked at the scientific literature, and then had an "aha!" moment: these two long-established religious views are not mutually exclusive. They are both correct. The secret lies in our cognitive duality.

When Peter wrote his book, mental duality was explained by the split between left-brain and right-brain, so that was his starting point. Like

Consciousness: A New Slant on an Old Conundrum

Zoltan Torey, Jill Taylor, and L.M. Archer, Novak took split-brain research as fact, and he went from there to build a theory. Peter, like so many others, has grasped the importance of our inherent duality, and he has applied his insights to cast light on the conundrum of life and death. He calls his insight *Division Theory*.

One of the fascinating ideas within Division Theory is that the concepts of "soul" and "spirit" correspond historically with our two minds. In my own terminology, this would translate to mean that the soul is the allocentric mind, and the spirit is the egocentric mind. Here is how Novak differentiates the two:

> Although the two terms tend to be equated today, "soul" is not properly the same as "spirit." The blurring of distinctions between these two is, historically speaking, something of a recent development; it used to be understood that they referred to two completely separate substances. In the ancient Hebrew of the Old Testament, just as in modern English, there was one term for "soul" and a different one for "spirit." The Old Testament consistently distinguishes the two; they were referred to as having completely different attributes. Souls were regularly referred to as "feeling" [experiencing] this or that; a spirit, however, was always "doing" or "thinking," but never "feeling" anything. The soul was thought capable of dying or experiencing death, but the spirit was never referred to as having died. After death, the spirit was said to "return to God," while the soul generally wound up in She'ol, the Jewish version of hell. ~ *The Division of Consciousness; The Secret Afterlife of the Human Psyche, Peter Novak, 1997.*

The premise of Division Theory is that, at death, our two minds "suffer" a different fate—they "die" differently. The soul's death, according to Novak's logic, is the explanation for heaven and hell. The death of the spirit, he tells us, explains reincarnation. When this viewpoint is fully fleshed out, according to Novak, thousands of years of confusion suddenly become very clear:

> Any truly satisfactory answer would need to resolve the ancient debate between our two primary traditions, the heaven-hell scenario

of the West and the reincarnation scenario of the East. It might also be expected to account for the contradictory findings coming in from today's near-death and hypnotic regression research. It would need to explain the worldwide reports of ghost phenomenon, the bizarre death-religion of ancient Egypt, and, in fact, the majority of humanity's beliefs about death around the world and throughout time. Perhaps it would even shed light on that unique event deemed responsible for launching the Christian era, history's only report-ed occurrence of a man rising from the dead. ~ *The Division of Consciousness; The Secret Afterlife of the Human Psyche, Peter Novak, 1997.*

Religions speak of the body, the soul, and the spirit.[2] The physical body clearly does not survive death, but what if, Novak postulates, the soul and the spirit actually do survive the loss of the body (the vehicle these two minds inhabited during a lifespan)? What would death be like for the soul? What would death be like for the spirit?

Novak uses the unconscious (soul), and the conscious (spirit), to make his case. He has read a lot of psychoanalytical perspectives and is comfortable equating the unconscious mind with the soul and the conscious mind with the spirit. So, what happens to the allocentric mind (the soul, the uncon-scious) after death? What does it "look like" or "feel like" when this mind can no longer perceive using a body as a vehicle for having experiences? The senses are gone, both internal and external—there is no vision, no hearing, and no proprioception. The flow of reality stops. The background mind—which the body used for navigation—fades into the void. There is no earth to walk upon, no domain that this mind was designed to exist within. Being separate from its egocentric twin, there is no-where and no-way to manifest forms or extract relevance—existence after death, for the allocentric mind, is all potential and probability; there is no substance, no hope of physical objectivity.

Likewise, what happens to the egocentric mind when the very concept of form, pattern, ego-and-other, and language disappear? What happens when the background disappears and there is no canvas upon which to manifest? How does an egocentric mind function when there is no center of perception and no surrounding atmosphere to extract meaning from? The egocentric mind has lost connection with the background. Nothing exists as a background, so no physical reality can arise. There is also no one to talk with since language is gone and other sentient creatures have vanished.

According to Novak, the soul has stored memories of experiences—all the experiences of a lifetime. These survive death and, therefore, the allocentric mind is trapped in a dream-like nightmare that repeats and repeats with no chance to gain new experiences. If there are a lot of sweet, loving, charitable memories, then the soul has arrived in a kind of heaven where good memories run in a never-ending loop. If you have been a bad dude on earth, and have stored lots of violent, cruel, selfish memories, then that is the hell you made, brother—expect to watch your cruelness replay for eternity (according to this theory).

However, the egocentric mind (the conscious mind, the ego, the spirit) would float free, but with amnesia. All the memories of a lifetime were stored in the allocentric mind (says Novak), so the ego is cut off from memories—including the memory that held the personality and ego. The dead egocentric mind has no identity, no knowledge, and no bearings. It is doomed to wander aimlessly, like the Scarecrow in the Wizard of Oz, in search of a brain. Reincarnation, however, saves the lost ego and it gets to rediscover the joy of being in a new body back on earth.

Unfortunately, my thinking comes to the exact opposite conclusion to what Novak has so eloquently expressed. If the two minds do, indeed, somehow survive death, I see the soul as reincarnating and the ego as heaven-bound. Also, they would both have their own memory.

Skeptics, materialists, and atheists, of course, find this whole discussion absurd. Western religions, especially those that reject the esoteric wisdom traditions, remain hostile to reincarnation. Eastern religions, meanwhile, cling to their belief in reincarnation. Still, Peter Novak's revelation that our cognitive duality logically suggests two kinds of death is satisfying, especially for dual-process theory enthusiasts like myself.

There is one more perspective that I need to share before I leave this discussion. The philosophy of Biocentrism (Lanza, 2016), holds that consciousness is pervasive and fundamental in the universe. The consciousness that human beings know is just a manifestation from an original background consciousness. This primal background consciousness contains potential information, memory, and contains a force which can result in the manifestation of information. Therefore, our consciousness is a figure in the ground of eternal consciousness. As such, human consciousness can never be separate from the universal consciousness—sentient consciousness is just a bulging out from the mother consciousness. Therefore, our separate existence is an illusion, and so is death. There is no birth and death.

A materialist could use the same kind of rationale. Instead of talking about souls and spirits, a materialist would simply say that almost the whole of existence is empty. There are a few rare entities, call them atoms or quarks. There is a manifestation of patterns from this field of emptiness, quarks, and energy—a bulging outward of the background field that gives the illusion of separateness. Death is a dissolving back into the universal field.

I will leave the discussion there. In my heart, I feel that neither reincarnation, nor heaven and hell, are valid—they both sound rather dreadful, and not at all anything to look forward to.[3] I hope there is something beyond Eastern and Western "habits of mind," something that transcends both soul and spirit, empty fields and quarks, something not spatial and not temporal, something that our inherent blindness is not allowing us to perceive. Therefore, I hope there is "somewhere" all sentient creatures go after death; a "place" where we find peace, love, joy, and clarity. If not, allow me to say thank you. What an amazing journey; what a miracle is this life. My ego has no regrets if it all ends here; and my soul has always accepted its role and fate.

Notes

(1) **Theory of mind** is a specific psychology concept. We know that we have a mind because we have sensations, thoughts, emotions, and beliefs—we call that process "mind." Because of our ability to communicate, we also know that other people have minds that are similar to ours, and yet somehow quite different. We assume that other people have minds even though there is no way *to be* them, so there is no way to be sure that their minds are the same as our own. We are only able to postulate a theory of mind. I often use the term, as I do in this book, in a more general sense to mean the "the study of how minds work."

(2) **Religions speak of the body, the soul, and the spirit**, a trinity. This fits nicely with my dual-process theory. The soul is the allocentric mind. The spirit is the egocentric mind. The body is the vessel that contains the two minds. There is another way to view the trinity as spirit, soul, and energy—the "Holy Ghost" in Catholicism.

(3) **Personally, I don't know if there is an afterlife.** It seems that I am in a purgatory of sorts, between belief and faith. I have no comforting words to share. We have two minds so it makes sense that they would each suggest a different scenario for death.

I have come to see that eternity and infinity are just concepts that result from having two minds. There is no way to know if these absolutes are real, or if they are perceptual masks that blind us to a greater reality. I will discuss eternity and infinity later in the book.

Four

The host who has no ego is here.
We look into each other's eyes.

~ "Dark Sweetness," *A Year with Rumi*, 2006.

The Host Who Has No Ego is Here

"What do you have for take-out, Surge? I'm in a hurry."

"No problem, Dutch. We have a cold and impersonal soup especially made for people in too much of a hurry to taste their food. It isn't quite done enough for us to care whether it tastes good or not. We use our short-order cook, a dull guy who is always in a hurry and preoccupied. We call it our "So What, I'm Busy, Soup."

"It's tasteless?"

"You won't notice."

"Can you add some salt?"

"We can add too much salt, or too little salt."

"How about just-right salt?"

"No. That would be against the laws of short-order cooking. By the way, what chapter are you working on in your book?"

"It's the chapter where I define duality and consciousness."

"Oh. That changes everything. I have just the meal for you."

"Can you make it quick?"

"Sure. We served it a few moments ago. You owe us thirty-eight dollars."

"I didn't get a meal, Surge, and you know it. What are you talking about?"

"Your consciousness has yet to register the meal. Given the speed of human cognition, you might not realize that you are full for another couple hours. That's not our fault."

"Oh, no, you don't, Surge. This is vegan insanity."

"Yes, of course, thank you. We will put it on your bill. And thanks for stopping by The Third-Eye-Watching Vegan Restaurant and Center for GPS direct-to-stomach-meal delivery."

A Strange Creature

Stars go out to graze in the night sky pasture
in the same way that animals love the ground.

~ "A Frog Deep in the Presence,"
A Year with Rumi, 2006.

The environment was mute until it gave birth to a strange creature, one that could reflect on its own body and mind, a creature that could look at the surroundings into which it had been born and could move about that environment and explore it with intent and curiosity. This ability to look back at nature, to witness a body separate from its domain, was the birth of egocentric consciousness, and from its inception there was a division between the seer and the seen.

When egocentric consciousness appeared on the evolutionary stage, a basic duality arose. Long moments of not being awake to "self-and-domain" alternated with fleeting moments of "self-remembering," as the philosopher and author P. D. Ouspensky called this glimpse of wakefulness.

Then, after years of staring in wonder at the origin of itself, trying to deal with the angst and the mystery of being awake in what seemed an unaware universe, another strange creature appeared. The ego sensed another mental "being" living inside the same body.

The mirror "image" looking back at us isn't an image; it is another mind, another "us." That is what dual-process theory asserts: two minds are staring at each other inside the same body—ironically, they validate each other's existence.

As I stressed and here repeat, the twins are mutually exclusive; they oscillate, they take turns, but they are never together on stage at the same time. However, the egocentric mind has evolved the concept of a twin, and so it can operate from that theory—with never a true "knowing". The allocentric mind is purely experiential; it is the background whole-body mind, and so "having a twin" is accepted unconsciously—there is no need to debate. I will go ahead and personify each twin for the sake of discussion, but keep in mind that they never manifest simultaneously.

Pretending to be Alone

I was walking home one evening after a meditation session at a neighborhood Zen center. It was summer and the sun was still high in a clear blue sky. The temperature was a perfect 70 degrees. I felt centered and at peace.

Suddenly, I had the urge to throw my ego into the sun. I stopped, reached inside my body, and I impulsively hurled my ego into the solar furnace. It was a moment of experimentation, an evoked experience. The feeling that followed, of emptiness, didn't last long, and I knew that my ego had only taken an imaginary trip to the sun. However, for a moment I felt that Doug Baldwin had actually been cremated and no longer existed. I imagined the end of my ego's self-effacing nature, the end of my systems-obsessed personality, the end of my many identities—as a retired teacher, writer, father, brother, son, widower—all my history, including all my adventure stores, were now gone. There was no past and no future without the ego. This imaginary experience was short-lived, yet the memory of that event has stayed with me.

I tell this story to illustrate how we pretend to be alone, as personalities with a history and a future. Yet when, in a flash, that personality evaporates, a silent entity still remains. Who is this empty creature with no voice? Who is this nameless, timeless "nothing?" This emptiness, this twin that shares "our" body is the background from which the ego manifests. This empty self is hidden by the verbal, in-your-face, old story teller called the ego. What is left after the egocentric mind dies is the allocentric mind. The allocentric mind is the self (a part of a universal soul) out of which the ego and all its concepts and artifacts manifest.

After a long period of evolution, after language appeared, egocentric consciousness "sensed" this allocentric sister. The egocentric mind could logically comprehend the existence of a twin, even without the ability to directly perceive the twin. Perhaps, the ego speculated, Mother Nature had, indeed, given birth to twins.

This awareness, finding a lost twin, should have been a joyous moment. However, the twins soon discovered how irritatingly different they were from each other. For a while, it was convenient to ignore the evidence of the contrary sibling; each mind managed the feat of "pretending to be alone" for centuries. But nature kept evolving. A time came—the time we live in now—when the twins could no longer pretend they were alone in existence.

They had to turn and face each other with honesty and courage. But more than that, the twins—in our dangerous age—now need to become best friends, and they need to provide a support system for each other. Together they can make the world a better place. Separately, their lack of communication and cooperation is preventing that better world from materializing.

We have two minds and two kinds of consciousness. If we don't start with that perspective, we fail to understand the pervasiveness of cultural duality, and we fail to comprehend how we work. We also fail to understanding why we can't agree on a definition of consciousness—because there are two of them, and each has a complex evolutionary history. Furthermore, each kind of consciousness has a strict job to do that is opposite to the purpose and intention of its twin.

Did a Higher Intelligence Make All This?

Consciousness stepped out of the order of nature and faced it.

~ *On the Problem of Empathy*, Edith Stein, 1964.

Debates between spiritual spokesmen, like Deepak Chopra, and rational philosophers, like Daniel Dennet, come down to this basic disagreement: Did our reality arise from a universal consciousness, call it "God" or whatever you want, or is our reality, our consciousness, a by-product of the evolution of matter? In other words, did we arise from spiritual vapors or did we arise from mindless matter? Or are they the same thing, like water as gas or water as ice? The spiritual perspective postulates benevolence and intelligence behind our reality, while the scientific perspective assumes that the harsh brutality of the fiery universe gave birth to a savage competition within nature, out of which fragile benevolence and flimsy intelligence are trying to emerge.

I think perhaps this debate is largely between the allocentric and egocentric twins stating their positions. The answer to the ultimate question is currently unknowable. Theoretical physicist Lawrence Krauss said in his book *A Universe from Nothing* (2102)*:* "What is the difference between arguing in favor of an eternally existing creator versus an eternally existing universe without one?" In a similar statement, in his book *The Taboo of Subjectivity* (2000), Alan Wallace, an expert on Tibetan Buddhism, says that ". . . there is little to distinguish religious ignorance from scientific ignorance." Therefore, the mystery will remain a mystery—and that is a good situation, because we need to be on a quest, to ask questions, to go on exploring. Some deep part of us refuses to arrive at ultimate conclusions.

The justification for either position, religious or spiritual, comes primarily from systems-oriented males who enjoy creating and debating orthodoxies—few women have historically participated in the debates. Women, of course, could be more involved in the debates, and brilliant modern writers like Susan Sontag and Simone de Beauvoir have weighed in on the big questions. But overall, women tend to be spiritual writers with a keen eye on relationships; they are not so much interested in esoteric debates. Females tend to comprehend the allocentric mind and be advocates for empathy and relationships. The reality, of course, is a continuum, with individuals falling somewhere along the bell curve.

I am well aware of the caution needed when weighing in on gender issues. I am talking more about yin/yang, masculine/feminine, than I am about sexual orientation. There is much more to say about the difference between the genders—and we are not served by ignoring the evidence—which I will pick up at a later time. But I do find a comment made by philosopher Jiddu Krishnamurti to be a greater truth. Krishnamurti said that all divisions come from the ego, including the concept of "male" and "female." These egocentric categorical distinctions detract from the greater understanding and appreciation of our common humanity.

Despite our inability to reason out whether creation came from Godly intelligence or Godless matter, both sides enjoy the debate and so it goes on; each side has evidence for their position, and each is paradoxically right— one side arguing that the allocentric mind is correct, and the other side stating with equal passion that the egocentric mind is correct. Each side is actually insisting that their mode of thinking, or their way of paying attention, is the correct avenue for eventually arriving at the "right" answer. The egocentric mind uses its only tool reductionism (deductive reasoning, dividing the whole into parts) to arrive at "truth," while the allocentric mind uses its only tool, holism (inductive reasoning, the ability to merge parts into wholes) to arrive at "truth."

As I tried to understand the philosophers and psychologists, each with their separate theory of consciousness, I was often bewildered and irritated at the abstractions, at the ever new vocabulary for the same concepts, and at the use of familiar words in totally strange ways. I finally realized that if I understood the underlying *initial assumption* of the "great thinker," I could at least start with an outline for comprehending. Struggling with Rudolf Steiner, Owen Barfield, Deepak Chopra, Daniel Dennet, and Carl Jung—to give five examples—was a lot easier when I realized which side of the God-fence these guys were on.

If you are on the God-Yes (the universe is inherently intelligent and ultimately compassionate) side of the equation, then you can believe in absolutes. If one absolute is true (God), then that absolute could logically give birth to other absolutes. A universal structure could be built that was true and dependable, ultimately based on intelligence and on a post-God logic that the world is basically intelligent. When you are dying there is a light at the end of a tunnel, which you pass through on your way to meet God.

If, however, you are on the God-No side of the equation, then the notion of absolutes is always questionable, never dependable, and always reducible. Everything is nested and sub-divided in the God-No worldview. A favorite "joke" of the rationalists is that when you are dying, if you see a light at the end of a tunnel, it is probably a train and you need to take evasive action if you want to go on living. That, at least, would be a logical reaction.

Two of my favorite writers were on the God-Yes side of the fence. Owen Barfield and Rudolf Steiner are brilliant Christians who started with the assumption that the universe is an intelligent, compassionate, God-ordained creation. But Barfield and Steiner were also philosophers; they did not shy away from abstractions and complexity. Deepak Chopra is also on the God-Yes side of the fence. His perspective, however, is not set around an anthropomorphic God, but is built on the assumption that consciousness is inherent within the background of existence. Chopra writes with great eloquence and clarity. Daniel Dennet, who at times has debated with Chopra in public settings, is a brilliant modern philosopher who is solidly in the No-God camp. He is proudly a part of the "Four Horsemen of the New Atheism," alongside Richard Dawkins, Sam Harris, and Christopher Hitchens. These thinkers reject the idea that a Christian God or Hindu cosmic intelligence preceded the material world that we perceive.

Psychoanalyst Carl Jung, another brilliant mind, straddled the fence between God-Yes and God-No. His internal voice struggled to comprehend the paradox of having contrary minds. Jung was a rational doctor, using the practical tools of medicine, yet he nurtured patients who refused to fit the egocentric mind's understanding of how the world must work. Primarily, Jung struggled to understand synchronicity, which appears to be a kind of acausal intelligence—something that refuses to fit with egocentric reason. Jung also used quantum mechanics as a possible hard-wired platform for understanding non-local, acausal "reality." Somehow, Jung mused, the answer to our duality must begin within this quantum world.

Perhaps all the players in this debate find themselves from time-to-time confused, sometimes drifting towards God-No, other times drifting towards God-Yes. This makes sense because each of the debaters has two inherent minds in constant conflict. The philosopher Georg Hegel, like Jung, straddled the fence between God-Yes and God-No. Hegel, now that he is dead and can't defend himself, has been taken hostage by both camps who wish to use his ideas to support their own. Hegel is so abstract and hard to translate that both sides can interpret his "logic" to support their contrary claims.

Consciousness: What it is Not and what it Might Be

Hegel is lumped with the idealists (Plato, Aristotle, Plotinus, Leibniz, Spinoza, and Kant) who believe in the evolution of individual and cultural consciousness. Hegel saw a constant struggle going on inside minds and inside cultures. He saw dualities. After he died, two camps developed, each taking opposing sides—the egocentric mind versus its allocentric twin. And, of course, like all the philosophers and psychologists who study the mind, Hegel changed his views as he matured.

As far as I can tell, every God-Yes and God-No advocate dies before they finish formulating their theories. Death, of course, doesn't care which side of the fence you were on during your tenure on earth, or how elegant your theory. As the future speeds forward, new evidence and new thinking eventually make everyone's musings obsolete, no matter how brilliant they seemed in the era when they spoke the truth. But that's okay. Thinking is fun. We each add our unique voice to the discussion. Our egocentric mind is driven to explore and to figure things out, so the debate is lively and invigorating, and it will go on. "Great thinkers" will continue the debate about the ultimate beginning of our realty, age after age.

However, the God-Yes versus God-No debate is about origins, and the implications that arise if we accept one premise or the other. *The debate is not about whether anything began.* That we are here in this conundrum is as-sumed by both sides in the debate. Here we are. What is going on?

As we look at "what is going on," we find two sides to every question. We continue to debate about various issues with the same emotional ferocity that fuels the God-debate. We "rip each other's throats out" over whether it is best to be a liberal or a conservative, or whether reason trumps empa-thy, or empathy trumps reason. Whatever the issue under consideration, we seem compelled to "take a side," as if there were only two sides, one right, and the other wrong. Why is that?

The answer that I find, after exploring duality and consciousness, is that human beings are on a spiral path of development. We are far more *individually* complex than we realize. Our debates are less about finding truth and more about explaining—or unconsciously demonstrating—how individual minds work. We have dual-cognition, two minds, two kinds of consciousness, so we live with conflicting, mirror-image perspectives. The battle within each of us plays out in physical reality as we attempt to communicate with others.

We can see this spiral of development clearly when we look at aging and maturation. Life forms have a beginning, a life span, and an ending. The pathway from beginning to end has a predictable, unerring trajectory. No lifeform escapes this pathway from birth to death. So physically there is no debate: we are born, develop, and then we die.

But is our ability to think also developmental? Do we also follow a cognitive spiral pathway that unfolds parallel to maturation? Do we get ever wiser, ever more intelligent, and ever more emotionally evolved as we mature? I believe we do, and I believe that all human beings are inevitably located somewhere on the spiral path.

Another way to say this is that all human beings are evolving physically, emotionally, intellectually, and spiritually. The speed of development may vary, and where we are along the spiral of development is dependent on physical age in each of these areas, and on the intensity of life experiences, but we are all somewhere along the path. This realization can help us understand why we are so often in conflict.

We carry our pathway of development into the lecture hall where the debates take place. What we have to say, our position, our side of the fence, depends upon where we are along the spiral of mental development. When two individuals are far apart cognitively, emotionally, and spiritually, they will be in conflict—the farther apart, the more they will disagree and the hotter will be the emotions. How we understand abstractions like "God," or "Love," or "Justice," depend on where we are along the spiral of development. Seven people at a debate might very well have seven different definitions of God, Love, and Justice. Not much is accomplished if we fail to take the spiral of development into consideration.

Therefore, there are two reasons we are in this vicious debating cycle. One is our failure to understand the spiral of development—we do not pause to consider that we are each at a different place along the spiral—and second, we fail to see that we have dual cognition. We are conflicted within ourselves—not often realizing it—and we fail to see that others are likewise conflicted. Our two minds are moving along a spiral of development, side-by-side, without being able to comprehend each other.

I have discussed this spiral of development extensively in my book *The Confusion Caused by Being Your Own Twin*. I won't elaborate further in this book. My goal here is to show you how evolution caused two kinds of consciousness. There is a reason that we behold duality everywhere we turn.

I will conclude this chapter with a look at the characteristics of consciousness— I call the next section *21 Puzzle Pieces*. It is a review of what we know about consciousness at this stage in our understanding. It is based on the assumption that dual-process theory is valid.

Consciousness, 21 Puzzle Pieces:
What Consciousness is Not and what it Might Be

Consciousness is a word worn thin by a million tongues. Depending on the figure of speech chosen it is a state of being, a substance, a process, a place, an epiphenomenon, an emergent aspect of matter, or the only reality. ~ *Psychology: The Science of Mental Life, George Armitage Miller, 1962.*

I join a long and rapidly growing list of people who are exploring consciousness. There is no shortage of books, articles, podcasts, and online presentations about our cognition. It is easy just to repeat what others have said, or to reframe the problem in the language of a specific discipline. It is also very easy to become confused by contradictions and obscure terminology. My perspective is based on three decades of teaching "navigationally disabled" individuals. Therefore, much of my knowledge is drawn from experience, which later was gradually refined through a study of the literature. The seed ideas that eventually found their way into this book are based on over 50 years of synthesizing these ideas into a coherent theory.

I started with a supposition: what if it was actually true that we have two minds? As I followed the yellow brick road to see where this seed idea was going, I came to various conceptual landmarks, each of which had a message. I collected these concepts, mused over the messages, and continued on down the road to the next landmark.

After years of reflection and study, I came to the conclusion that our debate about consciousness is framed by assumptions. For example, if we begin with the premise that consciousness results from language, we can build a convincing case. If we begin with the assumption that consciousness is about degrees of wakefulness, we can perhaps "prove" that case. If we say that consciousness is an off-line neurological system confined within the body, we find ample supportive evidence. If we say that consciousness evolved to enable socialization and communication, we can prove, more or less, that case. If we assume that "knowing we know" is the definition of consciousness, then we might also prove this supposition. However, all these assumptions, even when they contain substantial insights, will not be complete unless dual-process theory is understood and incorporated into the gestalt.

The following pages contain a summary of assumptions and conclusions regarding consciousness that I recorded on my journey. These conclusions are controversial; they are also filtered through my own cognition. The list contains more about what consciousness is not rather than what it is. Therefore, this is a kind of history of assumptions that have been used to gain insights into our mental functioning. Some of these assumptions I keep in a holding space that I revisit from time to time because I find them to be plausible and compelling—they are strange puzzle pieces that someday will fit together.

1. ***Consciousness is not a hard problem*. On the other hand, *Consciousness is a hard problem*.**

Is the study of consciousness so difficult, so full of paradox, mystery, and static, that it can never be fully solved? Some philosophers think so. Here is UC Berkeley Professor Alva Noë's opinion:

> We now confront a paradox. Science views its subject matter coolly, dispassionately, rationally. Science takes up the detached attitude to things. But from the detached standpoint, it turns out, it is not possible even to bring the mind of another into focus [to be empathetic]. From the detached standpoint, there is only behavior and physiology; there is no mind. So it would seem that a science of the mind is impossible. And mind itself is something paradoxical; it is a feature of our nature that cannot be made an object for natural science. ~ *Out Of Our Heads, Alva Noë, 2009.*

In other words, ***we can never study consciousness without using consciousness;*** consciousness *is not* a variable that can be set aside or isolated for study. A researcher's mind must always influence what he or she is studying and reporting. Scientific research is questionable and results are suspicious, especially when consciousness sets out to study consciousness. Mathematician and philosopher Kurt Gödel developed a famous theorem which can be used to proclaim that *consciousness cannot prove or disprove itself*—or, as Gödel's theory asserts: no consistent theory can prove its own consistency. I will discuss this in greater detail in Chapter Seven.

Furthermore, because we have two minds that alternate, we never settle within one mind. Cognitive oscillation is an invariant; it cannot be ruled out as an influence. In other words, a scientist studying consciousness is one moment using allocentric processing and the next moment using egocentric processing—without being aware of this constant oscillation. The fluctuating mind of the researcher impacts any study of consciousness.

In addition, space is expanding. We are on the edge of the expanding wave—we are located where the universe is creating space. One moment and the next moment seem to follow, but in each new moment space is new—it is as if we are constantly reborn into some never-before-encountered universe—we are never in the same river twice. A scientist studying consciousness is using an ever-fresh cognition moment-to-moment, as are the subjects of research. Therefore, considering everything discussed in the paragraphs above, consciousness is, indeed, a hard nut to crack.

One of our two minds, the allocentric mind, lives in a land of unmanifested potential; there is nothing for mathematics to get hold of when examining the allocentric mind because there is nothing to observe, no patterns, nothing that stands still. Only when something manifests from the background can it then be measured, compared, and catalogued. In other words, manifestation and unmanifestation are mutually exclusive. The manifest world of the egocentric mind *can* be proved or disproved. However, the unmanifest world of the allocentric mind *cannot* be proved or disproved. It is a case of either/or and never-both-at-the-same-time. Understanding consciousness is hard, especially when there are two mutually exclusive kinds of cognition.

There was a debate that began in the 1990s about duality and consciousness. The debate centered on the ideas of philosopher David Chalmers who coined the phrase "the hard problem." Other cognitive scientists and philosophers challenged Chalmers, saying that there was no hard problem in the study of consciousness—everything could be explained by science. One reductionist, Zoltan Torey, in his book *The Conscious Mind* (2014), puts Chalmers' position like this:

> To illustrate the point that experience and conscious experience are different phenomenon, Chalmers (1996) speaks of a hypothetical twin of his, who has the identical experience and the identical response to this experience as Chalmers himself but, unlike Chalmers, is not conscious of it. ~ *The Conscious Mind, Zoltan Torey, 2014.*

From my perspective, we *really do have* twin minds inside of us, one of which, (the allocentric mind) does not know that it knows. Chalmers' metaphor of the "hypothetical twin" is not a supposition; it is the actual situation.

Philosophers engaged in this debate about consciousness lacked an understanding of the twin minds—they did not have the knowledge base to understand the physiology, anatomy, and evolutionary unfolding of the two minds. Anti-reductionists used the term "Qualia," a term coined by David Chalmers, to mean "experiences." They could see that "experiences" could not be explained by the egocentric mind. From a dual-process perspective, only the allocentric mind can have experiences. The egocentric mind does not have experiences; that is not its job, not its mandate, not what it was designed to do. The egocentric mind is blind to experiences; it does not comprehend or deal with qualia. Instead, the egocentric mind is entirely about pattern recognition. It builds language, meaning, long-term memory, and associations around patterns. Only the allocentric mind can have experiences.

Confusion results partly because the ego believes it is located in the head. The ego has a notion of cognition that is not embodied—not in the whole body. The ego doesn't need a body to "play its mental games in the cortex." The ego is brain-bound and happily so. The ego also believes that it alone exists—that there is no allocentric mind, just some subservient, sub-processing thing lurking in the shadows and not worthy of serious attention. However, the allocentric mind cannot exist without being embodied. The allocentric mind only functions as a body immersed within and undifferentiated from a domain.

The body, the allocentric mind, has experiences. The egocentric mind has thoughts. Qualia are varieties of experiences. These experiences are inaccessible to the head-bound egocentric mind. It is only the egocentric mind that finds itself with a hard problem because it has no way to experience. There are also two kinds of meaning, one hard and the other not so much; both kinds of meaning relate back to navigation:

> . . . we have dual navigational life purposes: first, to *explore and figure out, because that is the evolutionary job of the ego,* and second, *the job of the self is to experience moments as they arise along our journey:* to soak up the awe, terror, joy, grief, angst, and peace of being alive moment-to-moment. ~ *The Confusion Caused by Being Your Own Twin, Doug Baldwin, 2017.*

Of course, if there is a higher intelligence behind the universe, if space and time are both illusions, and if a reality exists beyond what we are capable of comprehending with our limited cognition, then we cannot know about greater meaning.

2. ***Consciousness is not the voice in your head.*** Take that voice away (good luck) and for a few moments you realize that there is a peaceful, centered, non-verbal background self—the allocentric mind—that still operates behind the chattering monkey mind. Most of what that incessant internal voice does is worry, fret, complain, justify, and rant. It doesn't seem all that good at

problem solving. The quality of our internal thinking is suspect. As I discussed earlier, philosopher of mind Zoltan Torey argues that the voice in our head is a proprioceptive phenomenon. Many forms of meditation, for example, begin with the tongue firmly against the roof of the mouth, the lips consciously stilled, and the jaw muscles consciously relaxed. This slows down the inner voice because it partially inhibits proprioceptive feedback to the motor-driven language centers.

3. For similar reasons, ***consciousness is not language***. The confusion as to whether or not other creatures have consciousness is caused by confusing language with consciousness. You know full well that your cat and dog have consciousness because they communicate with you in their own way. However, what you don't realize is that other creatures have consciousness because they also evolved navigation through dual processing; they have allocentric minds and egocentric minds just as we do. Of course, our sophisticated language ability has pushed our dual minds well beyond what other earth-creatures have evolved. We have more sophisticated dual processors, especially our egocentric mind. So it is no wonder that we would conclude that language is what sets us apart, and it does egocentrically (but this is only part of the story). Here is the poet Rumi with a suggestion:

> *Close the language-door*
> *And open the love-window.*
> *The moon won't use the door,*
> *only the window.*

~ "THERE IS SOME KISS WE WANT," *A YEAR WITH RUMI,* 2006.

4. The issue of communication, however, is not the same as the issue of language. There is a silent allocentric communication that involves the immediate reading of body language, and the reading of environments. This is a non-verbal awareness and it is full of relevance in each moment. Spoken communication is egocentric:

it requires intention, posturing, and is goal-directed. In contrast, ***Allocentric consciousness is a relationship between the organism and the surround***, including signaling and mirroring other organisms. Psychologists have coined the term "embodiment," which is an acknowledgement that the whole body is one sensory unit. *The brain has no purpose without the body.* The body itself, from an allocentric perspective, can be thought of as one huge brain. Allocentric communication is non-verbal—it involves facial gestures, body mannerisms, postural signaling, moans, grunts, laughing, weeping, hugs, and kisses. Contrary to this, the egocentric mind is generated mostly in the neocortex of the brain, and its job involves the management of patterns. The egocentric brain isolates and extracts patterns, gives names to patterns, and associates concepts with patterns.

5. ***Consciousness is not the vision system, even though it is deeply affected by vision.*** This may seem obvious, but research about consciousness sometimes, often subtly, assumes that consciousness is entirely about vision. The visual system is very complex but when you remove it (which happens with blindness) consciousness continues.

Furthermore, since the senses can never be dissected out from their coherent synchronicity, there really are *no* individual sensory processing systems—simply sensory *portals*. According to dual-process theory, there are just two ways to pay attention; all the energy patterns that impact the body are immediately combined to serve either of these two processing networks, egocentric or allocentric.

Although consciousness is not all about vision, it is, nevertheless, deeply affected and guided by what the eyes take in. Because the eyes are quantum processors, they decode light patterns based, at least partially, on the dual-nature of the quantum world. Therefore, there are two vision systems, one to process wave dynamics and the other to process particle dynamics—this is my supposition, and is discussed in later chapters. The result of this processing duality is that we have two minds governed by quantum mechanics. I will make a leap of logic here and make these two further speculations: *wave frequency is related to consciousness*, and since frequency is related to energy, so is *consciousness related to levels of energy*.

Research in 2014 pinpointed a frequency of 40 cycles a second as being instrumental in creating consciousness:

> In 2014, European researchers reported results (in the journal *Nature Neuroscience*) of their investigation into how "higher-order consciousness"—abstract thinking and reflexivity—is generated by electrical currents called gamma waves. The researchers fired low-voltage currents through test subjects' frontal lobes to mimic the gamma band in an effort to induce self-awareness in unconscious patients. It worked. The dreams experienced by the test subjects started to become lucid. The researchers concluded that conscious awareness is induced at electrical currents pulsing at 40 cycles per second. It all strongly implies that the subjective experience arises at least in part because of electrical stimulation. ~ *Beyond Biocentrism, Robert Lanza, 2016.*

Consciousness corresponds to frequency states (electrical pulses). For example, we are more aware, and more awake, as we process higher and higher frequencies. Red light (low frequency) makes us sleepy, and gets us ready for bed. Blue light (high frequency) wakes us up in the morning and peaks our energy levels when the sun is at its highest point in the sky. Beta waves register when we are awake, but gamma waves register in monks when they meditate using higher states of consciousness—the monks can activate gamma waves. Therefore, *consciousness is a registering of energy changes.* This "registering of energy change" is a whole-body phenomenon—it is not just in the brain as the study above suggests.

Over time, I began to wonder if our dual cognition wasn't simply a division between our two most powerful senses, vision and hearing. Vision is spatial, a scene-generator, while hearing is temporal, a time-generator. I realize that consciousness is more complex than this simple differentiation, yet vision is the most significant and dominating system for allocentric processing, and vision can be activated by different wave states—like alpha and gamma waves. Likewise, hearing is the dominant sense for egocentric processing—so much so that we might make a case for hearing as the basis for egocentricity, just as vision is the basis for allocentricity.[1]

6. ***Consciousness is not a steady-state entity.*** It is a dynamic process. Nobel Prize winner Dr. Gerald Edelman agrees:

> The evidence . . . reveals that the process of consciousness is a dynamic accomplishment of the distributed activities of populations of neurons in many different areas of the brain. That an area may be essential or necessary for consciousness does not mean it is sufficient. Furthermore, a given neuron may contribute to conscious activity at one moment and not the next. ~ *Wider than the sky, the phenomenal gift of consciousness, Gerald Edelman, 2004.*

In other words, the idea of anatomical lobes in the brain, associated with specific functions, is an outdated idea. For example, the occipital lobe has been historically called the visual cortex. This is no longer completely tenable—it has to be modified. Vision, like all the senses, is a distributed neural phenomenon. Consciousness, unlike vision, has never been associated with a specific brain lobe, and it is even more neurologically distributed than visual processing.

Furthermore, whatever consciousness turns out to be, it is never without interruptions. It is extremely hard to stay conscious—focused on one thing, for example, or on the surround alone—for even a couple seconds. We cannot control the fundamental alternation between inhibition and excitation. Our mind flies all over the place: we day-dream, rant, nap, zone out, look, listen, fidget, concentrate, and so on.

Despite this inherent chaos, we somehow manage to have a co-herent image of scenes that give reality an invariance. Therefore, consciousness "hangs together," no matter what is oscillating or how wildly our attention wanders. Using dual-process theory, it is the egocentric mind that is ranting, daydreaming, and zoning out. The allocentric mind is the invariant that holds scenes and reality together.

When monks meditate they try to maintain allocentric aware-ness. The egocentric mind rants and raves, as it does in all humans, but the monks just observe this chatter, and then they return their focus to the stable allocentric mind. The monks know that within the human mind "nobody is in charge of the house." In other words, the ego rants, daydreams, reasons and emotes, in a random, uncon-trolled manner, unless some aspect of the mind steps up to monitor and redirect these errant energies. There may be a prefrontal execu-tive in charge of consciousness, but this executive is gone from the office more often than not.[2] It takes intention, a meditative practice, and discipline to harness the human mind.

Using a virtual reality model, we can say that **consciousness is something we create**, moment-to-moment, as needed. However, if no intention and no discipline are employed, then what the mind creates may not be stable or valuable.

7. **Consciousness is cyclical.** There is a diurnal and seasonal regularity to consciousness. From deep sleep, we awaken through the twilight zone of theta-consciousness, then through the relaxed state of alpha-consciousness, and then into the beta-wave conscious state we call "reality." There is a high state, called gamma-consciousness, which few people experience, except for monks and gurus. At the end of the day, the process is reversed. We leave beta-consciousness, slip back into alpha, then theta, then into the deep stages of sleep where we think of ourselves—probably incorrectly—as unconscious. This cycle repeats throughout our entire sojourn on the planet. The earth

also seems to go through states of consciousness, sleeping through the winter months, awakening in the spring, blossoming through the summer, and harvesting in the fall. As the earth goes through these long cycles, our personal conscious is correspondingly altered. This earthly cycling affects the whole body, and thus impacts the allocentric mind directly.

8. ***Consciousness is not the same at every developmental age.*** What a two-year old knows is not the same as what a three-year old knows. Whatever consciousness is, it changes with human development. As the brain and body rapidly grow during the first years of life, so also does consciousness evolve. On the other end of the life cycle, as the brain and body age, states of consciousness become less coherent and reliable. Research on consciousness often addresses the consciousness of an adult, roughly from age 25 to 50, a range of consciousness where there is a plateau in the velocity of change. We have to keep in mind that conscious is a process with a growth and decline cycle, similar and parallel to the changes to the body as it ages.

9. ***Consciousness is not the same in every species.*** There is an evolutionary component to consciousness. In a famous 1974 essay "What is it like to be a Bat?" Philosopher Thomas Nagel argued that only a bat knows what "bat-ness" is like. Each creature has some sense of its existence; each is living inside a mental shell that is different from any other organism. Whatever *human* consciousness is, it is unique to humanity. Taking this logic a shaky step further, perhaps **no two humans have the same consciousness**; only Karen knows what "Karen-ness" is like; only the blind psychologist and philosopher of mind Zoltan Torey knew what "Zoltan-ness" was like. A central thesis of philosopher David Chalmers' thinking is that even non-living "stuff" has a sense of knowing what it is like *to be*: a quark may "know" what it is like to be a quark, or a water molecule may "know" what it is like to be a molecule. Chalmers' calls this panpsychism. The confusion over whether animals have consciousness is really a failure to differentiate the egocentric mind from the allocentric mind. Animals do not talk, and they do not

have an inner voice. They do not have a highly evolved egocentric mind. "Language" is what animal's lack, not consciousness. Also, the allocentric minds of animals are presumably as evolved as in humans. Certainly, the same neurological substrates are found in animals and humans. The 2012 Cambridge Declaration on Consciousness stated that:

> Convergent evidence indicates that non-human animals have the neuroanatomical, neurochemical, and neurophysiological substrates of conscious states along with the capacity to exhibit intentional behaviors. Consequently, the weight of evidence indicates that humans are not unique in possessing the neurological substrates that generate consciousness. Nonhuman animals, including all mammals and birds, and many other creatures, including octopuses, also possess these neurological substrates." ~ *The Cambridge Declaration on Consciousness was written by Philip Low and edited by Jaak Panksepp, Diana Reiss, David Edelman, Bruno Van, 2012.*

Plants can also be said to have allocentric consciousness. They react directly to changes in the environment—they have allocentric experiences. This reactivity is inherent in their genetic makeup. Plants react to gravy, wind, sun direction, insect invasion, to water availability and to nutrients—to name a few of the variables.

10. ***Consciousness is not the result of biochemical processes***; or rather, the biochemical perspective is just one layer of understanding. A greater understanding has to account for quantum behavior, for coherent frequencies that bind all the body processes into a synchronicity.
11. ***Consciousness is both innate and learned***. It depends on whether we are talking about allocentric consciousness, which is mostly innate, or egocentric consciousness, which is mostly learned.
12. ***Consciousness is evolving. We are not the end result of evolution***, only the latest version. Even our children and grandchildren may

have a more advanced version of consciousness than we do. There are also levels of cultural consciousness, and this is evolving as well. Cultures contain a mixture of individuals with varying levels of consciousness. We have within our DNA the history of the evolution of consciousness and we "float" within these levels, one moment behaving like a Neanderthal at the office party, the next evening lecturing on quantum physics. Owen Barfield reminds us that consciousness is evolving as part of nature; nature itself is evolving—in a way, nature is "learning" how to evolve consciousness.

13. ***Consciousness is not a single brain state.*** Scientists know that mental states vary with recorded brain waves. For example, the relaxed alpha rhythm state is not the same as the hyper-alert gamma state. If consciousness has to do with levels of wakefulness, then it is a multiple and fluctuating phenomenon—it manifests in different ways at different times.

14. ***Allocentric Consciousness is a whole-body phenomenon that cannot be divorced from the environment.*** The cyclical brain rhythms mentioned above are a result of changes in the environment. For example, we have circadian and diurnal rhythms that originate outside of us; these continually alter our degree of wakefulness. "Environment" is just a term that refers to what we are exposed to moment-to-moment. The domain in which we live and navigate is not neutral; there is a continual reciprocal effect. In other words, as we impact the environment, the environment impacts us.

Remember, this is an allocentric perspective. The egocentric mind *can* exist as an island apart from the immediacy of the environment. Indeed, our advanced egocentric mind is the most significant reason that human beings evolved a level of consciousness beyond that of any other creature. We use our egocentric mind to understand the environmental situation in very sophisticated ways, beyond the capacity of other land-based creatures.

15. ***Allocentric Consciousness is not in the brain.*** Egocentric consciousness does feel as if it is in the head because the eyes and ears are

located there; egocentric processing starts with the head-bound senses. However, allocentric consciousness involves the entire nervous system; it uses the internal whole-body senses, including: proprioception, kinesthesia, the vestibular system, and photo-receptivity. "Mind," for the allocentric system, is a total-body phenomenon, and it cannot be dissected out from the environment. After further consideration, it is also true that the egocentric mind is connected to the whole-body and acts in harmony with the external head sensors, so it also is a whole-body phenomenon—even though it thinks it isn't!

Unfortunately, the assumption in neuroscience, the starting point for much research, is often that consciousness is entirely a brain phenomenon, entirely egocentric:

> The fundamental assumption of much work on the neuroscience of consciousness is that consciousness is, well, a neuroscientific phenomenon. It happens inside us, in the brain.
>
> All scientific theories rest on assumptions. It is important that these assumptions be true. [However,] . . . this starting assumption . . . is badly mistaken. Consciousness does not happen in the brain. That's why we have been unable to come up with a good explanation of its neural basis. ~ *Out Of Our Heads, Alva Noë, 2009.*

What Professor Noë is referring to is the allocentric mind. He is pointing out correctly that what we call mind extends beyond the brain, beyond the body, to include the environment—brain, body, and environment are one entity, and "mind" cannot be dissected out.

16. **Consciousness expands or contracts to fit the space that it is in**. The environment determines what there is for the individual to participate in. For example, in a familiar space, in your home or work place, consciousness settles into a familiar, almost automatic, repetitive routine—a center of gravity occurs wherein the level of consciousness is at a comfortable place. The people sharing

space with you within any environment, with their differing levels of consciousness, deeply affect your consciousness, as does the presence of other sentient creatures. Going into nature, being outside away from walls and routines naturally expands consciousness. Furthermore, the more confining the space, the more intense and chaotic the frequency state. The bigger the space—like in nature, without artificial walls—the more frequency slows down, balances out, and is less chaotic. We calm down when consciousness is allowed to expand.

This understanding has huge consequences for some of our institutions. For example, the prison population is confined to small cells, even to windowless solitary confinement. Consciousness is strangled in these conditions—there is little hope for cognitive evolution under these conditions. Unfortunately, the same is true for our children in schools. Most of the time, students are confined to small desks and small classrooms, which are often overcrowded and windowless. One of our primary goals in education ought to be the expansion of consciousness in our school population. However, the very structure of the school environment is a negative influence on the mental growth of kids.

17. ***Consciousness is not one entity***, it is an alternating duality. From a navigational perspective, and based on my pervious discussion, it follows logically that there must be two fundamental kinds of consciousness, allocentric and egocentric. These are attentional processing systems, so consciousness is a system of dual attention. You cannot decide that there is only egocentricity and call that consciousness, or that there is only allocentricity and call that consciousness. You can't hide from the twin that resides in the same brain with "you." You also cannot control the "light-speed" switching rhythm that binds the two systems into the illusion of oneness. We can only use one of the conscious systems at a time; we can be conscious allocentrically or egocentrically, but not both at once. Another way to say this is that we are either "being" or

we are "becoming," but we cannot do both simultaneously. Rapid alternation between the two gives the illusion of seamlessness, but we cannot perceive all-at-once the same time as we perceive one-thing-at-a-time. We can be aware of time egocentricity, or we can be aware of space allocentricity. We can be awake to others and to objects, or we can be awake to the surroundings. The ego can know, or the self can experience. We can believe, or we can have faith. We are stuck with this mental paradox.

18. ***Consciousness is related to purposeful movement.*** From an evolutionary perspective, it is clear that we are the top of the cognitive order—in land-based environments—because of the sophisticated way we understand and use purposeful movement and navigation. In an aquatic environment, dolphins seem to be the top of the cognitive order with their large and complex brains. The more an animal is able to control self-movement—the more ably they are able to navigate—the more conscious they are. In other words, there is a correlation between consciousness and navigation.

19. ***Consciousness is not a noun***, not an entity, not some object we might hold in our hand after successful neurosurgery. ***Consciousness is a verb (twin verbs)***, a set of processes that results in attentive (egocentric) and aware (allocentric) behaviors.

20. ***Consciousness is not "knowing that we know."*** Most of the time we are neither aware nor attentive. We don't know that "we know," or "don't know," and most of the time we don't care whether we are awake or not. We are a bundle of needs, desires, appetites, and habits. We are autopilot-creatures almost all the time. There are fleeting moments when a witness shows up who is aware of knowing. Most of the time, however, the witness-program is not running. We are fed—controlled, driven, ruled—by thirst and hunger for things and relationships. We are sleepwalkers most of the time.

21. **Consciousness needs a face.** When my daughter Anna was five, she announced to her surprised parents that she had decided to be a vegetarian, even though no one else in our family was a

vegetarian. She explained that she didn't want to eat anything that had eyes. What she was saying, in her five-year old way, was that she was taking a moral position. She refused to eat anything that had once been conscious. If it once had a face, she refused to eat its dead remains.

Many human beings, at a gut level, feel something similar. We have a hard time assigning consciousness to a rock or a cloud. However, we are okay assigning a crude consciousness to bugs or snakes, with their weakly constructed faces. But for cats and rats and elephants, and grumpy Aunt Mildred, we are pretty much in agreement that we can issue these creatures a semblance of consciousness. Flash any series of human faces on a screen, from any culture on earth, and we will all agree that these faces have degrees of sophisticated consciousness—you can "see it in their eyes," we might say. When eyes are dulled by disability, however, we are less generous in assigning consciousness. This has led to abuses of disabled people over the years. The face is a primary "clue" for the existence or non-existence of consciousness.

The review above has been mostly from a Western cultural perspective. Eastern philosophers and psychologists have evolved different perspectives and assumptions which I will discuss in Chapter Six.

I Have Some Bad News for You

"Hey, Surge."

"Hello, Dutch. You look perplexed."

"Yes, I need some succulent food for thought."

"We specialize in the feeding of sentient minds. We have pantries filled with thoughtful words. What bothers your tiny duality?"

"I saw a documentary about dragonflies. They evolved before the dinosaurs. They were on the earth 300 million years before humans arrived. Dragonflies perplex me."

"Yes, the tragic and magnificent dragonfly. You can tell by looking at them they are aliens. Dragonflies are vicious predators that are eaten by other vicious predators. If they survive, they die of old age after only a few weeks. Everything is born, everything dies. Everything eats everything else. The jungle has laws. That's not your fault. What's troubling you?"

"Essentially, dragonflies seem to have stopped evolving. The design that worked millions of years ago is still being used today. Dragonflies didn't evolve to be ever-wiser and ever more intelligent. They just kept doing the same repetitive behaviors over and over again. I look around at all the creatures on earth, and I see the same failure of evolution to go beyond the same set of routine movements. It's pretty boring stuff, if you ask me, Surge."

"No one asked you, Dutch. And a good thing, too."

"So, okay, what bothers me is how humans keep evolving while everything else in nature, like paramecium, fairy flies, and bacteria are essentially stuck in neutral."

"I have some bad news for you, Dutch. Most humans don't believe in evolution. They want to be left alone to do their 'dragonfly thing.' They want to drive mindlessly to Tim Hortons every morning to get their donuts. Most human beings don't want to be more intelligent. They don't want more wisdom. They might give lip-service to compassion, but they act as cruelly as their mood dictates. And they don't give a rat's ass about levels of consciousness."

"We are the King of the Food Chain, Surge. Shouldn't we rule our kingdom from a high throne of consciousness?"

"What makes you think that human beings aren't just savage killers like the dragonflies? Humans also murder in order to eat. They kill pigs, cows, sheep, chickens, turkeys, woodchucks, deer, and so on; then they eat the dead carcasses with delicate sauces. Human beings spray the earth with pesticides that mass murder insects and single-cell organisms. Human beings yank plants from the soil without saying sorry; they eat their carrots raw. Human beings hunt down wild animals and kill them for sport. Human beings execute each other. Human beings mass murder their fellow beings in war after war."

"I get it, Surge. That's enough."

"The dragonfly is a tiny predator that kills one-on-one; it kills, eats, and reproduces. In contrast, human beings are massively impersonal killers with the capacity to end all life on the planet. These are the highly evolved creatures of which you speak? The spiritual guys you champion? During happy hour, we like to watch old film clips of the human rise to dominance on planet earth. As far as we can tell, human beings are just murderous masters of arrogance. You are not the pinnacle of evolution that you suppose. God is fond of human beings, but he has made quadrillions of other beings that are more fascinating and lovable than you brutes. Just saying."

"Thanks for the rant, Surge. However, you might have mentioned your opinion of humanity before, at other meals, while I was trying to gulp down my gourmet gruel. But you aren't answering my question. Why are there evolving levels of human consciousness and not evolving levels of dragonfly consciousness? Why, if paramecia pioneered proprioception, didn't they evolve a spiritual allocentric mind? Why did human beings develop spirituality and empathy? No other creatures did so. Spirituality is contrary to a world where everything eats everything else, where creatures battle to the death for scarce resources. Why did we alone evolve benevolence?"

"Who are you to say that other sentient creatures didn't evolve ever greater degrees of consciousness? That isn't even close to being true. Of course, consciousness actually *is* evolving in other creatures. Evolution is a wave that goes on and on. The universe and everything in it is evolving ever more complex kinds of consciousness. So maybe dragonflies really *are* getting more sophisticated, more intelligent, and wiser. The fact that their external form didn't change is not proof that their internal processing didn't evolve. Didn't the human body plan stay pretty much the same for 200,000 years, while the brain and nervous system evolved?"

"Okay, Surge, let me try this again. Dragonflies move with a purpose, right? So they have both allocentric and egocentric cognitive systems. Why then did they fail to get more sophisticated? Why didn't each of their minds evolve to be more human-like? Dragonflies are still to this day savage killers; there is no evolved kindness, no empathy, and no ability to think. A dragonfly will even eat another dragonfly without remorse or awareness. Hey, Fred ate mom. Nature worked for over 300 million years to craft modern dragonflies and this is the result: a mind-less eating machine with a successful reproductive strategy. Modern humans have been on the scene for less than 200,000 years. Why did our dual navigational apparatus evolve wisdom and intelligence but other creatures, with their own dual navigational mechanisms, fail to evolve a thirst for exploration and a need to be kind? What am I missing, Surge?"

"Slow down, Dutch. Let's take this one meal at a time."

"That's fine; I'm just asking what happened, Surge. Why didn't empathy evolve in other creatures? Why didn't intelligence blossom?"

"Intelligence has evolved and so has empathy in many creatures—I guess you didn't have pets when you were growing up. Anyone who has a family dog knows how wonderfully empathetic they can be. By the way, what makes you think that human beings are so good at empathy? Most human beings can't even spell the word. Not only do most of them fail to conceptualize empathy, they also cannot comprehend how it works—many are baffled by the emotion and are incapable of feeling it. Masculinity, for example, is relatively blind to empathy compared to the feminine perspective."

"Fair enough."

"Here's your answer: alphabet soup."

"Finally, something to eat that has the potential for meaning."

"We serve a hearty meal of Alphabetic Vegetables floating in a rich background broth—which we call Rich Background Broth. This meaningful meal is served with Parenthetical Potatoes roasted and then arranged poetically in a Rumi Bowl. Dessert is included with the soup."

"What's the dessert?"

"Language-Laden Bread Pudding, Punctuated with Period Raisins."

"Okay, so it's about language, that's what you are saying? That it was language that transformed allocentric and egocentric processing and forged dual consciousness?"

"Well, of course. About 100,000 years ago, language appeared and then the evolution of the two human minds took off like a rocket-powered chariot. Language is the magic elixir that altered the allocentric system of awareness and the egocentric system of attention; language made them into two minds, then into two kinds of consciousness. Blame it on the words."

"So language is the reason we have consciousness, like the linguists have been saying."

"No. language was a necessary catalyst. Language would have been irrelevant had there not been two anatomical and physiological systems ready and waiting to evolve. Language altered the mouth, tongue, lips, and vocal cords; the whole facial structure was transformed by the evolution of speech. Then proprioception moved language inward to become the voice in your head. That became the ego, the personality. Thus was born proprioceptive egocentric consciousness."

Notes

(1) **Vision as allocentric and hearing as egocentric.** Consciousness is a complex puzzle with pieces of similar "colors and shapes"—it is easy to get frustrated and confused. Vision is overwhelmingly allocentric, and hearing is overwhelmingly egocentric, yet each system contributes to both minds as we go about our daily existence. I have in other places stated that proprioception is the dominant evolutionary contributor to both the allocentric mind and the egocentric mind. The notion that vision is the dominant sense is an egocentric perspective. We take our visual and auditory experiences for granted and pretty much fail to comprehend proprioception's fundamental role.

(2) **There may be a prefrontal executive in charge of consciousness,** but this executive is gone from the office more often than not. I am speaking here of the rational egocentric mind. Most of the time, we are sleep walking, and not at all rational or thoughtful—we are neither attentive nor aware. In other words, we are perceptually blind most of the time. Self-awareness, the watcher, is a highly cognitive skill that operates infrequently in most humans. The witness, or watcher, is a highly refined proprioceptive ability that can be manifested through intention and a meditation practice.

Five

No better love than love with no object.
No more satisfying work
than work with no purpose.
If you could give up tricks and cleverness,
that would be the cleverest trick.

~ "No Better Love," *A Year with Rumi*, 2006.

Emptiness Pudding

"I'm not very hungry today, Surge. What do you have for people who are full?"

"Full of what? That's my question. We have the perfect non-meal for satiated beings like you. We offer several kinds of Ersatz-Puddings, sweet lumps of nothing that are especially made for over-stuffed turkey minds.

"I came for help, Surge, not ridicule."

"Sure. I can help. You obviously need some Buddhist Emptiness. On special today, we have Buddhist Emptiness Meatloaf. Do you want to try the Emptiness Meatloaf?"

"What's in it?"

"Nothing is in it. What part of emptiness don't you understand?"

"Well, it seems that if I am to purchase a meal, a meal ought to come with the meal. Just put it in a bowl, and put the bowl in front of me on the table. Not that your track record for actually feeding people is so stellar."

"It doesn't come in a bowl."

"Put it on a plate then."

"It's only served in air cups."

"There's nothing in the bowl, is that your idea of feeding me, Surge? You are going to pretend to put stuff on the table, but there won't be anything real to perceive, right?"

"You said you weren't hungry. Is that my fault?"

"Fine, bring me a huge bowl of the Buddhist Meatloaf with a gigantic portion of your famous Emptiness Gravy. What does this have to do with my book, by the way?"

"When you apply Deep Thought to the serious questions, what you discover is *nothing*."

"That's just swell, Surge. Thanks for nothing."

"No problem, Dutch. Our motto is: *Nothing* is too good for our customers."

"This is going nowhere fast, Surge."

"Yes, thank you. Nothing ventured, nothing gained."

"Whatever, Surge. So what should I do in the next chapter?"

"Talk about nothing. Talk about something."

"Okay. I'll begin by saying that the following chapter is about our two minds, Thing One and Thing Two—they are opposites, yet they are allies. How does that sound?"

"I'm proud of you, Dutch."

The Failure of Deep Thought

Show your talent.
Do something outside
time and space.

~ "INWARD SKY," *A YEAR WITH RUMI*, 2006.

Douglas Adams (1952-2001) wrote a wonderful science fiction spoof called *The Hitchhiker's Guide to the Galaxy* (1979). Adams tells us that long ago there existed a race of hyper-intelligent-pan-dimensional beings (mice) that, like us, were frightened by infinities and eternities. Consequently, they built a super computer called Deep Thought to answer the question "What is the answer to Life, The Universe, and Everything?" After thinking it over for 7.5 million years, the computer concluded that the answer to All Things, Life, and The Universe was the number 42. This hilarious answer, of course, did not please anyone.

The master race (the rat race) unhappy with Deep Thought's failure to satisfy their passion to know ultimate answers, decided to build an even larger computer to answer the same questions. They called this new computer *Earth*. The self-replicating software used to program Earth was called *Life*.

To continue Adams' analogy, it has been close to four billion years since the creation of Computer Earth, which is still churning away. Life has tried all kinds of algorithms over the eons called "elephant," "snail," "amoeba," "tooth fairy," and so on. In other words, there have been many life-algorithms, none of which are satisfying. The most promising algorithm, however, is called *human being*. The Universe and Everything is hopeful that human beings can eventually figure it all out.

My dual-process theory fits nicely with the science fiction genius of Douglas Adams—Computer Earth has come up with a plan that uses dual-core processing units that operate separately but in parallel. These dual cognitive "computers" were constructed to solve two paradoxical conundrums: eternity/no-eternity and infinity/no-infinity. The first processor was called Ego, the second processor was called Self. The Ego-processing system was charged with solving the seemingly impossible problem of endless time, and the Self-processing system was charged with solving the equally bewildering and probably impossible problem of endless space.

Just this morning, Computer Earth spit out the most recent answer directly into my desktop computer in Saginaw, Michigan. I will share that answer with you now—well short of 7.5 million years. And no, the new answer is not 43. The latest answer to Everything, including Life and The Universe, is—drum roll, gasping of collective breath, cameras rolling, trumpets and flower girls—Ta da: Cookbook/Cooking. Evidently, just for the historical record, there is a God behind Life, The Universe, and Everything called Betty Crocker. But that is a story for later telling during happy hour at the Red Eye Café.

Evidently, there is a recipe for Life, The Universe, and Everything, a cookbook, but it just sits there until cooking happens—nothing manifests out of the recipe unless someone decides to make bread, so to speak.

This explains why Deep Thought was doomed to fail: it had been programmed to look for a single answer, not for two equal but opposite answers to the same question. Deep Thought was also searching for nouns; it had not been programmed to search for twin verbs. Deep Thought knew about concepts, meaning, and knowledge, but *it could not experience*. It knew about "cookbook" but not about the joy of cooking and the pleasure of eating. It could identify bread, talk about bread, post pictures of bread, write a blog about bread, create a Bread-World Facebook page, do a webpage bread-spread, post on wiki-bread, tweet bread wisdom every few minutes, build a Robo-chef to make endless types and quantities of bread, but Deep Thought could not *taste* bread; it was culinarily clueless. Deep thought was intelligent, but Deep Thought had no wisdom, no storehouse of experiences. Consequently, we can state unequivocally that, for a rich and full life, *Deep Thought— by itself—is insufficient without experience.*

In retrospect, even though nobody in the book cared for the answer "42" as the ultimate answer, Deep Thought did offer two helpful suggestions: First, build an organic super computer called Earth (check), and then, before you ask Earth any questions with the word "ultimate" in the sentence, clarify your thoughts—nonsense questions arrive at nonsense conclusions; abstract recipes bake abstract and tasteless cookies. In other words, using consciousness to understand consciousness is a tricky undertaking, which requires careful forethought.

It is also true, in retrospect, that finding the ultimate answer to eternity and infinity is a tad too frustrating for even an organic supercomputer called Earth. And there is, unfortunately, a dual curse that comes with dual processing:

- A mind that can create space, and therefore infinity, can go insane looking for final boundaries that can never be found.
- Likewise, a mind that can create time, and therefore eternity, can go nuts searching for conclusions that are forever and always beyond reach.

The cookbook of recipes is a background—the potential for bread. However, there can be no bread—nothing gets cooked—until something manifests in the egocentric mind. Thus, *infinity and eternity are metaphorical backgrounds out of which we manifest our reality.*

I don't know about you, but I find it strangely comforting that space perception and time perception are created by the brain. I don't have to get cranked-out-of-shape because I can't figure out infinity and eternity; neither of these "concepts" can be proven. The philosopher Immanuel Kant pointed out that we were stuck between an objective world that we can never perceive directly and a set of senses that are unreliable at best. He said that time and space are forms of consciousness. Arthur Schopenhauer, who came after Kant, also found it comforting that space and time are just illusory enough to let in the light of "spirituality."

Poet and songwriter Leonard Cohen (1934-2016) has a song lyric that says "there is a crack, a crack in everything; that's how the light gets in." Cohen, I think, is agreeing with Schopenhauer: If space and time are not real, then there is something on the "other side" of our reality; there is a crack in objectivity. Ironically, we are blind creatures unable to see a certain kind of "light." That unknown light streaming toward us has a calming power; some call it love.

The Ugly Paradox

Consider for a moment the consequences of our ability to cognitively conceive of and manufacture space and time. Such a creative brain would eventually, logically, come up with the notion of infinity and the notion of eternity. Within unlimited space, which is a product of the allocentric mind, it is impossible to find a center—egocentricity cannot survive, cannot be defined, within this kind of world. There is no possibility of one-thing-at-a-time when everything manifests at once. If you try to conceive of your essence inside this unlimited space you will discover what the Buddhists mean by emptiness, no-self, and the notion that we are all one. The logical conclusion for the allocentric mind is that "I" am nothing—there is no "ego." Personality is a construct.

Likewise, consider that the egocentric mind can represent and create time so completely that it can conceive of eternity, where there is no beginning, and no end. Eternity means there is no ultimate death, and there was never an original birth. There is no escape from having to exist, in some form, forever. Therefore, the big bang was not a great-beginning, and whether the universe comes to a cold dead halt or races off into cosmic vapor, it is only part of a series of happenings that go on forever. The concept of every-thing-at-once is an impossibility in the land of eternity where "now" is *always* followed by "later." In the land of eternity, there is no beginning, and there can be no end—yet here you are, smack in the "center" of the conundrum. The logical conclusion for the egocentric mind is "I am God"—or at the least: "I am a chosen one." When you try to add up the infinite-eternal numbers coming from each mind, you arrive at this equation: I am nothing, I am God. I am God, I am nothing—and so on, forever.

Somehow our two minds found a way to be coherent and synchronized, even though they are separate processing-universes. We are created to dually navigate—to explore, which is the ego's quest; and to experience, which is the journey of the self. To explore, to be egocentric, we need space in which to move about. To experience, to be allocentric, we need time to do so. Consequently, the twins, the egocentric mind and the allocentric mind, depend on each other; they cannot exist apart. The figure needs a ground, and the ground needs its figures.

Notice that *Heisenberg's Uncertainty Principle* (from quantum mathematics) reveals the same paradox: you cannot measure eternity at the same time you measure infinity. They cannot both exist at the same time. You have to alternate—one moment existing in eternity, where there is no infinity, and the next moment dwelling in infinity, where there is no eternity. Here is another way to say this: You can live in time, or you can live in space, but not both at the same time. You cannot exist as a purely time-based creature at the same time as you exist as a purely space-based creature. The fact that we do, indeed, seem to live in space-time is a perceptual illusion caused by a fundamental frequency oscillation—on/off: now one, now the other.

I don't know about you, but I get tired of making the bed every morning—the idea of having to make my bed everyday forever blows out all my circuits. As a kid, I would almost go insane realizing that something could never die—and that something might be me. I used to fall on my knees and quiver in terror. I would repeat over and over, "Please, God, I don't want to know; please! I don't want to know!" Woody Allen once said that he lived in fear that God might love the Ice Capades, so that every day for eternity all the saved souls would have to watch Dorothy Hamill do double axles again and again—for eternity. Her show was called Frozen in Time. There is no exit door, no way to say "Please God! Please, no more Ice Capades!" (You can check into the "Hotel California" but you can never leave).

I actually felt a lot better when I realized that my dual minds were creating this eternity-infinity stuff out of quantum thin air. This "eternity-insanity," this "infinity-madness," this bipolar genetic heritage, is just neural-wetware made from repeating spatial/temporal quantum-algorithms. It is very comforting (to me) to realize that beyond this illusory sensory wall is something far more wonderful and far stranger than our two crafty minds can fathom. We have no choice—as the legendary rock band *The Beatles* told us—except to "Let it be, let it be, there will be an answer, let it be."

As we switch between our two minds there is a subtle overlap: One mind arises as the other mind fades away. This gap is like a sine wave peaking or bottoming out. In the gap, infinity and eternity overlap. This gap is a sacred place where mystics and saints, seers and savants, and quantum physicists hang out—here, in the gap, is Einstein's space-time. Finding the gap, the overlap, is a goal during meditation. There are secrets to consciousness hidden in the gaps.[1]

The idea that there are no boundaries, no endings, and no walls that say "end of the universe, life, and all things," seems impossible for a creature witnessing its own developmental cycle. This strange creature is wasting away from old age and the dissipation of all its molecules back into the vapor from which it came. There is an ugly paradox, this bad news that we are not privileged to exist either in time *or* space, never to share in either infinity or eternity—we are just the flash of light that appears when flint is cracked together—there is a flash of light (our lives) and then eternal-infinite darkness (nothing). That is what the egocentric mind makes of our dilemma.

The Devil—a creation of the egocentric mind—is happy when our reasoning ends in depression. Eternally bad news is the ultimate birthday gift for His Evil Highness. However, if we use our allocentric minds, especially as we experience the gaps—the emptiness at the point of overlap—we can feel the void being filled with love, peace, joy, wisdom, and mindfulness. In the gap, dark collides with light, blindness collides with awareness, and despair collides with joy.

We seem to exist in an ugly, unfathomable paradox. Yet to be alive—to have experiences and to share experience with others—makes our mutual journey worth the cost of the ticket. You are on board the paradoxical, conundrum-heavy, freedom train. You might as well enjoy the ride—your parents paid good money for your ticket.

Two Contrary Ways to Be Conscious

Science investigates; religion interprets. Science gives man knowledge, which is power; religion gives man wisdom, which is control. Science deals mainly with facts; religion deals mainly with values. *The two are not rivals. ~ Rev. Dr. Martin Luther King Jr. from "A Tough Mind and a Tender Heart," August 30, 1959 (italics mine).*

Let's look now at how different our two minds have become after millions of years of evolution. I will discuss the two minds as if they are aware of each other—which they are not. I am simply making a case that there are two contrary ways to be conscious. Almost all students of consciousness know about our duality—it's no big news for the experts—but stay with me as I reiterate the obvious. Perhaps my perspective is just odd enough to spark some helpful associations.

Allocentric "perception" is a process for being *aware of immediate experience within a specific location* (domain). The allocentric mind constantly *adapts in the moment* to ever-changing circumstances. The location is very important because the environment must be understood if navigation is to be possible and fluid. Space is a background, a stage where we play the role of human being on planet earth. When we wake up in the morning, for example, we have faith that the floor will still be there to walk upon, that the walls will be where we left them, and that each room we enter will be familiar. The allocentric mind is "satisfied" with simply existing and with the faith-based evidence that space is reliable and apparently infinite.

The allocentric mind is that entity which builds, moment-by-moment, an invariant background. It allows us to cognitively measure relative movement—our own movement and the movement of others. The allocentric mind, therefore, builds *frames of spatial reference*. What we call experience is a set of movements that takes place against a relatively stable background. The allocentric mind neurologically builds this background and keeps it invariant.

The allocentric mind will not debate about reality because it cannot think, so it cannot debate. Therefore, it has no doubts and no beliefs to defend. There is no ego inside the allocentric mind. There is no personality, no sense of being separate from the environment. There is nothing in the allocentric mind that has opinions, so there is no reactivity when beliefs threaten the ego. The allocentric mind simply accepts what is happening in each moment. Allocentric consciousness has no reductionism and no time-based tool like the scientific method for problem solving. For the allocentric mind, there are no problems to solve. Allocentrically, we only understand by having experiences. For the allocentric mind there is no such thing as evolution or progress. It is only within our most recent stage of evolution— especially with the birth of language—that nature is getting around to dividing existence into time segments engineered by an ego.

In contrast, the egocentric mind is appalled at the allocentric line of thinking and accepting things as they are without judgment, since there is no thinking in it, only unsolvable, unexamined faith with nothing scientific to prove or disprove. Of course, deductive logic, reductionism, and time perception are not negative attributes; they are great leaps forward for mankind. Likewise, belief is not the Devil. We do need belief structures to function. We just need to realize that belief evolves. One day, long ago, we were sure that the earth was flat. The next day we were astonished to learn that the earth was a blue ball floating in outer space. Tomorrow we may discover that the universe is itself a sphere floating in some other dimension.

Whereas the allocentric mind is constantly adjusting, the egocentric mind has time to reflect. The egocentric mind is an off-line mind that can daydream, muse, reflect, ponder, review, and problem-solve because it creates a world called "time." The allocentric mind is an on-line mind because it is always active as it interfaces with an environment.

When the egocentric mind is active, the allocentric mind is relatively dormant. When the allocentric mind is flowing—adapting on-the-fly—the egocentric mind is unable to think. The ego is inhibited when the allocentric mind is active. We cannot be on-line the same time as we are off-line, even though the oscillation is so rapid that we perceive oneness.

There Is No Returning To the Days of Superstition

The rational (egocentric) mind has no intention of returning to the days when superstition did so much harm to humanity. There is a lot of ignorance on the Internet, Facebook, and in traditional media that is masquerading as science. This dissemination of false or incomplete communication is getting worse as the speed of communication accelerates. The logical tools used by the egocentric mind are essential for the evolution of a culture that is not dominated by shallow and unreasonable information.

The egocentric mind evolved all our technologies, booted ignorance from the throne, and created high civilizations—it brought order to a world that, left on its own, would dissolve back into the earth as dust. The egocentric mind knows that you have to create, fix, rebuild, innovate, organize, and work, otherwise the world will inevitably crumble—entropy wins out if life doesn't balance destruction with engineering.

The good-life for egocentric consciousness is to solve problems and to accumulate and catalog knowledge. Therefore, the egocentric mind does not accept the here-and-now as a finished product—there are problems to solve and progress to be made. Furthermore, and perhaps most importantly, egocentricity is the latest invention of nature. Evolution wants the ego to become stronger. There is a survival advantage to an organism with a sense of egocentricity. We are not going to drift backward in evolution, so try as we might to eliminate or dampen the ego, it is here to stay and it is the godhead of mental evolution—so says the ego!

In contrast, the allocentric mind's non-verbal response to the ego's self-important rant is to ignore it, to let it be. This makes the ego even more perturbed:

[The ego] has a deep fear of being nothing and is afraid of not having security, power, and possessions. It is thin-skinned and easily wounded, always eager to be recognized, easily discouraged against others, full of self-pity. There is almost constant fear—not a particular but a general fear—of being insecure or incapable, or some other vulnerability. And there is always avidity. I want to obtain, I want to change, I want to become.

> [The usual state of the ego] is negative, always reacting to people
> and events from a selfish, ego-centered point of view—what pleases
> or displeases me, what I like or what I do not like . . . The being, the
> whole being, is forgotten. ~ *The Reality of Being: The Fourth Way of
> Gurdjieff, Jeanne De Salzmann, 2011.*

De Salzmann's statement above that "the whole being is forgotten" is a
reference to the allocentric mind's attributes like love, peace, and joy,
which are opposite to the egocentric. The ego is dangerous, as the quote
above makes clear, and yet, ironically, the ego is a great leap forward in
evolution.

Two Different Jobs, Two Ways to Learn

The egocentric mind always has to pay attention *to something*, some object
of regard. So the ego attaches to things in the world. It attaches to beliefs, to
people, to locations, to habits, to pets, to emotions—good or bad. The ego
is never neutral—it is stuck like superglue to its ideological position. The ego
cannot understand non-attachment—that is not possible given its reason to
exist. *Its role is to attach.* The most powerful attachment that the egocentric
mind has developed is the attachment to objectivity. The ego believes that our
physical universe is filled with forms and that this is reality. There can be no
other reality for the ego—it is blindly attached to the physical. The allocentric mind, however, may have the capacity to pass through the veil of physical perception. Certainly, the world's wisdom traditions hold this faith-based
perspective.

Emotionally, the egocentric mind is appalled by the allocentric mind's
indifference to the past and future, and to the tacit assumption that everything is just as it was meant to be in the moment. To the egocentric mind
it looks hopelessly naive, unproductive, and downright foolish to live in the
moment when the future is so full of amazing potential and so dreadfully

dangerous at the same time. The past is dripping with the bloody evidence that living in the moment is outrageously and practically stupid. The past is strewn with the dead bodies of compassionate allocentric souls who failed to consider the past and future. Logic and evidence are required before the ego will start hugging its invisible twin.

The allocentric mind, however, knows that human beings do not learn or change simply because they acquire knowledge. Facts, systems, lists, plans, schemes, history, future projections—none of these egocentric machinations result in long-lasting biochemical change. No, only experience leads to change. This quote by author John Armstrong sums it up nicely (italics mine):

> . . . human nature is such that we have to learn by experience; it is only through our errors that we come to knowledge. Wisdom falls flat—appears to be merely a set of truisms or platitudes—unless grounded in experience. What we learn is not a set of beliefs; we acquire a set of capacities. *You cannot simply tell someone how to cope with life, how to bear responsibility or disappointment, how to love another person, and expect that will equip them to deal with such things. ~ Love, Life, Goethe, John Armstrong, 2007.*

Here is another way to say the same thing, from the Hindu guru and philosopher Bhagwan Shree Rajneesh (later called Osho):

> What is dead is dead; what has passed is past. The past has gone and the future has not yet come. The moment between the past and the future is the only thing that exists.

> The past is part of memory and the future is part of longing. Both are mental; they have no existence in themselves, they are human creations. If mankind did not exist on the earth there would be no past and no future. There would just be the present, the now, only now—without any passage of time, without any coming, any going. The meditative mind lives in the now—that is its only

existence. ~ *The Great Challenge; the Rajneesh Reader, Bhagwan Shree Rajneesh, 1982.*

Rajneesh is equating the meditative mind with the allocentric mind—they are the same. The only thing that exists, from an allocentric/meditative perspective, *is experience*. This is the core message of religion. There is an emotional, psychological, and physical heaviness to the past and future. However, the present moment has no weight; it feels lighter. From this awareness comes the possibility for what the Buddhists and Hindus call "enlightenment"—staying constantly in the moment, cultivating an ever-shrinking sense of weight and mass. The conceptualization of enlightenment is *a combination* of a lack of weight/mass (emptiness) combined with electromagnetic radiance.

Because the egocentric mind cannot *experience*, there is *no* possibility for ego enlightenment. Thus, enlightenment is an allocentric phenomenon, a spatial experience, rather than an egocentric time-based event. The egocentric mind can talk about space and conceptualize about space, and it can create mathematically eloquent symbolic renderings of space, but it is not aware, so it cannot experience space. The egocentric mind has no heart for space. Unfortunately, the egocentric mind must logically conclude that enlightenment cannot exist. Striving for enlightenment, for spiritual evolution, is not the job of the egocentric mind—it has other responsibilities that have great import for the well-being of humanity.

The egocentric mind knows only how *to do*. It goes from event to event, chore to chore, duty to duty, past to future. It needs recipes that describe how to proceed. It needs to know what it is supposed to do next. It doesn't seem to be happy without a project. It cannot just *be*. It cannot comprehend allocentric experience. Therefore, the ego cannot be, nor can it understand, the meditative mind. It cannot comprehend *being*. It also needs to verbally communicate, so it may be incapable of truly understanding silence. The egocentric mind exists in a cognitive off-line world that lacks humor, poetry, or any kind of artistry—perhaps with the exception of mathematical artistry.

On the other hand, the egocentric mind is the source of individualism, personal expression, and self-assertion. It came into town to kick up dust and make trouble—in its best moments, the ego came to make the world a better place. Get off your meditative butt—so the ego-hero tells us—and make a contribution or two. It is very hard to meditate or pray when the ego is in town.

The egocentric mind is on a hero's journey. The allocentric mind is on a saint's journey. A saint is aware of others; a saint is filled with empathy and with nurturing energy. The allocentric mind is the very opposite of the hero-on-a-quest. From the ego's perspective, saints tend to be roadblocks. Heroes find the saints annoying if not a little, or a lot, attractive. Saints find heroes to be arrogant, pushy, self-absorbed, and quite often very attractive. The twins are powerfully interested in each other, and they are puzzled by the other's beautiful gifts.

The egocentric mind is the home of beliefs, opinions, and a drive to do or to build. Ego is not neutral. It is good or bad, helpful or not. It cannot leave things alone; it has to decide something or solve something; it has to get results. The ego confuses language with emotion; this gives rise to flowery poetry one moment and declarations of war the next moment—all rational, of course, but without empathy. Ego, constructed around belief systems, cannot comprehend the allocentric experience called "faith," which the ego finds to be scientifically unsupported. The ego knows sympathy, but it cannot comprehend empathy, which is an allocentric ability.

Each Mind Has its Own Brand of Religion

Each mind also has its own brand of religion. Egocentric religion is based on rational belief systems. It is a religion that actually needs no god, or guardian angels, or any other person. Egocentric religions invoke deities, anthropomorphic objects-of-regard. These gods are egocentric manifestations, anthropomorphic projections. If the ego decides that there is a powerful god behind existence, then it builds this god in its own image. The ego's religion is, therefore, based on the individual as Hero-God—a human being who has become "all that he or she can be." God and the ego are outside Nature; both are manifestations out of some unknown cosmic background. When this egocentric mind dies, it goes to an Egocentric Heaven where it gets to keep its body, personality, and Enneagram number. Saints make the ego's bed each morning in Heaven and serve breakfast on the veranda everyday—for eternity (pancakes with eggs over-easy). The ego can survive perfectly well without a deity getting in the way because ego is its own deity.

The Egocentric God resides in a universe that is an impersonal force, an energy-matrix that has no way to care. The universe perceived by the Ego is without feeling; it is ruthless. Human beings have to evolve their own meaning for life. The Ego-God has set man adrift in the ruthless furnace of this world.

A healthy ego fights for justice in an imperfect, hostile universe. Ego evolved the rule of law, social structure, and the very idea of justice and injustice. The ego can decide to be good or bad, to care or not to care. The god that is created by this highly refined ego is a fighter who slays dragons and helps the good guys win battles (or football games).

The God of Allocentricity, however, is a wise and benevolent force behind all things. It is a universal, eternal, timeless Background-God that gives rise to all that manifests—a force that has empathy and is part of all that it brings into being. There is no need to examine this force or to doubt it. Our role in the allocentric universe is simply to experience the blessings of our manifestation. We are alive. We are part of the energy called God and that is enough. This Background-God is bastardized (belittled, made iconic) if we try to worship statues, or icons, or artifacts that are merely material manifestations—not the true Cosmic Self.

Meaning and Emotion

If there are two minds within us, then all human concepts can be divided into two general categories, allocentric and egocentric. Two of the most important and powerful concepts are "meaning" and "emotion." Therefore, my contention is that dual-process theory predicts that we would have two kinds of meaning and two kinds of emotion. I discuss this distinction in more detail in the companion book to this one, *The Confusion Caused by Being Your Own Twin*. I will simply make the following summary:

We know that navigation requires two mechanisms, so there are two ways to navigate through life. We are programed dually:

- To follow the mandate of the egocentric mind. *To explore,* to attain. To self-fulfill.
- To follow the mandate of the allocentric mind. *To experience.* To love.

We are designed to carry out two mandates. Our ego is driven to question, to explore, to solve problems, to repair and rebuild, to remain busy with tasks, duties, and commitments. That is its egocentric-job and it knows no other. The self (soul), however, is commanded simply to experience being alive. It has no tasks to accomplish, no duties, no obligations to fulfill.

Like meaning, emotion is related to relevance. Relevance for the ego is about satisfying needs for food, shelter, sex, and companionship. The ego is drawn toward and becomes "attached to" things that have positive relevance, and it is repulsed by things that are harmful and dangerous. However, allocentric emotion is a pure state that has neither attachment nor repulsion. The emotions of the moment are gratitude, appreciation, and loving-awareness. Therefore, like meaning, emotion is also dual. We are programmed dually:

- To have emotions which *are attached* to objects or to sentient others. The egocentric mind requires an object-of-regard. There is a seer (the ego) and a seen (the other) that is inherent and unchanging within the egocentric mind. The intensity of emotion toward the object-of-regard depends on positive or negative relevance.

- To have emotions which *are not attached* to objects or to sentient others. The allocentric mind has universal emotions that are tied to immediate experience.

"The Emotions" was the theme of the 10th Mind and Life conference held in Dharamsala, India, in March, 2000. Many of the world's leading authorities on emotion gathered at the Dalai Lama's headquarters in India to compare Buddhist perspectives with Western perspectives. Anthropologist and psychologist Paul Ekman was among the speakers; he and the Dalai Lama quickly developed a mutual respect and admiration for each other. Here is the Dalai Lama's summary of the gathering:

I learned a great deal from Paul [Ekman] about the latest scientific understanding of emotion. I understand that modern cognitive science draws distinctions between two principle categories of emotions—basic emotions and what some people refer to as "higher cognitive emotions." By "basic emotions," scientists mean those emotions which are thought to be universal and innate. As in Buddhist lists, the precise enumeration differs by researcher, but Ekman mentions as many as ten, including anger, fear, sadness, disgust, contempt, surprise, enjoyment, embarrassment, guilt, and shame. As in the Buddhist mental factors, each of these is seen as representing a family of feelings. By the "higher cognitive emotions," scientists mean a series of emotions that are also universal but whose expression is subject to considerable cultural variation. Examples include love, pride, and jealousy. Experimenters have observed that while basic emotions appear largely to be processed in the subcortical structures of the brain, the higher cognitive emotions are associated more with the neocortex. ~ *The Universe in a Single Atom, The Convergence of Science and Spirituality, the 14th Dalai Lama, 2005.*

The experts, Eastern and Western, agree that emotions fall into two categories. From my perspective, these are clearly allocentric and egocentric. For the lay population, the man-in-the-street, the duality of our emotions can also be understood egocentrically, as pleasure, or allocentrically, as happiness. Pleasure results from attachments. Pleasures provide a momentary surge of endorphins that cause a temporary elation. Pleasures ebb and flow; they are time-based, egocentric, and circumstantial. However, *happiness* is a total-body stability, an allocentric background state that is calm and consistent. Happiness is an aura around the body that can expand to fill a classroom, a city block, a nation-state, and ultimately all of existence. Happiness is spatial, not temporal.

The Meaning of Meaning

The world only manifests as we pay attention to it. We also manifest each other through our relationships, when we pay attention to each other. That which has meaning for us is constantly generated by our consciousness. The more attentive we are egocentrically, and the more awake we are allocentrically, the more meaning we derive from our existence.

The most common name for the energy that gives rise to egocentric attention is *the will*. What the will unearths and studies is what the ego turns into meaning. On the other hand, allocentrically, there is a different kind of meaning. Owen Barfield and Samuel Taylor Coleridge designated a name for the force that gives rise to allocentric meaning. They called it *imagination*. *The will and the imagination are the twin energy systems that create our two kinds of meaning.*

Both the will and the imagination are on the hunt for meaning, but they are both in pursuit of a different kind of meaning—each defines meaning to suit their own needs. The will is in search of the ego's needs. Whatever has relevance for an individual ego is said to have meaning for that ego.

Consciousness: A New Slant on an Old Conundrum

Consequently, egocentric meaning equals relevance. The imagination, however, is not on the hunt for anything relevant to the ego. Allocentric imagination is about creating relationships with others and with nature. It is, actually, the sense of *being* Nature, *being one-with-all.* The imagination is employed when we are in a state of absorption back into Nature, when we have sunk into the gestalt and have surrendered all egocentricity. When we "ride the light wave," as Einstein did, when we *become the process*, rather than analyze, then we find allocentric meaning.

When Rudolf Steiner talked about the science of the occult—the science behind the hidden allocentric mind—he meant that we can use imagination (allocentrically) to perceive in a whole different way—unlike egocentric problem-solving. Owen Barfield tells us, for example, that to hone the skill of intuition (to use allocentric perception) there are stages, mental processes that we must follow. In other words, there is a science behind things like intuition and synchronicity, according to Steiner. There is a whole different kind of meaning that is derived from the spiritual science of allocentricity, to use my terms. Goethe understood this in the 1700s, and he created an alternative to the scientific method of reductionism. He allowed his allocentric awareness to sink into nature and to absorb meaning through intuition. Steiner took Goethe's method and showed how there was a science behind intuitive learning. Steiner went on to form the Waldorf schools using the science of intuition as the philosophical foundation.

Aristotle said something similar about 2500 years ago. He wrote about it in his work called *De Anima,* which means "On the Soul." Aristotle's definition of the soul includes both the egocentric and allocentric minds. He argued that all sentient creatures had a soul, but there were variations in the sophistication of these souls. Essentially, he said that a rational soul (the egocentric mind) only existed within the human species, which had willpower. All sentient creatures, however, also had a basic soul that used sensation and imagination to know the world and interact within the world—this was the allocentric self, or soul.

Second Creators of the World

As concepts and abstractions evolved within the egocentric mind, mankind moved further and further away from a feeling that human beings were *in* nature. To the egocentric mind, environments are spaces separate from the organisms traversing them. The concept of environment comes from egocentric perception, a separation of a person within a surround. The ego-centric mind divided the world into subjective and objective reality, the seer (perceiver) and the seen (perceived). As they evolved egos, human beings became separated from their surround. They became islands floating within an alien atmosphere.

This materialistic consciousness cannot logically discern anything other than a sensory-based reality and so concludes that nothing can logically be hidden. This ego is separate from the world, a water droplet apart from the ocean. Originally, however, nature included everything and extracted no forms from the whole. Thus nature, from an allocentric perspective, is the background out of which life emerged, and yet life-forms can never truly be separate from this background.

The allocentric mind knows that everything is connected; there are no parts, only the whole. Because it *knows* this, the allocentric mind has a genetic predisposition for clean air, clean water, for a healthy atmosphere. There is a love for plants, animals, for the stars, for our common journey. The poorly developed ego, however, cannot understand this perspective. The ego wants an expensive meal in a posh restaurant—with good wait-staff. The ego would sell the earth and buy a condo on Alpha Centauri—if it could get away with it. The ego would turn the entire planet into a golf course for the rich and famous.

Psychologist Carl Jung had a peak experience in Africa when he was about 50 years old, wherein he became aware of this objective-subject evolution. Here is his telling of that experience in his autobiography *Memories, Dreams, Reflections*:

Man is indispensable for the completion of creation. [Man] himself is the second creator of the world, who alone has given to the world its objective existence—without which, unheard, unseen, silently eating, giving birth, dying, heads nodding through hundreds of millions of years, it would have gone on in the profoundest night of non-being down to its unknown end. Human consciousness created objective existence and meaning, and man found his indispensable place in the great process of being. - *Memories, Dreams, Reflections, Carl Jung, 1963.*

I love Jung's statement that human beings are the second creators of the world. Collectively, each day, we co-create our existence. Jung also saw that we had two minds, and he saw how important it was for human beings to understand how these minds worked. In this quote, Jung is stressing the importance of understanding how our cognition works, especially the soul (the allocentric mind):

For Jung the study of the soul became a matter of grave historical importance, for, as he once said, the whole world hangs on a thread and that thread is the human psyche. It is vital we all become more familiar with it. - *Jung's Map of the Soul, Murray Stein, 1998.*

Jung could see that the quantum world—which was a new perspective in his lifetime—was important for understanding the psyche. Jung was friends with physicists Albert Einstein and Wolfgang Pauli. Jung was hungry to understand the quantum world because he could see that it was a game changer for the study of mind. For Jung, quantum physics destroyed our supposition that both time and space are absolutes (italics mine):

In man's original view of the world, as we find it among primitives, space and time have a very precarious existence. They become "fixed" concepts only in the course of his mental development, thanks largely to the introduction of measurement. In themselves, space and time consist of nothing. They are hypothesized concepts born of the discriminating activity of the conscious mind, and *they form the indispensable coordinates for describing the behavior of bodies in motion.* They are, therefore, essentially psychic in origin . . . ~ *Synchronicity, Science, and Soul-Making, Victor Mansfield, 1995.*

Jung was a scientist and a medical doctor. In this role he was a materialist. However, as a psychoanalyst he found himself face-to-face with some very unusual minds. The concept of synchronicity kept appearing in the stories and lives of his patients. I have written about Jung's perspective on synchronicity in my book *The Confusion Caused by Being Your Own Twin.* Here, I will only point out that trying to solve the strangeness of synchronicity leads the seeker to the allocentric mind.

Should We Believe or Should We Have Faith?

Two languages slowly evolved to explain the workings of our two minds. For example, "intelligence" is what egocentric minds strive toward. While the word "wisdom" is often used as a synonym for intelligence, wisdom is actually what allocentric minds strive toward.

Intelligence is the egocentric ability to solve problems. It is about patterns: noticing patterns, grouping patterns, extracting whole patterns from partial patterns, measuring patterns, naming patterns, turning patterns into mathematics, remembering and recalling patterns, and so on. Intelligence is about explicit meaning, about features, about figures, about systems-thinking and foreground analysis.

Wisdom, however, is about quality and process. It is allocentric, empathetic knowing. Wisdom is the sum total of personal experiences that have affected life directions. Wisdom is about implicit meaning and about correct actions based on experience. The root of the word wisdom is "to taste." Wise people taste the world, drink it in, *experience the moment*; there is no need for the allocentric mind to dissect experience using the logic of intelligence.

Likewise belief is an egocentric concept while faith is an allocentric concept; the two words are often used interchangeably, but they are quite different concepts arising from the two minds.

Thomas Aquinas, one of our most important medieval philosophers and theologians, said:

"To one who has faith, no explanation is necessary. To one without faith, no explanation is possible." ~ *This is probably a paraphrase. No source was found during an internet search.*

Here, Aquinas is differentiating our two minds. Faith is a non-verbal knowing—it comes from the allocentric mind. It cannot understand explanations; it doesn't care about explanations. In contrast, the ego requires proof, logic, and dialogue. It cannot comprehend silence, or faith, or total acceptance without reason.

The egocentric mind is constructed around the ego. The allocentric mind is constructed around the self. We have, therefore, two fundamental identities. The egocentric mind uses the word language to mean information exchange. However, the allocentric mind uses non-verbal communication, which is a form of animal-based data exchange. Data used for allocentric communicating is based in the present moment and is never novel—it is based on genetics and heredity. However, language, coming from the egocentric mind, *can* go into the future or back into the past and is almost always novel—it is based mostly on learned behaviors.

So we have intelligence contrasted with wisdom, belief opposite faith, and ego facing the self. We have mirror images that arise because we have two minds. I have given this division of language in our two minds considerable thought and have summarized my conclusions in the next section.

Who We Think We Are Versus Our "True" Self

The great religions, according to Franciscan Friar Richard Rohr[2], ask us to go beyond the ego, to find a more sophisticated and loving mind. This wonderful quote from Richard Rohr begins by pointing out Albert Einstein's awareness of our dual minds:

> I often use this line, a paraphrase of Albert Einstein: "No problem can be solved by the same consciousness that caused it." Unfortunately, we have been trying to solve almost all our problems with the very same mind that caused them, which is the calculating or dualistic mind. This egocentric mind usually reads everything in terms of short-term effect, in terms of what's in it for me and how I can look good. As long as you read reality from that small self, you're not going to see things in any new way. All the great religions taught a different way of seeing, a different perspective. This alternative vantage point is the contemplative or non-dual mind. It is what we usually mean by wisdom. *~ Richard Rohr's Daily Blog, "Alternative Consciousness," Monday, May 9, 2016*

This quote even uses the term "egocentric mind" and compares that to a different mind, a different way to perceive, which is contemplative and wise (allocentric). Below, I have compared the mind that thinks it is alone in the universe—the small self, as Rohr calls it, the egocentric mind—with the soulful mind, the larger self, the allocentric mind.

ONE

The Egocentric mind is who we think we are: It is what we usually mean when we say that we are conscious. The notion that consciousness can be defined as "knowing that we know" is strictly an egocentric perspective.

Most of the time, the world appears to us as if we are the center around which everything manifests. No matter where we are, we are the center of a global surround. Always being the center of attention is a rather powerful affirmation of our importance to the universe—or so it would appear.

Egocentric attention is a kind of flashlight that we can control—it illuminates windows through which we observe the world. For example, if we want to pay attention to our right hand, we shine the mental light on the hand; as a result, we perceive the hand. We can adjust the flashlight to see just one finger, or a finger nail, or a sliver at the tip of a finger. We can also widen the flashlight's beam and take in the wrist, forearm, elbow, and so on. We move our window of observation into the objective world and focus the beam on objects and pathways. We use our egocentric flashlight to illuminate what it is that we wish to study. We want to know:

> "What is this? What is happening here? What meaning does this have for me? What information should I collect and remember? How can I use this information?"

This kind of observation of the world through a wandering window of perception we call *paying attention*, although it is purely egocentric and not at all like allocentric *awareness*.

In terms of survival billions of years ago, it was important to know during the act of perception whether "localized change in the background" had meaning for the organism. Meaning eventually determined what we call behavior. Slowly, over time, localization of disturbance gave rise to external senses that could observe—pay attention to—the area of disturbance. The area of disturbance eventually took shape; features emerged from the background. Egocentric attention resulted in the understanding of invariance—patterns that repeat. The neocortex evolved with the primary purpose of finding, analyzing, and remembering invariants. From the extraction of meaning, the extraction of relevance, language, emotion, and interpersonal relationships gradually evolved.

The allocentric mind is who we vaguely sense that we truly are—our hidden (real) self—our soul. At its biological core, allocentricity is about monitoring movement in the surround. It is hyper-alert for anything out of the ordinary in the spatial world. It is hyper-sensitive to flow. Therefore, it provides the whole organism with a navigational tool for relative motion detection. On the other hand, it is happy with a steady state—it loves dependable rhythmic flow. The allocentric mind is the background steady-state out of which figures can be perceived to emerge. Neurologically, it processes all-at-once, in parallel. Therefore, the allocentric mind gives us a sense of oneness. This sense of being merged with nature is called the *self* to differentiate it from the *ego*, which is an object separate from the environment. This allocentric self, in the esoteric literature and in religious contexts, is often called the *soul*.

Using a map-like perspective—an overhead, bird's-eye view of a terrain—allocentric mind gives rise to a global sense of relationship. Every object and every creature within the domain has a location compared to everything else. This is a gestalt, a sense of the whole. We can change perspectives regarding the bird's-eye-view, making the representational maps ever larger or ever smaller to get a different perspective—we can move closer or back away. As we look from ever-wider perspectives, our spotlight becomes a floodlight and we discover something miraculous: we find that everything is connected, everything has meaning, and the whole of creation is a relationship.

The egocentric mind is the seer, not the scenery. The allocentric mind is the scenery, not the seer; it is un-manifest potential movement. The allocentric mind is that part of animal life that is animated. The *soul* is a name for the animated *self*. From another perspective, the soul permeates the whole of existence, while the self is a manifestation of soul—a small part of *soul*.

Poets, mystics, spiritual seekers, artists, neuroscientists, psychologists, linguists, historians, writers, philosophers, educators, and many others from many different professions, have identified this same duality, each from the perspective of their own discipline. The duality each discipline has unearthed is the same entity, even if the words used to describe it differ. We are split down the mental middle between logic and intuition, between science and art, between rationality and spirituality. We have a mental duality that is extensively documented in the literature of every field of study. We cannot ignore the evidence; it is unearthed at every research level in every discipline. I look at these various disciplines and discuss how they have embraced duality in the next book in this series called *The Confusion Caused by Being Your Own Twin*.

The poem below is another example of our duality from the poetry of Sufi mystic Jelaluddin Rumi who lived over 700 years ago. This poem differentiates the allocentric mind from the egocentric mind; it is a contrast between the intellect (the ego) and a hidden self:

Two Kinds of Intelligence

There are two kinds of intelligence: one acquired,
as a child in school memorizes facts and concepts
from books and from what the teacher says,
collecting information from the traditional sciences

as well as from the new sciences.
With such intelligence you rise in the world.
You get ranked ahead or behind others
in regard to your competence in retaining
information. You stroll with this intelligence
in and out of fields of knowledge, getting always
more marks on your preserving tablets.

There is another kind of tablet, one
already completed and preserved inside you:
a spring overflowing its spring-box; a freshness
in the center of the chest. The other intelligence
does not turn yellow or stagnate. It is fluid,
and it does not move from outside to inside
through the conduits of plumping-learning.

This second knowing is a fountainhead
from within you, moving out.

~ "Two Kinds of Intelligence,"
A Year with Rumi, 2006.

The above poem, translation by Coleman Barks, seeks to put an ancient idea into modern language. Rumi understood 700 years ago that we had two minds, and he knew the role of these two minds. The egocentric mind is easy to see; it is the mind that gets explicit knowledge from books and teachers, and from observations attained through the use of the external senses. However, the allocentric mind, which arises from the internal senses, is "in the center of the chest." It doesn't get explicit knowledge from outside; instead, it implicitly generates internal knowing and sends this forth from the inside to the outside.

Two

The egocentric mind is about "doing" rather than "being." The egocentric mind is constantly, incessantly, trying to accomplish something.

The allocentric mind is about "being," rather than "doing." It has no desire or need to accomplish anything.

Three

The egocentric mind judges; it has an ego-opinion. The egocentric mind decides what has "meaning" and what doesn't. Life is all about *you*, a first-person perspective. Emphasis is on independence and individuality—it is about individual rights. The egocentric mind *reacts* rather than participates.

The allocentric mind is non-judgmental; it has no ego, it accepts what is. It is a third-person (bird's-eye) point-of-view. The emphasis is on co-dependence, cooperation, and social equality. The allocentric mind *responds* rather than reacts. It participates within nature—it is an inseparable part of nature.

Four

The egocentric mind reasons deductively. It is the "rational" mind.

The allocentric mind reasons inductively. It is the non-rational mind and has been called the spiritual or empathetic mind.

We can either solve problems using deductive logic, or we can solve problems using inductive logic. This is nothing more than egocentric problem-solving versus allocentric problem-solving. If we accept dual-process theory, it makes sense that we would have two ways to use our different cognitive systems to understand and navigate through the world.

The description of deductive logic is a description of egocentric processing. Deduction moves from the general (background) to the specific (something manifests out of the surround). Deduction is a flow from a broad perspective to ever finer observations, a reductionism, taking the whole apart to reveal ever finer particles. Deductive logic starts from an open, expansive perspective and proceeds toward ever narrower points of view. From a quantum theory perspective, deduction moves from wave perception to particle perception—it collapses the wave function.

Contrary to deductive reasoning, the description of inductive logic is a description of allocentric processing. Induction moves from the very specific and broadens out towards an ever more expansive background—it absorbs objects back into the surround. Induction moves from specific observations and phenomenon to create a gestalt; it takes puzzle pieces and builds pictures. Induction is an open-ended, bottom-up approach that we call the creative process, or the innovative process. In terms of quantum theory, induction is the opposite of collapsing the wave function; it is the reformation of the wave-state when particles dissolve back into the surround.

This distinction between deductive and inductive reasoning gets confusing when we realize that an object in a scene can itself become a scene (a background) which contains smaller units. It is as if logic itself is nested (like Russian dolls). The study of logic has a long history and many subdivisions. This broad generalization is just an attempt to fit logic into my overall theory.

FIVE

What the egocentric system "knows" is called "belief." It is the rational mind. This rational mind assumes that it is separate from physical reality. The rational mind is a new entity in evolution, and as such gives rise to a mind/body problem—that must be solved

What the allocentric system "knows" is called "faith." It is the empathetic mind. This allocentric mind does not think, does not categorize, and does not analyze the past or future. It has a genetic inheritance that allows behavioral response in the moment. It learns from experience.

Although philosopher René Descartes popularized what became known as the mind/body problem, awareness of human duality had been around for a long time before Descartes. In his book *Consciousness and the Brain*, cognitive neuroscientist Stanislas Dehaene, gives a nice encapsulation of this history:

> The idea that the mind belongs to a separate realm, distinct from the body, was theorized early on, in major philosophical texts such as Plato's Phaedo (fourth century BC) and Thomas Aquinas's Summa Theological (1265-1274), a foundational text for the Christian view of the soul. But it was the French philosopher René Descartes (1596-1650) who explicitly stated what is now known as dualism: the thesis that the conscious mind is made of a non-material substance that eludes the normal laws of physics. ~ *Consciousness and the Brain, Stanislas Dehaene, 2014.*

The debate over duality often is about this perspective. Descartes simply acknowledged the strange divide between our mental world and the world out there. There seems to be objectivity—the stuff of the universe—and subjectivity, the stuff of the mind. The mind can do whatever it wants whenever it wants, at least through the use of thoughts and images. However, these mind creations apparently have little impact on the stuff of the universe;

stuff which just keeps plodding along through evolution on a trajectory of its own design—no matter what the mind is obsessing about.

Descartes decided that the vaporous internal mind was spiritual and akin to God; he arrived at this conclusion logically because, as a scientist and a reductionist himself, a man who saw the body as a kind of mechanical automaton, he could not bring himself to believe that human beings were mere machines without purpose. Descartes is one of the original Western thinkers to frame the God-versus-Matter debate.

The philosopher Emanuel Kant, of course, had a now famous duality-label for this internal versus external conundrum. Kant defined the stuff of the objective world as *phenomenon*—a word that has stuck—and contrasted it with the internal mental world of *noumenon*, a word that has not stuck; except in philosophical heads. Phenomena are what our senses make of objectivity. We are, according to Kant, badly limited by our fragile and not so dependable sensory systems. Therefore, we can never know the real objective world; we can only know how our minds have represented objectivity. So, here we have another duality: what we can know versus what we cannot know.

People who rail against duality are almost always speaking about Descartes's assertion that the mind is independent of biology and physics, that it is a mysterious entity that is somehow God-ordained and uniquely human. Dual-process theory is not the same thing as Descartes's mind/body perspective, and "duality" is more complex and pervasive than his viewpoint.

SIX

The egocentric mind is the source of the secular. It is the domain of the intellect. It is more about *training* and less about *education*. Training is a sequential, curriculum-driven list of procedures that must be followed to receive credit. Training is entirely egocentric.

The allocentric mind is the source of the spiritual. It is about *educating* the whole person and less about *training.* Education addresses both minds and seeks to integrate them.

SEVEN

The egocentric mind likes quantities and measurements. It is not so keen on quality, subjectivity, and other questionable subjects. It cannot do qualia. It cannot *experience.*

The allocentric mind is about the quality of experience; it doesn't measure or compare quantities.

EIGHT

The egocentric mind examines the world rather than experiencing the world. It perceives the world from outside inward.[3] The term used to explain this kind of perception is *attention.* The egocentric mind reacts indirectly, filtering all input through a process that searches for relevance and meaning. Willpower and intention are egocentric.

The allocentric mind experiences the world, rather than examining the world. It perceives the world from inside outward. The term used to explain this kind of perception is *awareness.* The allocentric mind responds in the moment to changes in the environment. There is no filtering for meaning and relevance. This response system is hardwired and genetically determined. No will power is involved.

NINE

The egocentric mind has a memory system for "knowledge." It stores mostly long-term memories.

The allocentric mind has a memory system for "experience." It stores long-term memories of experiences, but uses short-term memories while navigating (while moving).

Consciousness: A New Slant on an Old Conundrum

The Greeks had two words for knowledge. *Episteme* was information gathered by the rational mind; the objective stuff of the universe could be collected, examined, and cataloged. This gave birth to epistemology. *Gnosis* also meant knowledge, but it referred to experiential knowledge, immediate awareness. *Gnosis is* knowledge gained through the allocentric mind, and *episteme* is knowledge gained through the egocentric mind. Therefore, the Greeks knew 2500 years ago that there were two ways to know about the world.

Philosopher of mind Zoltan Torey (1929–2014) postulated, as have many others, that our minds were split into two by the advent of sophisticated language. He asserted that language was unique to human beings and it became internalized very recently in evolutionary history. Essentially, Torey provided an anatomical and physiological explanation for a kind of consciousness unique to the human species. He said that we have an on-line (allocentric) perception and an off-line (egocentric) perception. The information collected by the off-line world was episteme knowledge, while the information gathered by the on-line world was knowledge based on experience. Later in history, Gnosticism came to mean "esoteric knowledge"—knowledge that was not based on rational thought.

The on-line world is our awareness of what's "out there" in objectivity-land, what we experience with our whole self. The off-line world is our language-based, reflective mental life. Torey believed that the everyday dualities that define humanity are the result of the balance between on-line and off-line perceptions. For example, we have art/religion versus science, spirituality versus rationality, faith versus belief, and so on, because language has given labels, and then concepts, to the operation of the two systems, on-line and off-line.

Torey bypassed the debate about ultimate origins and said simply that our dualities can be explained biologically. However, what Torey and other researchers have done is to equate consciousness with the egocentric mind. This is no longer acceptable—from the perspective of dual-process theory—because the egocentric mind lacks empathy and has created a barren "reality" in its own image. We are more than creatures with rational capability; we are also poetic souls with the ability to be compassionate and empathetic.

TEN

The egocentric mind's neural pathways primarily originate in the temporal lobe. This is a controversial statement since processing in the brain flows from the back to the front—all processing is thought to begin in the occipital lobe. However, the egocentric mind is dominated mostly by the sense of hearing (language). Auditory processing predominates in the temporal cortex.

The allocentric mind's neural pathways primarily originate in the occipital lobe.

ELEVEN

The egocentric mind is predatory, aggressive, and competitive. It tends toward pessimism. It gives birth to patriarchal, hierarchical systems. It lacks humor and irony.

The allocentric mind is passive, not aggressive. It accepts what is. It is cooperative and non-violent, and it tends to be optimistic. Humor and irony bubble up out of the allocentric.

TWELVE

The egocentric mind *believes that life is a random accident of nature.* The egocentric mind gave birth to materialism, spiritual capitalism, and dogma.

The allocentric mind knows that life is not an accident; everything is connected and interdependent.

THIRTEEN

Sticking my neck out—ever wary of philosophers with sharp axes—I suggest that the entire reason for *the differentiation between ontology and epistemology is due to the distinction between allocentricity and egocentricity.*

Epistemology is egocentric; it is about brain processing—assigning names, concepts, and relevance to what the ontological mind sends it for analysis.

Ontology is the study of what immediate experience (the allocentric mind) tells us about existence and about our sense of being.

FOURTEEN

The egocentric mind sees the world organized around dominator hierarchies. A dominator hierarchy is a top-down, oppressive structure in which those at the top are privileged and everyone below is progressively less privileged—a class stratification system. This is a model, or a perspective, that is primarily patriarchal, using masculine energy to control the lower rungs of the hierarchy. Business and government use this system to maintain order and to make decisions. Dominator hierarchies are based on exclusion.

The allocentric mind sees the world organized around growth hierarchies. A growth (actualization) hierarchy is a natural outgrowth of evolution in which smaller units (holons) are combined to make larger units. Small systems coalesce to create larger systems. This is the way nature evolves—it is organic. Growth hierarchies are based on inclusion. Here is how integral philosopher Ken Wilber differentiates the two kinds of hierarchies:

> . . . it is absolutely central [to differentiate] dominator hierarchies and actualization hierarchies. Actualization (or growth) hierarchies are not exclusive and domineering, they are inclusive and integrating. With each of the levels of a dominator hierarchy, the higher the level, the more it can oppress and dominate (as with the caste system, or criminal organizations like the Mafia). With growth hierarchies (or "holarchies"), it's exactly the opposite. In a growth holarchy, the whole of each level becomes an included part of the whole of the next higher level—just as,

in evolution, a whole quark becomes part of an atom, a whole atom becomes part of a molecule, a whole molecule becomes part of a cell, a whole cell becomes part of an organism, and so on. Each level is a whole/part, what Koestler called a "holon." The ever-increasing inclusiveness—genuine inclusiveness—of holons and holarchies demonstrates a direction that is grounded in nature and that has been operative from the first moment of the Big Bang forward, a direction of self-organization through self-transcendence that is the primary drive of evolution itself. ~ *Trump and a Post-Truth World, Ken Wilber, an online document published 2017.*

Verbal Versus Mute

Language is in service of the unsayable. When it comes to comprehending God and the great mysteries of love and death, knowing has to be balanced by unknowing. Words can only point a finger toward the moon; they are not the moon or even its light. They are that by which we begin to see the moon and its light. ~ *"All Language Is Metaphor," January 11, 2017, Richard Rohr's Center for Action and Contemplation, a blog post.*

ONE

The egocentric mind is verbal, the well-spring for words. It has an auditory language. It can discuss what it perceives. It deals with explicit knowledge. The very notion of perception is egocentric. The ego learns and accumulates knowledge. However, it does not move through levels of consciousness but rather expands horizontally and expansively as knowledge is compiled. Egocentric knowledge can be articulated, codified, and accessed. It can be transmitted to others. Most forms of egocentric knowledge can be stored in various kinds of media. The information contained on computer databases and textbooks are examples of explicit knowledge.

The allocentric mind is mute, silent. It has no verbal means of expression. The self or soul experiences and transcends vertically through levels of consciousness as it evolves. It deals with implicit or tacit knowledge. Tacit knowledge—as opposed to formal explicit knowledge—is knowledge difficult to transfer to others by means of writing or verbalizing. Art (creativity) is the allocentric mind's mode of expression

Two

The egocentric mind gave rise to "internal dialogue," to thoughts, to the monkey mind, and to concepts. It created an inner language. This is called the off-line mind. It does not directly perceive while it is musing internally. The egocentric mind is not such a good listener; it is not empathetic.

The allocentric mind has no internal dialogue, no monkey mind. It is receptive, a good listener, an equal partner in the gestalt. It is an empathetic mind.

Three

The egocentric mind expresses its ego concretely and directly. Written and spoken language—expressed logically and succinctly—is the egocentric minds only means for communication.

The allocentric mind expresses its "self" metaphorically and with analogy. The allocentric mind communicates through the imagination; it uses intuition to connect with others. We would rather hear a story about an experience than we would listen to a list of logical rules or concentrate on understanding a theoretical perspective. The allocentric mind is ravenous for experience and communion.

Here is a quote from Deepak Chopra about the Hindu Vedas that could easily be about allocentric and egocentric minds—neither Chopra nor the Hindu Vedas, of course, use dual-process terminology:

> Those who know *It* speak it not, those who speak of *It* know it not. The mystery here is tied up in the word *It*. If *It* means some kind of revelation, then you may struggle all your life to join the elite who have *It* revealed to them. Enlightenment turns into something like a secret handshake. But if *It* means a real place that one can journey to, there is no need for frustration. You just find that place, without pointless words. "Don't talk about it, go!" seems like sensible advice. ~ *How To Know God, Deepak Chopra, 1998.*

The Vedas in the quote above are reflecting the awareness that the allocentric mind is silent and unknowable through the egocentric mind. If we try to understand the allocentric mind, using reason, it is impossible. If there is talk, logic, and analysis, then we know that the egocentric mind is at work. We also know that the egocentric mind cannot know the allocentric mind. So those who speak do not and cannot understand the allocentric mind. Those who understand allocentrically have no way to speak, no way to verbally explain. Ironically those who use their allocentric minds *just know*. They have faith. They communicate through the imagination. Creativity and innovation, as expressed in the arts, are allocentric avenues for sharing.

Figure Versus Ground

ONE

The Egocentric mind pulls figures out of the background. *It manifests objects* within the global surround. It creates the material world. These figures are patterns, environmental invariants and they are processed in the neocortex of the brain.

The allocentric mind processes flow, rather than freezing objects/ events. It is the background mind. It assembles black-and-white, flat spatial maps. It processes almost immediately, as a whole-body system, responding to the environment in the moment.

Two

The egocentric mind is under voluntary control. To a degree, we can manipulate our large skeletal muscles. We move toward or away from objects and people that attract us or repel us. Our attention works the same way: we can shine our spotlight on objects and people at various depths, even if they are moving. Compared to allocentric processing, the *egocentric mind processes quite slowly*.

The allocentric mind functions without voluntary control, automatically; it simply responds to immediate experiences. *It processes very fast.*

Three

The egocentric mind must perceive in regard to something or someone. That is why intentionality, for example, is egocentric: you only have intent in relation to an-object-of-regard or an action. *The egocentric mind processes serially,* one event at a time. It has a temporal and a hierarchical design.

The allocentric mind perceives spatial arrangements and has a spatial and nested perspective—rather than hierarchal perspective. The allocentric mind processes in relationship *with* something, rather than in *regard to* something. The allocentric mind cannot "pay attention," but it has vigilance—levels of conscious *awareness*. *The allocentric mind processes in parallel—all at once.*

Four

Hearing is the sense that makes the major contribution to egocentricity. However, hearing makes both an allocentric and an egocentric contribution to cognitive processing. Since egocentricity is mostly a temporal process—that begins in the temporal cortex—it makes sense that it is the major contributor to the egocentric mind. Indeed, *all the senses*—contribute to both allocentric and egocentric processing.

Vision is the sense that makes the major contribution to allocentricity. However, vision makes both an allocentric and an egocentric contribution to cognitive processing.

FIVE

The egocentric mind sees parts, serial events—time sequences. It is not holistic. There is no sense of the whole. Because of this sequential nature, egocentricity concludes that everything has a cause, or is determined. There can be no free will. *This perspective gave rise to the very notion of evolution.*

The allocentric mind is holistic; there is a "feeling" of being immersed in the whole. Nothing is caused or determined. Everything arises out of potential within every moment. There *is* free will. Innovation and creativity can be allowed to manifest or allowed to be called forth—channeled. Nothing evolves in this state; there is no history and no evolution.

SIX

The egocentric mind cannot perceive flow; it freezes the world into objects. It inhibits flow.

The allocentric mind is hypersensitive to change/movement. It is activated by flow. It inhibits egocentric processing. It resists anything that inhibits natural flow.

SEVEN

The egocentric mind creates what we mean by "time." The egocentric mind manufactures a temporal matrix we call time. It can even go beyond objective reality to create imaginary realms. It is perfectly possible that time has no meaning outside of human perception. The egocentric mind conceives of, and creates, eternity because it can just keep creating more and more imaginary time. Therefore, the egocentric mind is the time-mind. *The egocentric mind is what scientists call the "What is it? Who is it?" processing system.*

The allocentric mind manufactures what we mean by "space." *It is the "Where is it? Where am I?" processing system.* It conceives of, and creates, infinity because it can just keep creating more and more imaginary space.

Because we have a genetically endowed oscillation between perception of the ground and perception of features, we can never stay permanently in one state of attention or the other. When we flip from allo to ego to allo—and so on—we move from time to timelessness in a rhythmic beat. The same for space: we flip from spatial awareness to no sense of space, like breathing in and breathing out.

Notice that we can also say that this oscillation is between creation and destruction, building up and tearing down, entropy versus syntropy (negentropy). It can also be understood as an intermittent, alternating blindness—one moment blind to time, the next moment blind to space. We are remarkable creatures who exist in two universes that can never collide.

EIGHT

The egocentric mind doesn't seem to be aware of "space." How could it, given its totally opposite mandate? The egocentric mind can only handle things one-at-a-time, in a time sequence, a serial order. Its job is foreign to spatial attention, to everything-at-once perception. It likes closed systems and is resistant to outside perspectives. The egocentric mind is systems-oriented, analytical, and pattern-hungry. *It is all about content, and not at all about process.*

The allocentric mind doesn't seem to be aware of "time." How could it, given its totally opposite mandate? The allocentric mind can only handle everything-at-once; it processes spatial arrangements in parallel rather than in serial order. Its job is foreign to temporal attention, to one-at-a-time perception. It likes open systems and is resistant to rigid perspectives. The allocentric mind is not concerned with systems or analysis; it is not searching for invariant patterns. *It is entirely about process, and not at all about content.*

NINE

The egocentric mind uses correlation and causal logic to articulate temporal relationships. *The egocentric mind tends toward linear causality.* Because egocentricity is tied to time perception, the ego-mind is aware that one thing follows another. This linear way of perceiving is a Western habit-of-mind. It is actually an old paradigm. The new paradigm is allocentric, akin to the Buddhist understanding of dependent co-arising. In her quite unusual (and fascinating) book *Mutual Causality in Buddhism and General Systems Theory*, Joanna Macy writes:

> Words like synergy, feedback, causal loops, symbiosis have become current and useful. They suggest that events affect each other in a back-and-forth manner, creating circuits and networks of contingency where causes and effects interact reciprocally. They express a paradigm which challenges the assumptions about causation that have dominated Western culture for over two millennia ~ *Mutual Causality in Buddhism and General Systems Theory, Joanna Macy, 1991.*

Linear causality has been applied to brain processing so that we tend to say—in the West especially—that processing proceeds from peripheral end organs to central brain centers, or that processing proceeds from the back of the brain at the occipital lobe forward to the prefrontal cortex. This can be misleading because the brain is a network, a meshwork of entangled neural cells of multiple varieties. It is misleading to think of brain processing as simply linear.

The allocentric mind uses analogy and acausal logic. Allocentric, total-body processing is a highly complex, co-arising, co-dependent system. Causality makes no sense in a simultaneous, co-arising, propagation of events. This complex co-arising can be understood as spatial expansion—rather than a temporal series. If logic and rationality fail to convey the language of the allocentric mind, only stories and analogies have appeal.

Duality Versus Non-Duality

ONE

The egocentric mind is where duality comes from; it gives rise to the mind-body debate.

The allocentric mind is the source of non-duality. There *is no* confusion in the allocentric mind caused by language or the perception of time.

TWO

The egocentric mind feels (is sure) that it is the only mind. For the egocentric mind there is, and cannot be, a twin, a hidden mind—the egocentric mind denies the allocentric mind. Yet, ironically, egocentricity is the mother of duality. Therefore, the egocentric mind can accept duality, and can conceive of—and eventually theorize—a twin, but it can never *actually perceive* the twin. It is forever in the shadow of doubt—hence, the origin of the *shadow self.*

The allocentric mind is a hidden mind (from the ego's perspective). The allocentric mind cannot perceive its own existence. It doesn't "perceive" and it doesn't "think." *There is no sense of "ego or other"* that is separate from the whole. Allocentricity is the mother of unity. This mind cannot know duality.

219

Defining the Two Minds

Psychologists like Sigmund Freud and Carl Jung popularized the idea that we had a conscious mind and an unconscious mind. In other words, we had an egocentric mind that was awake, and we had a hidden mind that the conscious ego could not perceive. The conscious mind, according to psychoanalytical theories, was not as powerful, nor as conscious as it supposed itself to be. The unconscious mind was seen as a vast well of underground activity and influence. You could converse with others or have an internal dialogue using the conscious mind, but the unconscious mind was non-verbal—it "spoke" in archetypes, mythologies, metaphors, and symbols. The unconscious mind was a behind-the-scenes dreamland out of which thoughts, emotions, and images could bubble up to the surface to inform the conscious mind. Here is how author Colin Wilson describes the nature of the conflict:

> The real conflict in man is not between "body and spirit," but between different levels of the mind—what Freud might have called conscious and subconscious. He only knows himself as a consciousness; but everyday consciousness is like the last quarter of the moon, a mere fragment. In certain moments of deep insight, the full moon suddenly appears and man experiences a curious revelation, knowledge of his true identity. ~ *Poetry and Mysticism, Colin Wilson, 1969.*

What Colin Wilson is talking about here is the few-and-far-between moments when the egocentric mind is deeply inhibited, allowing the allocentric mind to be fully expressed. We think we are egos, that egocentricity is our true essence, and that our personalities are who we actually are. In reality, we are mostly asleep, on autopilot. Rarely are we rational beings.

I believe that Freud's conscious and unconscious minds are another name for the duality that all seekers discover within their own domain of expertise; the psychoanalysts rediscovered the egocentric and allocentric minds and gave them yet another set of names.

THREE

The egocentric mind takes whole puzzles apart; it reduces the world to ever-smaller puzzle pieces; it is the source of reductionist science.

The allocentric mind assembles puzzle pieces into gestalts. It provides the framework, a foundation, for whole pictures.

FOUR

The egocentric mind is the particle, from a quantum perspective, not a wave. Egocentricity collapses the wave function.

The allocentric mind is a wave, not a particle. Allocentricity restores the wave function.

Notes

(1) Secrets to consciousness are hidden in the gap. The "gap" is a key idea that comes up in various contexts—the Buddhists, for example, speak of the importance of finding the space between the in-breath and the out-breath. We often get lost in the gaps—mindfulness is lost as we lapse into a dream state. Or, to the contrary, when we are daydreaming, a gap can return us to mindfulness.

(2) Richard Rohr is a delightfully complex Catholic priest who moves freely from Christianity to Buddhism. Rohr's daily blog has been an inspiration to me for many years. The great religions, according to Rohr, ask us to go beyond the ego, to find a more sophisticated and loving mind. Rohr sees the two minds of mankind and he writes eloquently about our internal cognitive divide. There is a Perennial Tradition, Rohr tells us, that is evident throughout history and is at the core of every religious practice across the globe.

(3) Perceiving the world inwardly or outwardly. The allocentric mind is served by proprioception, by the internal senses. Therefore, the allocentric mind responds to information arising from within the body—from inward "perception" comes outward responses. Contrary to this, the egocentric mind is served by the external senses which gather information and funnel it *into* the body—from outward perception to inward reflection. This line of thinking can become confusing when we compare *processing* as opposed to *perceiving.* The egocentric mind *processes* inwardly and reacts outwardly. This means that egocentric *processing* flows from inwards to outwards, the exact opposite of what happens during perception. Likewise, the allocentric mind processes immediate outward environmental impacts and then makes immediate bodily adjusts—so it *processes* outwardly to inwardly. To remain clear as we dialogue, we must be sure whether we are discussing perception or processing.

Six

Sometimes
organization and computation
become absurd.

~ "The Miracle-Signs," *A Year with Rumi*, 2006.

Levels of Consciousness Stew

"Welcome back to The Third-Eye-Watching Vegan Restaurant and Center for Contemplating Palate."

"Thank you, Surge. What's your special today?"

"Today is very special, indeed; today we have Levels of Consciousness Stew."

"What else?"

"That's it. The cook went home with the flu."

"Oh, great. The cook made stew with a drippy nose."

"I assure you that the stew is delicious, although every level of palate has a different opinion about taste. Nothing we can do about that. We just serve the consciousness-combo and see how people react."

"Okay, then. I'll have the Levels of Consciousness Stew."

"That's a good choice. I had a bowl last week and then passed the bar exam."

"Really!"

"No, not really. That was a joke. I'm a quantum bio-roboticist from a planet unknown to your egocentric mind. Are you interested in quantum biology?"

"What's in the stew?"

"We have Cream of Mindfulness, Sticky Toffee Intention, Marinated Joy, and Roasted Red Pepper Wisdom. Tibetan monks make it in sacred caves and then ship it over here giftwrapped in dark matter. Sometimes it has a little conundrum residue when it arrives, but we scrape that off."

"No vegetables?"

"Nothing but veggies, I assure you. It's a very complex vegan recipe, spicy yet subtle, satisfying but with hints of angst. Most people leave satisfied and mildly confused, although a few others depart befuddled. We get all kinds in here."

"Okay, fine. Bring me the soup."

"That's a good choice, as I said. Man needs a little adventure, otherwise, he might not try other levels of consciousness and there-in discover layers of previously unknown fields of freedom. Hey, I should write that down—in my poetry notebook. Where did I leave my poetry notes?"

"Write it on a napkin. That worked for other great thinkers. However, I'm not so sure that unwieldy sentence is poetry, Surge. I am just being honest."

"Actually, I just remembered that Zorba the Greek said the same thing many years ago: 'A Man needs a little madness, or else . . . he never dares cut the rope and be free.' I think that's what I was trying to say. Except Zorba was a patriarchal male. He forgot to say that 'women also need a little madness and rope cutting.' Of course, the madness of a woman and the madness of a man are often different. Each sex goes bonkers in a unique way. That's a discussion for later. But you get the idea: escape from the chains."

"I am pretty good at being irrational, Surge, so I get what you are saying."

"That's good. So, I've been thinking about your book, and I am troubled that you are missing a large and important piece."

"I see. I appreciate the attention. Thank you."

"It's about Socrates. Know thyself, he told us. But what do you think? I mean, about Socrates. What did he mean when he told us we should know about our *self*?"

"I guess we are supposed to figure out how our minds work. That's what I am trying to do in my book."

"Yes. Of course, we are proud that you are on this mission. But Socrates didn't say "Know thy ego." And he didn't say, Know thy Enneagram numbers, or thy Myers-Briggs score, or know thy astrological personality. He said know thy *self*."

"Aha! I get it! That means the Greeks, just like Buddha and Lao-Tzu, knew about the hidden self at about the same time in history—2500 years ago! Know thy allocentric mind!"

"There you go; glad to help. Actually, just to be clear, the Eastern sages told their disciples, "You are that." They required the aspiring gurus to make "I am that" part of their essence. Of course, we both know that the teacher was actually saying 'Be the background.' Allow yourself to become nature. Ride the wave of expanding space (be the wave). Sink back into the ever-present process, the ever-present origin. That is really what Socrates was getting at: be the soul, not the personality."

"Well, that is quite helpful. Thank you."

"Sure, we are here *to not serve*. However, there is more that troubles me. As your guardian-angel waiter, I gather each evening with the helpful staff at

The Third-Eye-Watching Vegan Restaurant and Center for Contemplating Palate to muse about your book. We think perhaps you are struggling because the allocentric mind is automatic and silent. To *be all that you can be*—right now—requires an enlightened allocentric mind. The highest forms of soul-consciousness arise within allocentric beings. But this doesn't make sense if that higher mind is unconscious of even its own *self*. The soul that doesn't know about souls, the religion of non-religion, is a seriously obscure vapor—"Serendipity-Soul Soup" is what we call it in the kitchen. That's the part you don't get. But I can help. I did my minor in non-religion."

"Oh, Lord. I am not surprised."

"Just for the record, are you happy being so unevolved? I don't mean to be cruel, but are you really okay with such a minor degree of wakefulness? How long are you intending to pitch in the minor leagues? Maybe you should move out of the minor leagues and get a real job, something with mindfulness and health insurance. That would make your dad happy."

"Fine, okay, and no, I am not happy being unhappy and unevolved. I would like to be all that I can be—right now. What did you say I was on the Enneagram?"

"You are a hopelessly optimistic, enthusiastic, overly-energetic, "Seven" on the kahuna scale. Sugar droplets form on your forehead instead of sweat. But I don't believe I mentioned the Enneagram."

"I prefer adventure to despair. Maybe the world has more than one thing on the menu—just a suggestion."

"No offense taken. Here's your stew. Eat the whole thing, like a big boy, and then get back to me. We do great desserts here."

———

"Well, that was, indeed, satisfying, Surge. I feel inexplicably happy and paradoxically full. Thanks so much. What's for dessert?"

"We have only one dessert today, but it is strangely delicious."

"I'll have that. What do you call it?"

"We call it "The Sweet Inner Meaning that is Strangely Delicious Chocolate Pie."

"That sounds intriguing. What's in it?"

"I am glad you want to know. It has a solid base of Goethe Sage, with a slightly bitter hint of Steiner Thyme. We fold in small chunks of Caramelized Imagination. The result on the palate is both meaningful and yet not meaningful at all. It is the perfect balance between journey and arrival. Most people stagger from the table with eyes aglow and a sweet knowing smile."

"Oh, Lord. That sounds really romantic! I want some!"

"Very well, but I should warn you that coming so soon after the Levels of Consciousness culinary experience, the jolt of Caramelized Imagination cuts your rope, so to speak, and results in a madness of sorts—at least that is how the world will see you as you walk about with that silly grin on your face."

"Would you do me a favor?"

"Sure."

"Would you speak plain English prose from time to time, please? Fancy food phrases are fun, but what exactly are you talking about?"

"I am talking about the connection between your two minds. What is the force that moves attention around? Who drives the chariot that keeps the two minds pulling in unison? What is the force that sucks the ego back into the mother-soul? What is the force that manifests figures from the background? You found the evidence, but not the mystery that moves the universe."

"You know, I get that. And I *am* troubled. But is it my job, really, to figure out what moves the universe? I've got laundry stacked up in the basement, dishes in the sink, and, anyway, I am a retired teacher—my philosophical brain-organs never fully developed."

"You already know there is a fundamental oscillation; that's where the force is—your will is just riding the waves of alternating current. You don't need to name the nameless, unknowable whizzbang. Your problem is human and solvable."

"Okay, then. *Meaning* bothers me. Every time I find the word in a sentence, my airwaves constrict."

"Have some more dessert—on the house—then walk around awhile with that grin on your face and that paradox in your heart. Then, as the Buddhists say, "sit on it.""

"Thanks for the splendid meal."

"My pleasure. Before you go any farther with your musing, have a look at Buddhism and sitting—it might help you evolve. I'll catch you later on down the road."

The Blending of Eastern and Western Thought

There is a deep need to combine the wisdom of spiritual traditions with the intelligence of secular traditions. In other words, the allocentric mind, as expressed through global religions, and the egocentric mind, as expressed through science, must come together to create a healthy world. The most articulate spokesperson for this worldview is the 14th Dalai Lama, Tenzin Gyatso:

> Though there are areas of life and knowledge outside the domain of science, I have noticed that many people hold an assumption that the scientific view of the world should be the basis for all knowledge and all that is knowable. This is scientific materialism. Although I am not aware of a school of thought that explicitly propounds

this notion, it seems to be a common unexamined presupposition. This view upholds a belief in an objective world, independent of the contingency of its observers. It assumes that the data being analyzed within an experiment are independent of the preconceptions, perceptions, and experience of the scientist analyzing them.

Underlying this view is the assumption that, in the final analysis, matter, as it can be described by physics and as it is governed by the laws of physics, is all there is. Accordingly, this view would uphold that psychology can be reduced to biology, biology to chemistry, and chemistry to physics. My concern here is not so much to argue against this reductionist position (although I myself do not share it) but to draw attention to a vitally important point: that these ideas do not constitute scientific knowledge; rather they represent a philosophical, in fact a metaphysical, position. The view that all aspects of reality can be reduced to matter and its various particles is, to my mind, as much a metaphysical position as the view that an organizing intelligence created and controls reality. ~ *The Universe in a Single Atom, by His Holiness, the Dalai Lama, 2005.*

Historically, science has questioned the validity of religion, and religion has questioned the validity of science, despite the fact, as the Dalai Lama points out in the quote above, both science and religion fall within the domain of metaphysical philosophy.[1] Both approaches to understanding the properties and origins of existence are correct. But how could that be? How can opposites both be correct? As I argue in this book, it is because we have two mutually exclusive minds. Religion represents –speaks for or through—the allocentric mind, while science represents and speaks from the perspective of the egocentric mind.

Buddhists consider our essential nature to be good and pure. I call this essential nature the allocentric mind. As the egocentric mind has rapidly evolved with the rise of language, especially internal dialogue, there has been a powerful masking of our essential nature. A core understanding of Buddhism is that the often-negative, self-obsessed nature of the ego can be counterbalanced through meditation. Essentially, through a meditative discipline, the ego can be calmed and removed from the forefront of the mind—the mind can be "emptied of ego." In this egoless state, the allocentric mind can be directly experienced:

> [Within Buddhist tradition] there is a widespread understanding . . . of the mind's capacity for transformation from a negative state to a tranquil and wholesome purity . . . the essential nature of the mind is pure and its defilements are removable through meditative purification. ~ *The Universe in a Single Atom, by His Holiness, the Dalai Lama, 2005.*

As we blend the findings of Buddhism with the findings of neuroscientists and quantum physicists, there is mounting evidence that both directions have led us to the same conclusion: we have two minds. After 2500 years of thinking about the mind, two fundamental practices emerged in Buddhism. These essentially relate to our two mutually exclusive minds, the egocentric and the allocentric:

> In Theravada Vipassana meditation, "meditation involves practicing both "focused attention" and "open monitoring." These terms, although derived from traditional Buddhist meditative vocabularies, were recently coined by scientists and contemplative scholars in order to delineate the specific kinds of mental processes involved in various Buddhist and non-Buddhist meditation practices, ranging from Vipassana to Yoga to Zen. ~ *Waking, Dreaming, Being, Evan Thompson, 2014.*

Focused-attention meditation is used to steady and control egocentric energy. It is actually a method for developing the egocentric mind. Contrary to this, "open monitoring" meditation is used to develop the allocentric mind—the mind is emptied of chatter, and the ego quieted.

The irony for me is that both Eastern Buddhists and Western neuroscientists have the evidence in front of them concerning our inherent duality, yet fail to entirely grasp the conclusion.[2] I have read enough from both perspectives to realize that they do, indeed, support the awareness of oscillating attention and they both hold that purposeful movement is a key to a greater understanding of human nature. However, we have stopped short of saying outright that evidence from both Eastern and Western perspectives evolved into two minds and two kinds of consciousness. This chapter explores the blending of Eastern and Western thought and its relationship to dual-process theory.

The Arrogant Little Sea Squirt

One morning I awoke to find my cat, Napoleon, standing on my back, head-butting me. Napoleon insists that I get up and fed him his shredded turkey with cheese. I don't really care to wake up from a deep sleep just because he wants me to, and shoving him off one side of the bed only results in his leaping back on the bed from the other side—his Napoleonic idea of compromise. On this particular morning, however, when I was trapped between wakefulness and sleep, a little voice in my head whispered "Arrogant Little Sea Squirt." My eyes popped open and I immediately rushed downstairs to open a can of shredded turkey and cheese. Something in the twilight of consciousness wanted me to elaborate on the Sea Squirt.

Recall from Chapter One that the young sea squirt is a fish that decides to become a plant. One day it is swimming around the sea, enjoying the varying views and then, abruptly, it grows roots and sticks itself to a rock; for the rest of its life. For an animal, human beings especially, this is an act of treason. The arrogant little sea twerp absorbs its brain, eyes, fins, and nervous system for lunch—kind of an in-your-face gesture to the animal kingdom. As a plant, it no longer needs to move, so it no longer needs a navigational nervous system. Over time, it also must transform its musculature, eventually absorbing that as well. It goes from being bilateral to being symmetrical. The arrogant little sea squirt becomes a brainless colony of plant cells at the mercy of ocean tides.

What I realized that morning, while scooping cat food into Napoleon's bowl, was that the sea squirt was onto something important. Its act of transformation was not only a biological statement, it was also a metaphor. I could, I thought, use the "sea squirt message" to further the discussion of the allocentric mind. It turns out that human beings can choose to be "plant-like" any time they care to, and the feeling of being plant-like is wonderful, joyful, and peaceful. The sea squirt gave up a world of struggle and stress for a world of acceptance (surrender) and peace.

We call being plant-like "meditation" (or "contemplation" in Western esoteric practice). Eastern cultures have known for centuries the value of being plant-like, but our rational Western scientific perspective demands anatomical and physiological evidence for the benefits of meditation. If science cannot measure the process in the lab, then it isn't considered valid. Therefore, scientists are now measuring what happens when we sit still, like a tree. They are asking what consequences arise when we are deep in meditation or contemplation.

At the turn of the century, Western researchers leaped with enthusiasm onto the meditation bandwagon. Over the last several decades, scientists have been busy jacking Buddhist monks into the mothernet: running them through MRI scans, pasting electrodes to their bald scalps, taking blood samples, and measuring changes in their pupil sizes. Meditation has become popular in modern Western culture; more and more articles are exploring the benefits of being plant-like. For example, the lead article in the November, 2014, issue of *Scientific American Magazine*, by Matthieu Ricard, Antoine Lutz, Richard J. Davidson, was called "The Neuroscience of Meditation; how it changes the brain, boosting focus and easing stress." The scientists doing this research were inspired by their direct conversations with Buddhist philosophers and religious leaders, most notably The 14th Dalai Lama. Notice, in the title of the article, the phrase *meditation . . . changes the brain*. We are talking about a very powerful way to actually alter neural networks, to change biochemistry through intention.

Consciousness: A New Slant on an Old Conundrum

The 14th Dalai Lama, Tenzin Gyatso, suggested decades ago that scientists and meditators collaborate. Although he has had a strong interest in science since his boyhood and has been dialoguing with scientists for many years, he began in earnest the quest to blend Buddhism with neuroscience in 1987. That's when the biannual Mind and Life conferences started—and continue to this day. The Dalai Lama believes that human beings, if we are to evolve, need both the intellect of science and the wisdom of Buddhism:

> The great benefit of science is that it can contribute tremendously to the alleviation of suffering at the physical level, but it is only through the cultivation of the qualities of the human heart and the transformation of our attitudes that we can begin to address and overcome our mental suffering. In other words, the enhancement of fundamental human values is indispensable to our basic quest for happiness. Therefore, from the perspective of human well-being, science and spirituality are not unrelated. We need both, since the alleviation of suffering must take place at both the physical and the psychological levels. *~ The Universe in a Single Atom, by His Holiness, the Dalai Lama, 2005.*

But how do you blend a spiritual practice with a scientific practice? These are contrary orthodoxies; they are, in essence, the allocentric and the egocentric minds trying to work together for the benefit of all sentient creatures—it seems hopeless at first glance:

> Science deals with matter, Buddhism with mind. Science is the hardware, Buddhism the software. Science is rationalist, Buddhism experiential. Science is quantitative, Buddhist qualitative. Science is conventional, Buddhism contemplative. Science advances us materially, Buddhism advances us spiritually. *~ "Buddhism and Science: On the Nature of the Dialogue." Buddhism and Science: Breaking New Ground, Ed. Alan B, Wallace, 2003.*

What the Mind and Life conferences are exploring, from a dual-process perspective, is the relationship between a world created by the egocentric mind—the scientific world—and a world created by the allocentric mind—the world of spirituality that is at the core of the wisdom tradition:

> In the year 2000, because of the efforts of monks and scientists, a new sub-discipline of neuroscience was created called "contemplative neuroscience." Since this new arena of study began "more than 100 monastics and lay practitioners of Buddhism and a large number of beginning meditators have participated in scientific experiments at the University of Wisconsin-Madison and at . . . 19 other universities. ~ *"The Neuroscience of Meditation; How it Changes the Brain, Boosting Focus and Easing Stress." Scientific American, Matthieu Ricard, Antoine Lutz, Richard J. Davidson, November, 2014.*

The *Scientific American* article identified three kinds of Buddhist meditation: focused-attention; mindfulness meditation; and compassion and loving-kindness meditation. Focused-attention calms the mind and seeks to hone the ability to stay attentive even though the mind naturally wanders—it is egocentric contemplation. Mindfulness meditation is different; it is an awareness of the surround, being still, vigilant in the moment, staying awake despite endless distractions. The attention is not focused on an object-of-regard; it is diffuse—this is an allocentric meditation. Likewise, compassion and loving-kindness meditations are also allocentric; they seek to develop empathy and altruism towards all sentient creatures.

I like Dzogchen Ponlop's distinction in his book *Rebel Buddha* (2010) between calm-abiding meditation (shamatha) and insight meditation (vipashyana). During calm-abiding meditation—using the allocentric mind—we are meeting a friend, our egocentric mind, for tea. Our friend is troubled and needs a sympathetic ear. Our role is not to interrupt what our friend is saying, but to listen with compassion. We get to know our friend by observing what he or she is thinking, saying, and feeling. In contrast to calm-abiding meditation, insight meditation—using the egocentric mind—comes only after we master the art of befriending our anxious ego. Insight meditation cultivates a deeper friendship, where profound concerns can be explored with honesty and with greater focus. Another name for insight meditation is analytical or discerning meditation, which I discuss later in this chapter.

Scientists use brain studies in their effort to see what is happening neurologically in these different states of meditation. What they find is that widely separated areas of the brain form networks specific to different meditative behaviors. To do the focused-attention meditation, for example, requires four sets of brain regions: one set activates when attention wanders, a second set becomes aware that it has been distracted, a third set resets intention, and a fourth set resumes focused-attention. The sets of brain regions necessary to do focused-attention meditation are, I suggest, parts of a larger whole, the egocentric mind. Allocentric meditations like mindfulness, calm-abiding, and empathy-awareness require a whole different set of brain regions.

The definition used for mindfulness meditation in the *Scientific American* article is a match with allocentric processing: awareness of the gestalt as opposed to focal-attention:

> In our Wisconsin lab, we have studied experienced practitioners while they perform an advanced form of mindfulness meditation called open presence. In open presence, sometimes called pure awareness, the mind is calm and relaxed, not focused on anything in particular yet vividly clear, free from excitation or dullness. *The meditator observes and is open to experience* [my italics] without making any attempt to interpret, change, reject or ignore painful sensation. ~ *"The Neuroscience of Meditation; How it Changes the Brain, Boosting Focus and Easing Stress." Scientific American, Matthieu Ricard, Antoine Lutz, Richard J. Davidson, November, 2014.*

Meditation on compassion and kindness is also allocentric. The ego must be dissolved so that the practitioner can emotionally *become* the other person. That is the very definition of empathy. For everyone who seeks a non-violent, loving world—rather than the current mess—this meditation is vital.

Meditation benefits health care workers, teachers, and others who run the risk of emotional burnout linked to the distress experienced from a deeply empathetic reaction to another person's plight. Practitioners develop the ability, after many hours of practice, to share the emotional space of others without being emotionally overwhelmed themselves. They are able to feel the emotional pain of others, to convey deep compassion for others, without being affected by the pain.

According to the Buddhist contemplative tradition from which this practice is derived, compassion, far from leading to distress and discouragement, reinforces an inner balance, strength of mind, and a courageous determination to help those who suffer.

The *Scientific American* article also looked at meditation as a "doorway to consciousness." Using mental capabilities, meditators can change their brain frequencies. These different levels of brain-wave activity correspond to different levels of consciousness,[3] to different states of awareness. Monks can transition to different levels of awareness simply by using the power of internal intention:

> We found that these long-term Buddhist practitioners were able, at will, to sustain a particular EEG pattern. Specifically, it is called high-amplitude gamma-band oscillations and phase synchrony at between 25 to 42 hertz. ~ *"The Neuroscience of Meditation; how it changes the brain, boosting focus and easing stress." ~ "The Neuroscience of Meditation; How it Changes the Brain, Boosting Focus and Easing Stress." Scientific American, Matthieu Ricard, Antoine Lutz, Richard J. Davidson, November, 2014.*

What meditators are doing when they are employing their egocentric attention system is not "calming the mind"—although that happens anyway—they are raising the whole-body frequency to a higher state. Sitting still, with eyes closed, shuts down much of the body's sensory input. Consequently, energy builds up within the meditator and this energy reserve can be used to deepen understanding and awareness:

> It acts like a particle accelerator in nuclear physics; the particles go faster and faster until they are able to split atoms. In meditation practice, we build up the energy of awareness until it grows powerful enough to see entirely different levels of reality . . . Such increasing momentum comes from continuity of awareness and the periodic effort to return to the primary object of our meditation. ~ Insight Meditation, The Practice of Freedom, Joseph Goldstein 1993.

Buddhist practitioners are not dropping into the theta state, as do psychics who tap into a twilight world, nor are they settling into the relaxed state of peace that occurs during alpha wave activation. Instead, meditators are generating high gamma-wave frequencies, a state of awareness that few humans even know exist. I suppose that as the monks fine-tune their consciousness, they are able to explore the worlds that each level of consciousness opens into—each frequency domain is a portal that allows specific kinds of energy and information to enter the mind.

Overall, meditation can be divided into two practices: either the meditator can use the allocentric mind to be aware all-at-once, or the meditator can narrow and focus attention using the egocentric mind to attend to one-thing-at-a-time. Here is how professor of philosophy and Buddhist scholar Evan Thompson explains this practice in his book Waking, Dreaming, Being:

> In the yogic traditions, meditation trains both the ability to sustain attention on a single object and the ability to be openly aware of the entire field of experience without selecting or suppressing anything that arises. In both modes of meditation—focused attention (or one-pointed concentration) and open awareness—one

learns to monitor specific qualities of experience, such as moment-to-moment fluctuations of attention and emotion, that are difficult for the restless mind to see. ~ *Waking, Dreaming, Being, Evan Thompson, 2014.*

———

I t is crucially important that science and Buddhism continue this co-operation. Collaboration is especially important for scientists and for Western civilization. Buddhism has a clear and simple mission: to alleviate suffering and to promote happiness for all sentient creatures. Science does not have such a simple and clear mission; it seems, instead, to be a run-away cement truck with a robotic driver, a driver with lots of brains, but no heart and no spiritual awareness. Science is a "let-loose monster." Science *must* embrace Buddhism, and the spiritual practices that come from the wisdom tradition, if for no other reason than to accept and promote the alleviation of suffering and the promotion of happiness.

Science is busy proving to itself that meditation can lead to changes in the brain (and mind) that have beneficial effects on individuals and on whole cultures—and thus the world. Speaking of brain plasticity and findings from MRI studies, the Dalia Lama stated the following:

> At the Mind and Life conference in Dharamsala in 2004, I learned of the growing sub-discipline of neuroscience dealing with . . . "brain plasticity." This phenomenon suggests to me that traits that were assumed to be fixed—such as personality, disposition, even moods—are not permanent, and that mental exercises or changes in the environment can affect these traits. Already experiments have shown that experienced meditators have more activity in the left frontal lobe, the part of the brain associated with positive emotions, such as happiness, joy, and contentment. These findings imply that

happiness is something we can cultivate deliberately through mental training that affects the brain. ~ *The Universe in a Single Atom, the 14th Dalia Lama, 2005.*

Meditation is a prescription for a non-violent, loving universe. It is a method for crafting a benevolent, empathetic cognition. *It is a way to actually change the biochemistry of the nervous system.*

When asked what his religion was, the 14th Dalai Lama said that his religion was simple; *his religion was kindness.* Science must also state that its ultimate purpose is kindness.

Why Did Buddha Sit Under a Tree?

Pascal famously remarked that all the evil in the world comes from our inability to sit quietly in a room. It is activity that causes all our problems. ~ *Love, Life, Goethe, John Armstrong, 2007.*

After centuries of exploring the powers of meditation, Buddhist monks and Hindu gurus arrived at a common beginning: first, you sit down and plant yourself on the earth; second, you close your eyes; third, you pay attention to breathing. If you do only these three things for a few minutes every day you will discover that you are happier, more at peace, and you have become a kinder person. How could this possibly be?

This common beginning for meditation can be understood in a slightly different way: first, stop all gross muscular movement, second, induce an artificial blindness (or low vision), and third, *become conscious* of your breathing in and breathing out. These three purposeful behaviors are the physiological foundations of meditation, the beginning of a long journey toward self-knowledge. Ironically, blindness, immobility, and not-breathing—the gap between in and out breaths—appear to be critical to the development of human consciousness.

Peace and happiness *are felt* because meditation inhibits the egocentric mind and allows the allocentric mind to express itself. This section is an exploration of meditation and how it is used to strengthen the allocentric mind.

Why do we sit doing nothing? Because sitting with "nothing" is a state of being that is foreign to the egocentric mind. It is a state far removed from *doing*. The answer, told to us again and again, is that we meditate to understand what it means *to be* rather than *to do*. In navigational terms, we sit to activate our allocentric mind. We cool down our hyperactive egocentric consciousness. We learn to *experience* rather than to *accomplish*. We can't comprehend spirituality unless we experience being spiritual.

Spirituality is not serious business or a goal. The word "spirit" has a practical history; it meant "breath" in ancient languages. Breath is not a goal or a business to be managed. It is easy to add the quest for spirituality on our egocentric to-do list and thus miss the whole point of meditation. I like this quote from Hindu guru Bhagwan Shree Rajneesh:

> A serious person cannot be religious. And all religious people are so serious! It seems as if only diseased people with long faces become religious. But meditation is not something that is a "must," it is something absolutely purposeless; it is something whose end is intrinsic to it. There is nothing to be achieved by it or through it—it cannot be made a means. ~ *The Great Challenge: A Rajneesh Reader, Bhagwan Shree Rajneesh, 1982.*

Many people who become interested in meditation are interested in attaining something; meditation, for these people, is used as a means to attain egocentric goals. They may be interested in silence, in achieving a non-tense state of mind—they may be interested in anything—but they are not simply interested in meditation as such, so they cannot be open to it:

> Meditation comes only to those who are interested in meditation as an end in itself. Silence comes: that is another thing. Peace comes: that is another thing. The Divine comes: that is another thing.

These are consequences, by-products; they cannot be longed for because that very longing creates tension. ~ *The Great Challenge: A Rajneesh Reader, Bhagwan Shree Rajneesh, 1982.*

In other words, the only goal allowed for the egocentric mind is meditation itself. The ego will get in the way if it is allowed to make a list of reasons to meditate.

In navigational terminology, meditation is a "returning to our roots," to the days before we were animals—even beyond our plant-like ancestry—to a time when there was no meaning and when there were no brains or modern nervous systems to control conscious movement. Meditation is a total return to allocentric existence, a pure experience, a return to nature. So we sit in silence, very still, and, like the sea squirt, we metaphorically ingest our brains and spinal cords. That is the first step in the process of meditation: just sit there.

Karma and Sitting

The concept called "karma" becomes important when we realize that it is primarily about purposeful movement, and as such is relevant to dual-process theory.

Here is how the Dalai Lama defines Karma:

Literally, *karma* means "action" and refers to the intentional acts of sentient beings. ~ *The Universe in a Single Atom, by His Holiness, the Dalai Lama, 2005.*

"Action" is movement and "intention" is purpose. Purposeful movement requires both our minds, *intention* arising from egocentricity, and *movement* arising from allocentric activity. Karma is another name for purposeful movement.

From the perspective of science, the brain and nervous system evolved to enable navigation. To move with a purpose across a domain required a foundational mind—the allocentric mind—that held the background gestalt, and a second egocentric mind that manifested the forms of our reality. In other words, karma requires two minds. Here is the Dalai Lama explaining the Buddhist perspective:

> As to what might be the mechanism through which karma plays a causal role in the evolution of sentience, I find helpful some of the explanations given in the Vajrayana traditions, often referred to by modern writers as esoteric Buddhism. According to the Guhyasamaja tantra, a principle tradition within Vajrayana Buddhism, at the fundamental level, no absolute division can be made between mind and matter. Matter in its subtlest form is *prana*, a vital energy which is inseparable from consciousness. These are two aspects of an indivisible reality. ~ *The Universe in a Single Atom, by His Holiness, the Dalai Lama, 2005.*

Therefore, karma requires "two aspects of an indivisible reality." From a dual-process perspective, these two aspects are background and foreground, figure and ground. However, it is often misrepresented. Here is the Dalai Lama again:

> Karma is not a transcendental unitary entity that acts like a god in a theistic system or a determined law by which a person's life is fated. ~ *The Universe in a Single Atom, by His Holiness, the Dalai Lama, 2005.*

The Buddhists say that everything is energy in various forms. The subtlest energy, prana, is a form of background out of which other less subtle, more solid energies can manifest. The solid forms are never free of the background, so they are appearances—shapes that the mind creates. Consciousness is a reflective ability of the mind to be aware of energy changes. We perceive levels of energy, which we name, conceptualize, compare, communicate, and employ. Because karma, by definition, requires intention, it is under our control. *We can alter the energy before it is expressed as communication or action.* This is a very serious responsibility;

we have to be careful about wielding this kind of power. The consequences of our decisions ripple through our communities and eventually impact the quality of life of all sentient creatures.

We manifest thoughts and behaviors from a background that is pure potential. If we have inner sophistication, an evolved spiritual mind that is under our control and watched over by a witness, then we can control the karma that we manifest; our actions can be compassionate and helpful—our karma can be positive. *We sit in mediation to practice generating good karma.*

We sit in stillness, not moving the large muscles of our body, as we experience our spiritual development. The purity and coherency of our inner mind is healthier when our intentions are positive and our self-control has developed through practice. A spiritual journey is an ongoing transformation. We clear the mind of clutter, bad habits of thought, and then we cultivate kinder and more positive thoughts:

> Intentions result in acts, which result in effects that condition the mind toward certain traits and propensities, all of which may give rise to further intentions and actions. The entire process is seen as an endless self-perpetuating dynamic. The chain reaction of interlocking causes and effects operates not only in individuals but also for groups and societies, not just in one lifetime but across many lifetimes. ~ *The Universe in a Single Atom, by His Holiness, the 14th Dalai Lama, 2005.*

I believe the Dalai Lama is saying that out of a background mind come endless manifestations *that we can watch, control, and filter.* We can think good thoughts. These become good intentions. Then our intentions result in right (good) actions. Right actions affect our individual lives, the lives of friends and neighbors, and then, through a chain reaction, the character of cultures. Therefore, we sit in meditation to craft a spiritual mind that knows how to evolve, hold, and transmit beneficial energy (beneficial karma). The spiritual mind radiates health inward and outward. With enough spiritual minds transmitting enough healthy energy, the whole planet can heal itself.

Looking, Sitting, Breathing

Why do we close our eyes during meditation? Why not leave the eyes open? Different kinds of meditations do, indeed, allow the eyes to be partially open or fixed on a single point in space. The important idea is that vision is brought under control during meditation.

We close the eyes—or reduce visual perception—because visual processing eventually must be stopped to allow the opening of inner perception[4], which results from the management of internal experience. Inner seeing is different from external seeing—inner vision seems as if it has a cosmic presence to it. To see within, we must manage the flow of outward images, so they don't mask or interfere with inner perception.

As soon as the eyes are closed, the retina stops sending streams of redundant images to the brain. Huge areas of brain real-estate, given over to processing vision, go quiet when the eyes are closed. This causes an immediate reduction in stress. Closing the eyes also activates brain waves called alpha rhythms, which are close to the coherent vibrational rhythms of the earth. Alpha waves generate a state of peacefulness, of being grounded.

Ironically, but understandably, the egocentric mind starts looking around for something to do when vision is absent. Suddenly, there are reserves of energy freed up for other activities, like inner perception. The beauty and power of our vision system comes at a cost. There is high velocity franticness to seeing, a stress-inducing superficiality that has to be silenced, temporarily blinded, before the spiritual portal can be opened. Therefore, the second step of meditation is that we close down visual processing. If that was the only reason to meditate, it would be sufficient on its own.

Consciousness: A New Slant on an Old Conundrum

But why do we need to *sit still* to meditate? Why do we need to stop moving? Why do we need to become more like a tree than like a human being? When we meditate, we "put down roots" and connect with the earth; we metaphorically absorb our brain and spinal cord. But why would we go against the very reason that brains and nervous systems were created? We are mandated to move with purpose, so why would we deliberately shut down the brain and nervous system?

I think the answer follows this logic: before the brain evolved, before the nervous system became sophisticated and complicated, there developed a primal system for survival within a specific domain (water for fish, air and the earth's surface for mammals). Organisms had to be completely compatible with the surrounding environment if they were to survive. There had to be a total oneness with the surround so that immediate and simultaneous adjustments could be made to environmental changes. If the concentration of light, nutrients, competition, and temperature went beyond certain delicate thresholds, an organism would perish. When we sit in total stillness, therefore, we approach this primitive allocentric state of oneness with our native habitat. This is a deep suppression of the egocentric mind. Notice also that sitting still, without using the large muscles of the body for movement, frees up a tremendous amount of energy.

Yoga masters and great mystics have been telling humanity for centuries that the goal of the spiritual seeker is *to be*. "Simply be," they repeat over and over again; to enter through the spiritual portal, we must *first just be*. The sages urged their disciples to live in the moment, to "be here now," to be mindful, to simply *experience life*. They used language and exercises in as many ways as they could to convey why it was important just to exist without purpose. However, language is so limited that they finally instructed their followers to simply sit. Close your eyes and sit there—that is the beginning. The sages knew that sitting quietly with eyes closed, not moving—not navigating—would be an experience with powerful consequences that words could not express.

Sitting still and closing down vision frees up tremendous brain power. There are other things that minds can do, and with so much freed up energy, the next levels of human consciousness can be probed and experienced. However, there can be little development in human consciousness in our overly active, image-bombarded modern world. Humanity will not evolve unless we learn to meditate.

Doing and seeing are so powerful that we become obsessed like a drug addict, or alcoholic, or a person with obsessive compulsive disorder. We feel like we will starve to death without our seeing and doing. Therefore, the practice of meditation involves slowly increasing the length of time we are able to sit sightless and just be, with no purposeful muscular actions. *We are bringing our major sensory-motor addictions under central control.* This is not easy, so we take it slowly, with discipline. Indeed, the very understanding of discipline, our habits of eating, drinking, and exercising, for example, are improved by meditation. The danger, of course, is that the egocentric mind will make meditation an addiction. Goals, intent, and purpose must be left out of the equation—there is nothing to "attain" during meditation.

Of course, the egocentric mind will not stop trying *to do*—that is this mind's biological mandate. Just as lungs breathe and hearts beat, the egocentric mind must think. Sitting quietly calms the egocentric mind but cannot eliminate it; we watch it butt in and then fade out, again and again.

Why do we concentrate on breathing? Here is our friend Rumi on the subject of breathing:

> *Be wary when you breathe.*
> *In every instant a new species rises*
> *in the chest—*
> *now a demon,*
> *now an angel,*
> *now a wild animal,*
> *now a friend.*

Consciousness: A New Slant on an Old Conundrum

The inner working of a human being
is a jungle.
Sometimes wolves dominate;
Sometimes wild hogs.

~ "THE INNER WORKINGS," *A YEAR WITH RUMI,* 2006.

Between each breath, the mind starts anew. The gap between breaths initially may enter in a state of calm-abiding and compassion, but may exit as a hateful warthog. Be wary when you breathe. You are switching between allocentric processing (out-breath) and egocentric processing (in-breath). Between the two minds, at the gap where the turn-around occurs, the egocentric inner voice will seize control and inhibit allocentric awareness. Daydreaming, sensory review, and attention to physical discomfort will keep the egocentric mind in control. The benefits of the allocentric mind—love, peace, joy, wisdom, mindfulness—will be blocked.

Spiritual teachers keep returning to the breath, imploring their followers to *consciously* breathe. Consciously breathing, not to be confused with the automatic breathing we do without awareness, has at least five objectives:

- First, *it gives the egocentric mind something to do.* Whatever the mental obsession-of-the-day is—and there will always be something to make the body feel stress and cause the mind to feverishly perseverate—awareness of breathing will inhibit the flow of words and the flood of associated destructive emotions. The key is to keep the focus on the breath—keep reminding the egocentric mind that its purpose while meditating is to remember to be aware of breathing, to direct *conscious* breathing.
- Second, *conscious breathing calms the body*, reduces stress, and sends peaceful alpha waves flooding through the brain and body. These calming waves ebb up and down the spine and trickle outward along nerve endings that reach the very tips of the extremities. These alpha waves are fundamental oscillations of the brain that are

associated with peaceful, serene, unstressed sensations. Alpha waves only occur—by definition—when the eyes are closed. When fully under the spell of alpha waves, breathing is full, even, and balanced.

- Third, we consciously breathe from the diaphragm. This *places attention not on the heart (emotions) or the brain (problem solving), but on the silent, nonverbal, and unemotional gut.* The belly goes out and then the belly goes inward; the lungs fill up and then empty; the diaphragm pushes down and the diaphragm pushes up. Concentration on the spiritual center of the body—at about the level of the navel—is the result of this breathing practice. Breathe-in through the nose and out through the mouth. Breathing out through the mouth powerfully inhibits egocentricity.

- Fourth, and very importantly, *we consciously breathe to remember to be mindful.* To be mindful is to "remember to be spiritual," to remember that we have a practice. Mindfulness is also the awareness system of the allocentric mind—a mind that has no voice, no verbal language, no opinions, no beliefs, no politics, no religion, no drive to do and solve. It just is. However, it does watch, it has awareness. It can witness. It can experience. It has a *self,* a soul, but not an ego. It is primal awareness.

- Fifth, we become aware that the in-breath is an egocentric action—the egocentric mind is operating while the allocentric mind is inhibited during the in-breath. The out-breath is the opposite: the allocentric mind is active and the egocentric mind is inhibited. Therefore, *during meditation we can become aware of the actual switching of our two minds.* As an example, think of the act of smelling a flower. We are gathering information with the in-breath, which is an egocentric action. This is a very focal activity; it is not at all holistic, not concerned with a gestalt. To smell—the act of gathering information—requires that the allocentric mind be inhibited.

My editor (and friend) Karen Horwath pointed out that in the wisdom tradition, each of us "is the face of God." We are manifestations of an essence. As we breathe in, we bring into ourselves the intelligence, the consciousness that is all around us. As we breathe out we release what we have learned back to "God." Breathing can be thought of as an exchange of information, a sharing.

We began our life outside the womb with a first breath. In a way, the first breath is our first audible communication that says "Here I am." The first breath has been understood as the first word: "In the beginning was the Word." The breath is a primal sound, evidence of life. With each breath we say, "I am here, I still live; I exist."

The gap between the in-breath and the out-breath is a potential space where the ego-mind and allo-mind change places; it is the turn-around place where dominance can be passed from one twin to the other. Meditating using OM—the audible exhalation of breath—helps the meditator navigate the gap so that the allocentric mind maintains awareness. The sound OM is followed until it reaches the gap and then the exchange occurs. Here is how Joachim-Ernst Berendt explains this phenomenon in his book *The World is Sound*:

> Speaking and meditating on OM is irresolvably connected with the correct method of breathing. OM "happens" while exhaling [while the allocentric mind is active]. The M has to vibrate for a long time—extending as much as possible into the space between exhaling and inhaling, which is the actual moment of emptiness and becoming one. OM signifies that point where "breath," becomes "word" and where "word" becomes "breath" . . . ~ *The World is Sound, Joachim-Ernst Berendt, 1991.*

Something becomes "word" when the egocentric mind takes control. Conscious breathing is a non-thinking, non-verbal, activity and it happens when the allocentric mind is dominant. In the gap, we exchange control between allocentric non-verbal awareness and egocentric attentive language—the "word."

We breathe to survive—we have to breathe—so if we mentally associate mindfulness with breathing we are constantly reminded to activate the witness, to be aware of self, to dwell within the allocentric mind. Therefore, we seek to be mindful in each breathing moment of our day. Breathing makes meditation a constant companion. Whatever we are doing, wherever we are, whatever the circumstances of any moment, we are breathing, and, therefore, we are meditating and being mindful. The Christian equivalent to this Eastern perspective is "prayer without ceasing."

The ego is dominant. It is so dominant, indeed, that most of humanity believes that this is who they are. When they seek to "know their self," they search for only one of the two minds, the egocentric mind. Ironically, this is because the purpose of this mind is to seek and to understand, as if it were the center of the universe. However, the second mind, the allocentric, exists neither to seek nor to understand. It exists to experience. When you sit still, with no purpose, no movement, you open a portal into the second mind. This is very complicated for the egocentric mind to comprehend because it doesn't believe there is a hidden mind to discover. Therefore, the ego sees no reason to go seeking for a miraculous twin with magical powers. It also isn't inclined to cooperate with a process that is meant to shut it off—meditation can feel like suicide to the ego.

Meditating with the Whole Body

Monks also use walking meditation to practice mindfulness. If we associate walking with being mindful, just as we associate breathing with mindfulness, then each step whispers "wake up!" Walking meditation is a technique that links purposeful movement with mindfulness. If we remember to be awake with each step we take, then we effectively meditate whenever we move our bodies. If we combine breathing and walking with mindfulness, we get a powerful system for being mindful at all times.

It is also important to understand that meditation is a whole-body activity. We tend to think that meditation occurs in a head-space. But this is false and leads to a purely egocentric meditation. The major senses, the eyes and ears, are located in the same spatial plane—eyes, nose, and ears are within a three-inch zone that encircles the head. So we receive most of our information from a perceptual field, a band of perception, at head level. For the eyes, this is an entirely forward perceptual system that monitors straight-ahead. Hearing is equally balanced on two sides and effectively is a global surround perception, a halo of sound around the body. The internal, whole-body senses—proprioception, kinesthesis, and the vestibular system—also must be stilled during meditation; all the senses are coherently affected. Consequently, when we meditate allocentrically, we do so with the whole body.

Therefore:

- Meditate not as if the mind is in the brain, but rather meditate as if the mind is in the whole body.
- Meditate as if the mind can stretch beyond the body—as indeed it can. The environment is also part of what we call mind—it is inseparable from the body.
- Meditate as if the mind includes all other conscious creatures. You, as a mind, are not alone. The mind that extends into the environment also includes other minds. Therefore, meditation includes the collective mind.
- If you meditate on self-love and self-health, you affect the larger mind and, consequently, spread love and healing to others. If others meditate on their well-being, your well-being is positively affected.

The egocentric mind and the allocentric mind are out-of-balance in our world. Meditation restores the balance. The unbalanced world lacks love, peace, joy, wisdom, and mindfulness—the attributes of allocentricity. The rebalanced world includes these positive qualities.

Modern human beings are overburdened and overly distracted because they are almost always in an egocentric mode of attention. Even the intention to meditate is overruled by obligations. Few of us have the time to become monks. Ironically, neither do the monks. They have another way to stay awake. Monks rarely sit and meditate. They are mostly active during their waking hours:

Georges Dreyfus, in his marvelous philosophical anthropology and biography, *The Sound of Two hands Clapping* (2003), explains that contrary to Western assumptions, most Tibetan monks meditate very little, being more involved in chores, in ritual performances for patrons, and in loud memorization and recitation of texts. ~ *The Bodhisattva's Brain, Owen Flanagan, 2011.*

Monks don't sit in traditional postures and meditate incessantly because they take their allocentric mindset—their mindfulness—with them as they move through their day, as they breathe and walk about. Meditation may begin with a deliberate time and place to practice allocentric awareness, but it can be practiced through slow, deliberate, aware movements, and with the constant in-and-out flow of the breath. Every moment of a life can be contemplative.

Spiritual traditions use various strategies for inhibiting the ego so that the allocentric mind can awaken. Yoga, for example, is a series of carefully designed postures that are held steady. These postures, especially as they are difficult to get into or to sustain, effectively stop the egocentric obsession with time-based problems. In the Mevlevi Sufi tradition, to give another example, a specific hypnotic dance ceremony is used to quiet the ego and induce different states of consciousness. This Sufi order is famous for their whirling dervishes. Tai chi is another tradition using slow controlled movements that take so much concentration it is impossible for the egocentric mind to function. Rudolf Steiner used a system of movements called Eurythmics, a performance art that required the egocentric mind to subside. Gurdjieff also used rhythmic movements and orchestrated dance routines to still egocentricity. Every kind of dance is spiritual therapy.

Music too, performed or listened to, releases the allocentric mind and inhibits the egocentric mind—the egocentric mind is overpowered by music. Whenever we "get into a zone," whether in sports, music, poetry, or art, we are experiencing allocentric consciousness, and simultaneously inhibiting the angst and hyperactivity of the egocentric mind.

Consciousness: A New Slant on an Old Conundrum

Meditators also study sleep states. During sleep the eyes are closed and the body is so motionless that it is temporarily paralyzed. Meditation during the wakeful state is difficult because the senses and the muscles want to do their daily job. However, during sleep it is natural that purposeful movement and sensations fade. Slow brain waves like theta and delta predominate during twilight sleep and deep sleep. This is an opportune time for the meditator to explore how the mind works.

Lucid dreaming[5] and hypnogogic dreaming[6] are also avenues for the exploration of the meditative mind. The dreaming mind is disassociated from objectivity; there is no seer, or seen. There is nothing to do and nowhere to go. The brain is disembodied from the physical form and from a physical domain. There is no environment to move about in and no body to move through that environment. Dream state mediation is an important part of Buddhist practice.

All brain waves fire synchronously in the brain, but states of consciousness are characterized by just one dominant frequency (others are still firing, but subliminally). It is perhaps the case that the alpha rhythm is *the* steady background frequency against which all others are measured. Furthermore, it may also be true that while one brain region is experiencing a dominant wave frequency, other areas are experiencing less dominant frequencies. Brain wave measurements tap into a complex, networked system that cannot be over-generalized. We are only beginning our pioneering journey into the realms of consciousness.

———

Buddha sat under a tree with body still, eyes closed, while breathing consciously because he was, in my opinion, experiencing the allocentric mind—with minimal interference from the egocentric mind. The ego seems to feed on the higher frequencies, probably because great energy is needed for manifestation to occur. However, the background mind is a resting place where we exist as unformed potential. Buddha was also cultivating what I call "faith," the opposite of belief. Faith has no ego. Sitting in silent meditation, we practice what it means to have faith.

When we sit in meditation, words cannot harm us. A person who simply knows and accepts has no need or desire for debate. The angry, opinionated monkey mind—the belief machine in our heads—cannot penetrate the meditative membrane wherein resides a calm person aspiring to faith. During meditation we are practicing *pure knowing* and *pure accepting*; rants and raves have no effect. Belief is ego-bound; people who have false or unkind beliefs about you can harm you. They rain their emotion-heavy words down on you almost as if you were a mere statue put there to absorb the shock waves. Those who know about faith, and practice being in their allocentric mind, leave you alone to be who you are meant to be.

You sit with faith until you realize that this faith is coming from your allocentric mind and this mind is already part of your body, part of the world, and part of the collective mind. Therefore, as the followers of the wisdom tradition have been telling us for centuries, there is no need to seek that which you already possess.

Let's look now at the consequences that arise when the allocentric mind becomes an equal partner with the egocentric mind.

Emptiness

An important concept (practice) in Buddhism is emptiness. This is a state of mind where pure awareness exists without egocentric intrusions, where "nothing" happens and nothing is intended. In this pure state, a portal opens that reveals what I will call "allocentric purities." I am out on a limb here since my own ability to sustain emptiness, or even to understand the full scope of the concept, is weak. And yet what I have glimpsed during meditation is the arising of certain essences that are closer to pure awareness and pure emotion.

Emptiness is related to energy. When the mind is free of words, free of thoughts, free of habits, free of immediate emotions, free of visual images, free of gross muscular movement, then a void is created. Into that void rushes energy that can be used for healing, for clear thinking, and for pure compassion. This is energy that can be used to intentionally alter the bio-chemistry of the brain and body. We have the power to create minds that are self-healing, peaceful, and loving. However, the egocentric mind suppresses this life-affirming, highly-beneficial energy—it drains the body of energy as it pursues its obsessive needs. The egocentric mind is like a wild animal. It has to be tamed if we want to cultivate kindness and reduce suffering. Here is how the Dalia Lama explains this untamed ego:

> The root of all this is whether our minds are tamed or not. If they are untamed, we commit various destructive actions, and in keeping with those destructive actions, disasters, unhappiness and so forth come about. If our minds are well tamed, then these things won't happen. So if misfortunes occur, we can't point our finger at the Buddha, nor can we put the blame on somebody else. Similarly, we can't say that our happiness came from someone else. All these things arise depending on whether our minds are tamed. When our minds are tamed, we engage in constructive actions, build up posi-tive force (merit), and happiness comes as a result of that. If we want to get rid of our problems, of our sufferings, then we have to work on our attitudes and tame our minds well. - *"Attitude-Training like the Rays of the Sun," the 14th Dalai Lama," http://www.holybooks.com/wp-content/uploads/A-Commentary-on-Attitude-Training-Like-the-Rays-of-the-Sun.pdf, May, 1985.*

To tame the mind we need a discipline, intention, and motivation. Meditation actually changes the biochemistry of the mind, so like a weight-lifter building muscles through practice, meditation builds neuronal con-nections through practice.

Emptiness is also deeply connected with emotions. Research now shows that the whole body has a response to environmental circumstances, to spe-cific experiences, *before* the information reaches the egocentric mind. This is a survival mechanism because the body must react to danger rather than analyze and categorize immediate danger. This means that the allocentric mind becomes aware and has a reaction long before the egocentric mind is able to pay attention and process. When the egocentric mind does eventually and inevitably receive the good or bad news, it labels, categories, and assigns meaning to what has already come into being. For example, if you make eye contact with a perfect stranger and "suddenly" fall in love (love at first sight), your whole body goes from a steady, routine state into a kind of good shock. If you are hiking in the woods, however, and come face-to-face with a snarling bear and her cubs, your whole body has another kind of shock—you go from a steady calm state to an emergency state. These whole-body allocentric changes occur *before* the egocentric mind gets the news.

The allocentric mind "knows" about the state of the world, the state of the body, and it "knows" what to do (and does it), long before the egocentric mind is able to weigh in with an opinion or a plan. Many of the emergencies that the body faces in the modern world are not caused by emergencies in nature. For example, if we are insulted at work, or overhear a disparaging remark about our appearance or personality, our whole body reacts as if a bear suddenly showed up in the doorway. This kind of false alarm is very common in the modern world—we live as if a bear leaps into view two or three times a day. By the end of a day, our whole body has gone in and out of shock several times, upsetting body chemistry, energy reserves, and mood.

Meditation can build "body-armor" that overrides or prevents shock. This building of psychological body armor comes from the egocentric mind. After reflection, the egocentric mind can make a plan and formulate an intention: *I will program my whole body—my allocentric mind—to react only when it is appropriate for survival*. The two minds are symbiotic, they affect each other.

Buddhists have a fundamental understanding called "downward causation." In a way, downward causation is the opposite of upward causation, which is what we normally think of when we say that we live in a causal universe. We know that moment-by-moment what manifests in our reality is caused by something that came before. Downward causation, however, means *that our egocentric mind can cause something to happen*; the egocentric mind can change/control the body through intention. The mind can also alter its own functions through intention.

Buddhists actually understand this as a *neurological* change; we can actually change the anatomy and physiology of our own brains and bodies—our own minds—through intention. *We* are a causal force, acting not only on our surround and on others, but also morphologically on our own essence. In one sense, this is obvious: if we establish a practice (a discipline) and stick to it, we alter our muscles, our memories, and our health. The Buddhists, however, are also talking about how our mind, our state of consciousness, can alter itself. For example, we can actually modify brain/body neurons and biochemistry by willing ourselves to be more joyful, more at peace, more loving, more aware, and so on. Neuroscientists use the term neuroplasticity to mean that the brain can self-repair. Evidently, neuroplasticity was discovered by Buddhists 2500 years ago.

In this book, I have stressed the distinction between the background allocentric mind and the foreground egocentric mind. Emptiness *is* the background mind. In both Christian and Muslim traditions, it is said that we should be *in the world* rather than *of the world*. This simply means that we should be the background—empty, formless, not egos. A spiritual mind is an allocentric mind; it is a quiet, empty well of possibility. If we go to this place of emptiness, we discover energies that are healing and loving.

When we are on a spiritual journey, or cultivating a spiritual life, we are deliberately crafting a mind that is loving, non-violent, calm, and clear. This takes a lot of work; it takes intention, discipline, and a practice. The concept of "mind" is understood to include the brain, the body, the surround, as well as the collective cognition of all sentient creatures. Therefore, our personal spiritual journey impacts all other sentient creatures and the environment.

The topics in the following "mantra" can be used to ready the mind for meditation. It is also a training strategy, a self-fulfilling practice, for evolving a spiritual mind—it is a practice that I use during meditation. I offer it here for consideration; it may not be how you craft your own practice.

The mantra repeats:

- "The mind is love,"
- "The mind is peace,"
- "The mind is joy,"
- "The mind is wisdom,"
- "The mind is awake,"
- "The mind is coherent,
- "The mind is thankful,"
- "The mind is enlightened."

This mantra sets an intention to be compassionate, awake, and thankful. At the same time, this mantra creates a mindset that allows the overheated, over-used, habitual egocentric mind a much-needed rest. I will spend the next few pages discussing the connection between emptiness and the allocentric attributes of this mantra.

Love Arises Through Emptiness

The love that comes through emptiness, and that is crafted through intention, has many forms. For example, there is *compassion*, which is love for all sentient creatures. This is a generalized love, a conceptualized, but felt, love for all creatures. Compassion holds the additional notion of empathy for the suffering of others—an awareness of suffering and a desire to take action to reduce or eliminate suffering..

Another kind of love is *empathy*, which is a bridging across space to "enter" another mind. To have empathy is *to become* another mind, to feel the emotions of another person without being consumed by the emotions of the other. Empathy is a deep connection, a powerful mind-to-mind touching. Empathy is a conscious, aware, merging of energy fields.

There is also *loving-kindness* which is a one-on-one awareness of shared experience, a genuine caring for *another person's presence*. Loving-kindness creates a holding space, a psychic bubble where you and another person are alone. Loving-kindness is a way to fill space with concern, caring, and sincerity. Loving-kindness is also a receptacle where you and another can share a safe, one-on-one connection that goes deeper than the superficiality of custom and habit. In this "holding space filled with love," you are able to allow and receive the love of another. Loving-kindness is love-sharing, a spatial relationship.

There is also *care-giving* love that seeks to nurture. This is the all-powerful love that we extend—like a bubble—over those we love. To have a child is to know this protective embracing love. This is a vigilant love that watches over another sentient being unceasingly. Good teachers expand the nurturing bubble over groups of children. Parents know this lifelong energy, as do children who care for their elderly parents. Those who care for suffering humanity use care-giving love.

There is also *loving-forgiveness*: self-forgiving and forgiving others. You were born human—that is not your fault. You arrived on earth through two parents. They were who they were—nothing you could have done about that, for good or not-so-good. You landed in a certain environment, at a certain place in time, and within certain spaces. You didn't control the variables—there is no guilt or grief that needs to be carried about—you can forgive God for the circumstances of your birth and the challenges of your life. You arrived with a personality, with desires that were unique—no other human exists or will ever exist that is you. However, all personalities are flawed. You might as well forgive yourself for your lack of kindness at times and for your blunders. You have only recently realized that *you can build* a spiritual mind. You are on a new journey with each passing moment. Let judgements, especially self-judgment, evaporate.

When we say "I love you," that is often an egocentric statement, an owning of an emotion, an acknowledgement of an important feeling, or need, or desire. It is also an attachment, because that is all the egocentric mind knows how to do—it attaches to others. This is okay—to use the egocentric mind—as long as we are aware that we are attaching. However, allocentrically, we could more appropriately say "we are within love." This characterizes a relationship, a shared space where two people are experiencing life and relationships together. There is wonderment and joy inside a space where two people are held in happiness.

Finally, and perhaps what the wisdom traditions most mean by "love," there is *a felt one-ness with the flow of existence. We are where the universe is expanding.* If you spread your arms and feel this realization, there is a rush of love as you surf the wave of existence. This *surfing the flow of existence* is the love of spirituality. Phrases like "God is love" or "God is light" refer to this surfing at the speed of creation.

Peace Arises Through Emptiness

The spiritual journey is a *deliberate changing of the mind.* From a chaotic, self-centered mind, we move toward a state of peace. However, through intention, we are also changing the quantum state of the mind. Through meditation, we craft a mind that radiates peacefulness within and outward. Peace, like love, is also a complex set of emotions and concepts.

Peace means *calm-abiding*, sitting inside an aura of satisfaction and acceptance. Calm-abiding is free of worry about past and future because it is located within and resides in the moment where time has no meaning. The world outside may be in a frantic, anxiety-ridden state, but calm abiding practice keeps the body relaxed and the mind steady and balanced.

Peace also means *total acceptance*: what is happening now is what is happening now. What happened in the past is over. What the future will bring, it will bring. There is no need to feel like we are wasting time.

Peace also means *relaxation*. It is an alpha wave state. It is a physical sensation of feeling rested for no reason—being okay with the moment so deeply that a gentle, healing flow washes over the body. This is a deeply physiological peace.

Peace is also *non-violence* toward other sentient creatures. It is an *intention* to be part of safe, happy, gentle neighborhoods, communities, and relationships. Peace is, therefore, a wish that all sentient creatures be free from suffering and that all sentient beings dwell in happiness. It is the conviction that this "building a non-violent world" needs to be a personal intention. Peaceful internal goodness constantly radiates outward from your peaceful essence; this starts a chain reaction that spreads outward forever.

There is a movement in some schools to send children to meditation rather than detention. This is a loving reaction to their acting out. Boys, especially, need to learn how to monitor and control their aggression. We have to teach all our children how to understand themselves and how to self-generate a sense of personal peace.

Dwelling in a state of peace is not a naïve head-in-the-sand approach to life. We can perceive and acknowledge the conflicts, ambiguity, and complexity that define life on earth. We have a choice. We can cooperate and cultivate peacefulness or we can contribute to unrest and violence toward others and toward our environment. We can meditate and promote peace, or we can stay within the ego and promote disharmony. Cultivating peace tames the ego and turns the egocentric mind towards the cultivating of social justice and community building.

Joy Arises Through Emptiness

Joy can come through emptiness if we wait for it to percolate upward into our essence. However, joy can also be an intention. We can deliberately craft a mind that is joyful. We know from the introspective heritage of Buddhism that this is true. We also know from the scientific understanding of neuroplasticity that the mind can be intentionally altered. Joy can be self-created through meditation and through intentionally created life-affirming moments.

The writer and philosopher Colin Wilson tells his readers about "absurd good news." This is a wonderful feeling that arises suddenly and for no apparent reason. You feel grateful and joyful at the same time. This blissful state of joyful awareness is within us, beneath all the guilt, dread, sadness, routines, and habits. Rumi calls it a gemstone—a yellow carnelian—that resides deep under our house of emotions and duties:

> *Tear down this house.*
> *A hundred thousand new houses can be built*
> *from the transparent yellow carnelian buried beneath it.*
> *The only way to get to that [sacred energy]*
> *is to do the work of demolition;*
> *dig beneath the foundation.*
>
> *If you wait and just let it happen,*
> *You will bite your hand and say,*
> *I did not do as I knew I should have.*

~ "The Pickaxe," *A Year with Rumi*, 2006.

Joyfulness seems to arise only when the ego is gone and when we are empty. Human beings keep searching for this state through drugs, yoga, exercise, fasting, and so on—it seems to be a pure form of happiness that is just beyond reach. Meditation on emptiness opens a portal that allows subtle energy to enter. This energy holds the seeds for the sprouting of joy. Absurd good news awaits us as we learn to dwell more often within our allocentric minds.

Wisdom Arises Through Emptiness

Wisdom begins with the intention to have experiences, to lead an interesting life, and to be awake when we experience life. Memories based on wisdom become the collected knowledge of the allocentric mind and, consequently, wisdom informs us of *right action*. Wisdom also means to have a practice, a discipline—a wise individual has a spiritual routine that continues throughout a lifetime.

Wisdom is a lifelong search for all the egocentric barriers that hide the allocentric mind. It is the collected awareness of how to bypass or transcend egocentric inhibitory forces. Wisdom is a search for the portals through which we discover what the allocentric mind is whispering. Here is Rumi again:

What matters is how quickly you do what your soul directs.

~"THREE TRAVELERS TELL THEIR DREAMS."
A YEAR WITH RUMI, 2006.

The root of the word "wisdom" means *to taste.* Knowing this helps us remember to "put ourselves out there" to have deliberate experiences. Wisdom is both a going with the flow—letting life happen—and the intention to craft a curious mind that has experiences.

Wisdom also means "to have good judgment." The problem is figuring out what constitutes good judgment, especially when faced with contrary needs and desires. Often the best judgment is not egocentric. Wisdom is allocentric and is about cultivating healthy relationships.

The wisdom tradition is a core that runs through all religious teachings; it binds them together regardless of their history or rituals. The wisdom tradition holds that there is a recurring spiritual understanding—generation after generation and culture after culture discover it—that transcends the power structures associated with institutionalized egocentric religion. This perennial wisdom is the awareness of our duality and of the need to resurrect the "lost" or "hidden" allocentric mind. Here is how Franciscan Friar Richard Rohr explains the Wisdom Tradition:

There have been many generations of sincere seekers who've gone through the same human journey and there is plenty of collective and common wisdom to be had. It is often called the "perennial tradition" or the "perennial philosophy" because it keeps recurring in different world religions with different metaphors and vocabulary. The foundational wisdom is much the same, although never exactly the same. As in the Trinity, spiritual unity is diversity loved and overcome, never mere uniformity. ~ *Richard Rohr's daily mediations, January 15, 2017.*

Here is a good succinct summary of the Perennial Tradition:

There is a Divine Reality underneath and inherent in the world of things.

There is in the human soul a natural capacity, similarity, and longing for this Divine Reality.

The final goal of all existence is union with Divine Reality. ~ *Richard Rohr's daily mediations, January 15, 2017.*

Mindfulness Arises Through Emptiness

Monks have shown us that we can intentionally change our brain waves. This is the same as changing our relative states of wakefulness. This, in turn, affects how we pay attention and what we pay attention to. The monks have also shown us that we have a Witness, a part of our mind that is aware in different states of wakefulness. For example, we can "stay awake" while deeply asleep in the delta wave state, we can watch the words flow out of our mouths in the beta state, and we can put our bodies into a relaxed state when alpha rhythms are predominant. We can learn to control brain waves through meditation and intention.

What this ability means for science is that the brain itself can be altered by willpower. We can create the hardwired brain of our choosing:

Buddhism has long had a theory of what in neuroscience is called the "plasticity of the brain." The Buddhist terms in which this concept is couched are radically different from those used by cognitive science, but what is significant is that both perceive consciousness as highly amenable to change. ~ *The Universe in a Single Atom, by His Holiness, the Dalai Lama, 2005.*

If our thoughts are peaceful, joyful, and loving, then we will actually create a brain that has these qualities. On the other hand, if our brains are molded by propaganda, by commercials, and by the rote habits of old cultures, then our minds will be made from without and by others.

When we change our own biochemistry, we do so in a way that seems limitless. Suppose we decide that we will cultivate a mind that is filled with compassion. There is no limit within the endless universe of consciousness, so our compassionate mind will grow in scope and power through practice and intention:

> [Buddhist scholar] Dharmakirti argues that the natural constraints on consciousness are far fewer and are removable, so that in principle it is possible for a mental quality like compassion to be developed to a limitless degree. In fact, for Dharmakirti, the greatness of the Buddha as a spiritual teacher lies not so much in his mastery of various fields of knowledge as in his having attained the perfection of boundless compassion for all beings. ~ *The Universe in a Single Atom, by His Holiness, the 14th Dalai Lama, 2005.*

Esoteric philosopher, P. D. Ouspensky, spoke of *self-remembering*. He meant that we need to keep in mind our practice, the discipline we are following, and that we need to be aware of when we are following the discipline and when we are not. To be mindful is to remember that we are on a spiritual journey, a consistent, lifelong practice to stay awake. Practicing with discipline leads to a high level of awareness:

> In its developed form, mindfulness brings about a highly refined sensitivity to everything that happens, however minute, in one's immediate vicinity and in one's mind. ~ *The Universe in a Single Atom, by His Holiness, the 14th Dalai Lama, 2005.*

Equanimity and Flow Arise Through Emptiness

We are bilateral creatures. The left side of us and the right side of us need to be in balance. The mental equivalent to this physiological balance is equanimity. The vibrational frequency of the body needs to be harmonious and coherent throughout—from the top of our head to the soles of our feet. This whole-body frequency-balance arises within meditators as they set their intention; it is part of the practice that stabilizes coherence. We can "will" our bodies to be coherent and healthy—up to a point, obviously.

Science has discovered that the universe is expanding. When we contemplate this understanding, we at first envision the edges of our universe where this expansion is supposedly occurring. However, this is a misconception. The expansion of the universe is happening through you—you are a part of it. *You* are expanding. This is what I mean by "the flow that arises through emptiness." We can "surf the wave" that is this expansion—we can feel it if we spread our arms apart, shut our eyes, and let go. All that exists in nature, in the universe we know, is surfing this same wave of expansion; it is a continual creation. The egocentric mind inhibits our sensing of this expansion; the ego grounds us, freezes us to a single location. Meditation is a means to drop the inhibitions of the ego and return us to sensing the flow.

Energy Arises Through Emptiness

When the ego is resting and the allocentric mind is released, a meditator is able to work on changing the biochemistry of the mind. This "working on the biochemistry of the mind" is the deliberate practice of transforming/ altering neurons, neural pathways, and neuronal networks. This discipline creates all of the qualities discussed above: love, peace, joy, wisdom, mindfulness, equanimity, and flow. The fuel used to do this transformation is energy in all forms, from the gross to the most subtle. This energy is made available when the eyes are closed and the body is stilled.

What we call a spiritual practice is actually this transformation of the mind. Self-generated, biochemical changes craft a brain and nervous system to be ever more loving, peaceful, joyful, balanced, and wise.

Enlightenment Arises Through Emptiness

Evidently, enlightenment is both a sense of weightlessness and a filling up with light—becoming light. The biochemistry of the mind becomes ever more refined during meditation, until a point is reached, a threshold, which is called enlightenment. This is my best guess. I am not an enlightened being—far from it. Like you, I am on a spiritual journey. However, I do have a meditation practice. In my best moments, when relating to others, I use what I discovered during loving-kindness meditation.

Loving-kindness is a method for communication. Using this meditation technique, it is possible to connect to an enlightened essence, or so it feels. During ordinary discourse, the practice of loving-kindness involves the building of an imaginary bridge that connects your own mind with the mind of another. Here is how it begins: imagine a bridge that spans from your third eye to the third eye of the person you are dialoguing with. Your essence, your soul, will make a journey across this bridge so that quality communication can occur.

As your soul departs through the portal of the third eye, it has lots of company. The ego has decided to accompany the soul on this adventure. The ego, as it is so accustomed to doing, wishes to be in charge of communication. Consequently, the ego has packed its bags and is departing with its personality, its beliefs, its knowledge-base, its associations, its preconceived notions, and with overstuffed bags filled with passionate emotions and judgmental opinions. This ego train is miles long, and is attached to the soul-engine as the journey begins.

About halfway across the bridge, the soul and the ego encounter a gate. This gate is impervious to egos. The ego cannot make the journey across the bridge to another mind. It must remain suspended in space between one body and another. It cannot survive in this rarified atmosphere— disembodied—so it evaporates, taking the entire rational, emotional train down with it. Consequently, when the soul reaches the other mind, it does so without the egocentric mind. Only the allocentric mind has made the journey successfully.

The allocentric mind has no voice; it is mute. It can only listen deeply to the other mind. It can only, in a sense, *become* the other mind. Therefore, it experiences empathy and becomes symbiotic with another soul.

In actual practice, of course, when two people try to communicate, this method is often interrupted by egocentric associations, opinions, and disruptive emotions. It takes practice for loving-kindness to work well. It is very difficult to remain purely focused, deeply listening, and absorbed in another's mind. The role of the egocentric mind during loving-kindness communication is just to ask short questions, and then disappear so the al-locentric mind can get back to deep listening and deep empathy.

This is a very different style of communication than most people are accustomed to. Usually, two egocentric minds square off and battle it out, hurling emotions, interrupting, posturing, hardly listening, going from as-sociation to association, falling into neural loops that rant and repeat, and all of it without a trace of empathy.

This state of being—in which the ego is left behind—is not hard to understand because we are very accustomed to feeling egoless. Consider what happens when we watch a good movie, or even when we become im-mersed in a video game, sporting event, or concert. The egocentric mind with its personality disappears. We become totally absorbed outside our ego-structure.

Consciousness: A New Slant on an Old Conundrum

During loving-kindness *meditation*—unlike when we are communicating—we are not building bridges into other minds, nor are we getting lost in one kind of social event or another. In loving-kindness meditation, we are building a bridge between our embodied self and a universal, enlightened soul. We don't bother egocentrically trying to conceptualize such an enlightened soul. We just begin the spiritual journey from our third eye into a world beyond space and time.

We drop the skeptical ego and head up into the third eye portal. We push whatever mind-buttons we can imagine, and then we envision a bridge being created that spans between our embodied self and the third-eye of God—or whatever you want to call the loving entity you will be visiting. When you step onto the bridge, you start the spiritual journey.

When you get to the other side of the bridge, "you" become a non-verbal, nameless entity, without a personality, without an ego, without emotional baggage. On the other side, you can only deeply listen. The ego may only intervene to ask short questions and then it must evaporate—until a new question is asked. This is a spiritual communications technique. Ask your questions respectfully, with compassion, and then just sit there waiting, deeply listening. A universal empathy will arise and answers will come—as long as the ego doesn't block the transmission.

Use loving-kindness during your daily discourses as you communicate with others, but especially use it to connect to what we might call *Divine Wisdom*. In other words, when imagining a bridge from your third eye, allow the bridge to expand beyond your space-time boundaries and connect to the *ever-present origin*—to use Jean Gebser's phrase. Meditate on a bridge that connects you via loving-kindness with something beyond the shell of everyday reality. Of course, the ego will reject the bridge metaphor. The ego will reject the idea of connection with anything beyond sensory reality. It will reject the process of transmission and reception, and especially the notion of a universal intelligence. This is just the ego doing its valuable job. That being said, build the bridge anyway, despite the ego, and go beyond physical reality.

Changing Mental States through Meditation

When Buddhist monks meditate they can move through levels of wakefulness, if that is their intention. They are then able to study various states of mind using the highly refined skills that are developed during a lifetime of practice. In the quote below, the Dalai Lama underscores this ability for the Western thinker:

> What occurs during meditative contemplation in a tradition such as Buddhism and what occurs during introspection in the ordinary sense are two quite different things. In the context of Buddhism, introspection is employed with careful attention to the dangers of extreme subjectivism—such as fantasies and delusions—and with the cultivation of a disciplined state of mind. Refinement of attention, in terms of stability and vividness, is a crucial preparation for the utilization of rigorous introspection, much as a telescope is crucial for the detailed examination of celestial phenomenon. Just as in science, there is a series of protocols and procedures which contemplative introspection must employ. Upon entering a laboratory, someone untrained in science would not know what to look at, would have no capacity to recognize when something is found; in the same way, an untrained mind will have no ability to apply the introspective focus on a chosen object and will fail to recognize when processes of the mind show themselves. Just like a trained scientist, a disciplined mind will have the knowledge of what to look for and the ability to recognize when discoveries are made. ~ *The Universe in a Single Atom, by His Holiness, the Dalai Lama, 2005.*

Buddhist monks and Hindu gurus have more than 2500 years of experience developing the mental tools for introspection. They have much to teach humanity about how minds work. It is heartening and powerful to realize that the final conclusion reached by these spiritual beings is that our goal on earth is to relieve suffering and increase happiness for all sentient creatures.

Consciousness: A New Slant on an Old Conundrum

Let's take a quick look at the levels of wakefulness that the Buddhists and Hindus have studied for centuries. These are the mental realms (as far as my own limited skills can reveal) that seem to match what Western scientists study when they look at frequency variations, especially brain waves.

When I was doing research for this book, I was surprised to discover that monks and gurus not only explore waking states of consciousness but also sleep-states. Talented meditators are "awake" as they drop into deep sleep. Buddhists take the study of the sleeping mind as seriously as they consider all other levels of alertness. Let's start at the sleep state with delta brain waves and slowly "wake up" as we associate Eastern introspection with Western science. Notice that if we can stay attentive during sleep, this means that different areas of our brains can be at different levels of wakefulness at the same time—asleep, and yet "awake and watching," is just one example.

As we drop into the various deep stages of sleep, we lose two abilities that are invariants in everyday reality: the physical environment becomes irrelevant, and our ability to move with a purpose becomes irrelevant. During sleep, there is no stable visual background image, and there are no stable objects to be seen—so there is no need for an egocentric seer. The brain, during sleep, has lost its reason to exist—i.e. to perform purposeful movements. The brain is disembodied during sleep; it has no proprioceptive sense of being an entity separate from a domain. In deep sleep, the body is actually paralyzed; there is no ability to move the large muscles. This state of disembodiment combined with the loss of a stable visual background creates a kind of mental "purity," a "plant-like understanding" of being alive. The mind has an inherent need for purposeful movement so evidently—during the dream state—it creates new worlds, unusual illusory domains. Frames of reference can also shift without restraints during sleep.

As we begin to wake up, we leave deep sleep—and the delta-waves that characterize the dream state—and we enter a theta-wave state of consciousness. Here we find a most bizarre and fascinating realm of reality. This theta-wave level of consciousness seems to be a transition, a struggle, between a reality that has no dependable domain and no body capable of affecting the domain, and our "normal" state of perceiving reality—characterized by alpha and beta-waves. Evan Thompson, in his book *Waking, Dreaming, Being* (2014), speaks of the hypnogogic state (another name for theta-wave consciousness) as a form of synesthesia. It is as if all brain processing is cross-wired during synesthesia so that thoughts can have colors, or locations can have a voice, poetry can write itself in the air, or landscapes can self-animate. Hypnogogia is where saints, seers, poets, and painters "travel" to receive messages and inspiration. This state of consciousness seems to be a portal through which "channeled information" flows. Theta-wave consciousness is a semi-dream state, a land of mystery. We have yet to deeply probe the depths and extent of this strange domain; it is a relatively unexplored frontier.

As I said above, when we fall asleep, the theta-wave zone of consciousness is called "hypnogogia." However, when we awaken from deep sleep in the morning, we again pass through theta consciousness, but now it is called hypnopompia. As far as I know, the process is the same, although I am not sure. More research needs to be done to compare the two transitional states as we fall asleep and as we awaken.

When we are not asleep, nor in the twilight zone between sleep and being awake, we exist in a frequency soup composed of alpha waves. This is a relaxed, peaceful, secure zone of comfort. Alpha waves fire when the eyes are closed. Just before entering our "normal" wakeful state, we pass through this alpha state of consciousness. We begin and end each day as we pass through this sea of calm. Brain wave systems that seek to entrain the mind have these alpha waves as a background. Alpha waves may also be a stable frequency that grounds the entire nervous system—a steady state, a kind of mental background, against which other frequency states manifest.[7]

After we pass from the alpha-wave state as we are waking up, we slowly enter a world dominated by beta waves. This is the zone we call "reality." When we say that we are awake, or conscious, we are referring to this beta-wave cognition. This state is often subdivided into low, middle, and high beta.[8] Low beta is a daydreaming, habitual, and routine state of awareness. Middle beta is our conversational world, more aware than low beta. High beta is a zone where cognition is clearer, more rational, a high problem-solving state.

Meditators who are very skilled at energy management can move from high beta to a gamma-wave state. Here is another zone of unusual power and mystery. Monks hang out in this world, and thus they have an understanding that is beyond what most people can appreciate—or imagine. The Buddhists have discovered many gradations within each level of consciousness, but science has yet to record and study these gradations in detail.

C an the mind connect to energies below delta and higher than gamma? I don't know, but many mystics and clairvoyants seem to dwell in realms that require subtle energies. I do believe that such subtle energies await scientific exploration, and I honestly feel that we have capabilities well beyond our everyday idea of reality. Understanding dual cognition (dual-process theory) will be an important conceptual underpinning as research is conducted.

There are seven chakras in the body according to Hindu practice, these energy nodes (centers) are located along a plane parallel to the spine. From the first chakra at the base of the spine to the crown chakra that is above the head, there is a gradual rise in frequency (energy). These chakras also correspond to the octaves of the musical scale. I wondered, as have many others, if the seven levels of the chakras and the octaves of music correspond to levels of wakefulness, to the wave-states of consciousness. Esoteric researchers also find the number seven to be significant. It is as if there is a biological or quantum reason that seven levels of frequency turn out to be significant for comprehending how our minds come into existence.

Discerning Meditation

Buddhists and Hindus did not have scientific evidence for two minds, but they found the duality through meditation and reflection; they discovered and carefully explored both minds. Through meditation, monks and gurus could see that the *ego* was dominating, hiding, and ignoring the *self.* Much of Eastern religious practice involves reawakening the dormant twin—the allocentric mind—without losing the goal-directed positive contribution of the ego. Buddhist practitioners know that there needs to be a balance between the allocentric mind that experiences, and the egocentric mind that actively explores:

> The critical balance we need to discover in meditation practice—and indeed in all aspects of life—is the equipoise between effort and surrender. On the surface these two qualities seems to contradict each other. How can we make effort, be purposeful, and at the same time surrender to what is happening, to the natural unfolding of our experience? Grasping this paradox is a decisive turning point in coming to understand the whole spiritual journey. - *Insight Meditation, Joseph Goldstein, 1993.*

The allocentric mind surrenders to the flow of experience. The egocentric mind works hard to unearth knowledge. These are not mutually exclusive, but they can get out of balance. Meditation is a strategy for maintaining equanimity—harmony, synchronicity, and balance.

Over two thousand years ago, and perhaps longer than that, after they reached the end of their mental journey, Buddhists and Hindus came to a startling realization. When they had totally deconstructed the mind, they were left with the unsettling awareness that the world we perceived through our senses was not really there—at least, not as we perceive it. Something basic about our reality was an illusion. They called this illusory awareness "Maya." Maya can also be translated to mean "delusion." We are fooling ourselves if we think that "sensing is believing." All the senses tell us, at best, partial or metaphorical truths.

Deeper meditation reveals the illusion of egocentricity itself. It is this ego-mind that emerges from the allocentric. It is impossible for "objects" to separate entirely from the background; the foreground is part of the background even as forms bulge into 3-D manifestation. We are just one entity, allocentric wholeness. Unbounded awareness comes from the allocentric mind, but when a form manifests it is possible to sharpen egocentric attention into a bounded focus. Meditation techniques can address the needs of one mind or the other but not both at once.

What, and how much, can be manifested from the allocentric background? If we look deeper and deeper and pull out ever more subtle "objects" from this unbounded potential, what might we find? What kind of sacred, spiritual, powerful wonderment might we cause to manifest from our own allocentric ability, if only we knew the pathways to follow? I will leave you with that question and challenge.

Notes

(1) **Metaphysics** deals with "first causes." Where did the world—the universe, objectivity, everything and anything—come from? What is reality? Does the world exist outside the mind? Is there a God, or many gods? Why are we here? How did we get here?

(2) **Buddhists and neuroscientists fail to entirely grasp dual-process theory**. I could be way off-base with this statement because my understanding of Buddhism is not sophisticated. Perhaps I should say that I have yet to come across a clearly stated discussion about the evolution of two minds from a Buddhist perspective. Dual-process theory falls under the broad umbrella of neuroscience, but again, I have not seen a clear indication that neuroscientists understand the significance and evolution of duality—especially as it relates to navigation.

(3) Different levels of brain-wave activity correspond to different levels of consciousness. There is a vast amount of research about brain waves. For example, the discipline of neurofeedback is the study of brain waves and associated states of consciousness—biofeedback techniques use various wave forms to induce different levels of consciousness. Usually five brain wave frequencies are identified—although each can be subdivided. Delta waves are present during deep sleep. Theta is a zone between sleep and being awake—a zone containing dream states, like deep daydreaming. Alpha waves show up when we are relaxed—these waves arrive as Theta waves decrease. Alpha waves give way to Beta waves which represent what we normally call "being awake and communicative." Gamma waves are hyper-alert states; few people except expert meditators can generate these high frequency waves.

(4) Inner perception is not a verbal process, not a self-dialogue. It is what flows into us when we are empty of external sensation. It is not like the act of looking at an object-of-regard, or listening to a sound source. It is whole-body awareness: becoming the surround—becoming nature, becoming the flow of existence. That being said, I don't want to give you the false impression that I am an authority on inner perception. I am not. I am a seeker, perhaps as you are a seeker. Knowledge I seem to have in abundance from reading many books and watching documentaries and *YouTube* clips; wisdom, I do not have in much abundance. I meditate to have the experience of meditation.

(5) **Lucid dreaming:** this means to be aware, or relatively awake, while dreaming. Watching and remembering your dreams is said to be a lucid phenomenon. The stage of sleep called Rapid Eye Movements (REM sleep) is correlated with lucid dreams.

(6) **Hypnogogic "dreaming:"** This is another name for theta-wave sleep. It is the zone between deep sleep and the relaxed alpha-wave state. Hypnogogic "awareness" occurs as we are falling asleep. The same or similar state of awareness occurs as we wake up. However, the waking theta state is called hypnopompic.

(7) **Alpha waves as a background frequency** for the whole mind. I got this information from a lecture presented at the Redwood Neuroscience Institute in California. A Scandinavian research team suggested that alpha waves were the "frequency glue" that made the brain coherent—all other frequencies were foreground to the foundational 7.83 Hz alpha wave frequency.

(8) **Beta waves have been further divided** into low, middle, and high beta. Low beta waves (12-15Hz) are reached when we are musing, letting our minds drift. Middle beta (15-22Hz) is synonymous with our problem-solving mind. High beta (22-38Hz) is a highly alert state that comes about when we are being challenged, or perceived to be. Each level requires ever more energy.

Section Two

THE HARD-EVIDENCE

Section Two

Seven

Quantum Science, Duality, and Consciousness

The sun rises,
but which way does the night go?
I have no more words.
Let the soul speak
with the silent articulation of a face.

~ "Beyond Love Stories," *A Year with Rumi*, 2006.

Einstein Jam

"Okay, Surge. I need a meal full of bravery."

"I see. And what is the occasion for your lack of courage?"

"Today I talk about the philosophical foundation for my ideas. Today I write about quantum theory."

"Yes, of course you do. Everyone who writes about consciousness must tip-toe through the mathematics."

"Please don't use the "M" word."

"Okay. Let me get this straight. You are going to enlighten the masses about quantum theory with no grasp of the "M" word. You want to speak with authority about the most complex ideas in modern times, and yet you have no authority whatsoever. Is that it?"

"You are making me unhappy, Surge. What's for breakfast?"

"We have the perfect coward's breakfast. Fools also like this dish. You should be delighted."

"I forget why I keep coming back here for my meals."

"For breakfast on this overcast morning, we offer a Quivering Soufflé cowering beneath a thick crust of Fried Dread. We drizzle Melted Grief over the Gutless Concoction, and then we lather a thick layer of Embarrassing Ersatz Chocolate Frosting over the whole Cowardly Mess."

"You have outdone your depressing self, Surge. That is sadly disgusting."

"It's *your* mood, Dutch. It's not my fault you were born with chicken genes."

"Just bring me some coffee and an order of toast and jam."

"You want the Heisenberg Jam?"

"Okay. Sure. What's that?"

"We aren't certain."

"You don't know what you are putting on my toast?"

"We also have Thomas Young Jam."

"Do you know what that is?"

"Well, it starts off as jam but then becomes both jam and not-so jam."

"I am sure you have an Einstein Jam, right?"

"Yes, of course, but its only Relatively Jam; it keeps moving off the toast."

"That's it! That's what I am trying to say, Surge. Everything is based on movement and no-movement."

"That's very clever. You might be the first person, after Einstein and 600 philosophers, to figure this out."

"Your own personal jam, Surge, is sarcastic and demeaning."

"I offer a teaching jam that does not suffer fools lightly. Do you want the coward's breakfast or not, Dutch? By the way, I remember reading that 'Dutch' means 'overly self-conscious nerd' in Swahili."

"No cowardly breakfast for this crusader. I came in for some courage."

"Start with dessert, then. We have Fortified Oats drenched in Killer-bee Honey with Sumo-wrestling Raisins Thrashing in Tiger's Milk. You must eat it aggressively and without remorse. Eat boldly with insane hunger. Don't stop eating until every soggy oat is vanquished and every swollen raisin is swallowed whole. Be brave. Remember that even special education teachers can have opinions; they can even exhibit spasms of sporadic logic. Who says you can't make a contribution? Damn the torpedoes. Full steam into the madness."

"Let me in Coach."

"Don't let those big linemen push you around, Dutch. Speak your mind. Hold your chin up and stop drooling."

"I'm not drooling."

"Say what you have to say, Dutch. Stick to your guns. Place your innocent neck in the guillotine and smile at the incredulous quantum physicists and the dumbstruck philosophers of logic. You have as much right as anyone to be uncertain, and yet unabashedly loquacious."

"You are such a strange kind of help, Surge."

"Yes, of course. Go get your helmet and mouth guard, soldier. Make us proud."

"Here I go Surge."

"Your mother is overjoyed at your faux courage. Don't embarrass her again."

"She's ninety-two so I better get this right."

"What's Chapter Seven about, Dutch? Maybe I can help."

"It's about science. In the beginning, I explore quantum physics through the eyes of Albert Einstein, Werner Heisenberg, and Thomas Young. Then I muse about metaphors of science—because a big part of our confusion lies with language. We use abstractions and metaphors as if they were solid entities instead of processes. Then I talk about the head-on collision between entropy and syntropy."

"That's a real snore, Dutch."

"No, it's not! It's very fascinating."

"Okay, but I am going to bring a pillow, just in case. Is that it, or are you going to claim to understand holography too?"

"You are giving me a belly ache, Surge."

"Good luck, Dutch. I'm going to call Karen and tell her you need emergency editing. You are trying to cram all of existence into one chapter."

"I can do this, Surge."

"You already did it, I saw the draft. God help us. The Titanic has sailed."

Mind-Hurting "Logic"

I am breathing too close
to this mirror's face.

~ "TOO HAPPY, YOU COULD NOT SLEEP LAST
NIGHT," *A YEAR WITH RUMI*, 2006.

Consciousness: A New Slant on an Old Conundrum

There is a cosmological logic that I am following, a set of assumptions that guides the dialogue about the origin of our two minds and two kinds of consciousness. This logic is fundamental to the dual-process theory of cognition proposed here. I feel it necessary to provide an outline of the logical flow that evolved our two minds, if for no other reason than to lay it bare for future dialogue. In a way, this is a recap of my main themes, but in another way, this is a fresh perspective. I often find it necessary to rethink and rewrite, even at the risk of redundancy.

Step One: We are born into a world of relative movement. Everything is in motion. Everything is moving relative to everything else. Nothing is absolutely still in our universe. That is how we found the world as we popped out of the womb. We don't need to ask why this is so, or what the origin of relative movement might be—relative movement is just a fundamental characteristic of our world. Yet there is something innate within us that *does* want to know if ever there was a time when movement did not exist. Was there a time when nothing moved?

Either there was a beginning to existence, in which no-movement gave way to movement, or there is a rhythm between movement and no-movement that has always existed. In either case, there is a fundamental oscillation in our universe between movement and no-movement (change versus no-change). Therefore, I suggest, as have many others, that *"movement" is the key to understanding our universe and ourselves, including mind and consciousness.*

Two thousand years ago, Aristotle proclaimed an "unmoved mover," a background entity, out of which movement manifested. Everything is now moving relative to everything else, and that, of course, is what Einstein said in his theory of relativity. Things may only appear to have ceased movement relative to an outside observer. The universe that we know is defined by what the physicists call conservation of momentum. This simply means that constant and relative motions are expected.

Step Two: "On" oscillates with "off." This is the same as saying that movement oscillates with no-movement. This fluctuation is constant and equal. It is a primal metronome. It is a "perpetual motion machine." There is no shut-off switch. It just keeps exponentially making more of itself. It is a self-replicator and a self-organizer; it is a process that duplicates itself without apparent end. Metaphorically, movement can be called "wave" and no-movement can be called "particle." This suggests a primal connection between quantum theory and movement.

Step Three: "On" cannot exist the same time as "off;" no-movement does not equal movement. We cannot measure movement the same time as we measure no-movement. This is the basis—a laymen's explanation—of the Heisenberg Uncertainty Principle. I will discuss this important principle below.

Step Four: The first three steps above are a template for creating living organisms that purposefully move. All navigational creatures have within them this oscillation, at many scales. *Movement/no-movement* is the fundamental frequency out of which all frequencies are created; we live in a fractal universe composed of repeating frequency patterns.

Step Five: The sensory systems of all creatures that purposefully move are built on the scaffolding of these "on-off frequency patterns." *Therefore, we would expect to find (in every creature capable of self-movement) one biological system for movement management and a second biological system for no-movement management.* Indeed, this is what we do find. One system deals *only* with movement (allocentric processing) while the other system deals *only* with non-movement (egocentric processing). These two systems are mutually exclusive. They cannot exist simultaneously. They must alternate.

Step Six: The allocentric mind manages "flow," which is another name for "movement." The egocentric mind "freezes" flow; it manages "no-movement." Egocentric processing is serial, one-thing-at-a-time. This temporal processing is primarily, but not exclusively, what the sense of hearing does. Spatial processing of flow is parallel, every-thing-at-once, and this is primarily, but not exclusively, what the vision system does. However, all the senses work together. They coordinate either in allocentric mode or egocentric mode.

Step Seven: After eons of evolution, two processing systems became separate minds. One mind is a whole-body, all-at-once system for having experiences (a self), and the other mind is hyper-focal, taking in the world one-thing-at-a-time—it gave rise to the ego. The world of the *self* is spatial; it is silent, and overwhelmingly visual. The world of the *ego* is vocal. It speaks outwardly and inwardly—it is overwhelmingly temporal and auditory. The basis of both of these types of consciousness is "memory of memory," which is a proprioceptive function. The source of consciousness, as I stated earlier, is "proprioceptive layers of memory." These layers of memory are fed information through two kinds of attention systems.

Step Eight: The two minds evolved their own kind of consciousness, which I call "self" and "ego". *Self* arises from the allocentric mind, and it manages movement-through-life, which we call "experience." The *ego* arises from the egocentric mind—it freezes life into events. The literature that tries to explain consciousness is a confusing mess, of course, often using *ego* and *self* as synonyms. The ego is the center of the universe no matter where it goes—it exists as a separate form in a world of objects. The self, however, is what remains after egocentricity is taken out of the equation. The self is not separate from the world around—it is "connected with everything that flows." Self is merged with nature; it is not a "form" separate from nature.

Synonyms for the allocentric mind include the soul, the self, the unconscious mind, and the subconscious mind; it is the background mind. Synonyms for the egocentric mind include the spirit, ego, rational mind, and conscious mind; it is the foreground mind.

Calling Einstein to the Witness Stand

In layman's terms, Einstein's famous relativity theory reveals that *there exists no privileged spatial frame* of reference—everything is moving relative to everything else. No one and nothing is the center of infinity. There cannot be a center to infinity. I suggest that Einstein's relativity theory is proof of allocentricity; it is a mathematical snapshot of the *self*. There is no center to the allocentric mind.

Einstein's *special theory* of relativity, again in layman's terms, reveals this: events that occur at the same time for one observer can occur at different times for another. This means there *can be individual temporal frames of reference*. Wherever you go, you are the center of the universe. Your body provides a temporal frame of reference in which you appear stationary relative to all the movement going on around you. All events happen with you as the creator of a personal time frame. Other human beings have their own temporal frame of reference as well, but it cannot be the same as yours. Again, I am stretching this mathematical theory to fit my proposition that we have two minds. I see Einstein's special theory of relativity as proof of egocentricity. The special theory of relativity is a mathematical snapshot of the ego. We are each the center of a temporal universe; we are stationary, and the world revolves around us no matter where we go.

Given the above perspectives, I suggest that Albert Einstein inadvertently "proved" that all creatures capable of self-movement have two minds: a spatial mind that manufactures infinite space, and a temporal mind that manufactures an eternity of time. These two minds have to be mutually exclusive.

Of course, my musings could fall under the broad category of flaky metaphysics. I am sure that several physicists had apoplectic seizures after reading the above interpretations of Einstein's and Heisenberg's famous equations, and half-a-dozen philosophers choked on their Tim Horton's donuts. I put "logic" in quotes in the title of this section to protect myself, but I fear there are too many ways to challenge each of my eight steps. In the esoteric world, however, where the dictum "As above, so below" is "God-ordained," this line of thinking—my "mind-hurting logic"—is not outrageous.

While I am hanging over the edge, I will add one more piece to this section (actually, I am repeating an insight I shared earlier). According to current cosmological thought, the universe is expanding. But where is this expansion taking place? We usually envision some far-away edge—the "end" of the universe—where this magic is happening. Instead, let's realize that *we* are the edge of the universe. The universe is being created "on-the-fly," and we are part of this cosmic expansion. *We* are the edge of the universe. *We* are "expanding" with the speed of light—every moment. This is "cosmic flow." Later in the discussion, we will find that "flow" is a fundamental characteristic of the allocentric sensory system.

As Above, So Below: Heisenberg's Uncertainty Principle

Heisenberg's mathematical formulas show a basic conundrum of existence: We cannot measure movement the very same time as we measure no-movement. Heisenberg's actual equations show that we cannot measure location at the very same moment we measured momentum. You can't measure the location of something that won't stop moving. Another perspective of the Uncertainty Principle is that energy and time are negatively correlated. The more we know about the energy of an event, the less we are able to say when it happened. Energy is flow; it is an allocentric process, another name for movement. Time is egocentric, another name for no-movement. We use different words, but they represent the same process, the same conundrum, the same negative correlation.

The more I thought about dual-process theory, the more Heisenberg's Uncertainty Principle kept popping into view. I seemed to see this theorem everywhere I turned my attention. It even showed up in the retina of the eye, in the attention system of creatures that purposefully move, in attentional oscillation, and in the heart of quantum theory. Therefore, it occurred to me that what Heisenberg found was a fundamental pattern that was repeating throughout evolution.

In the wisdom traditions, and in esoteric psychology, there is a fundamental law that states "As above, so below." Each tradition couches this differently; for example, Christians say "On earth, as it is in heaven." This insight is helpful when we look back in evolution to see how things evolved in the past, and when we look forward to see where evolution is going.

Esoteric cosmology says that our world is filled with correspondences. This simply means that we can see ever smaller or ever larger examples of the same principles. A correspondence is an example of "as above so below." For example, if we discover that things are made out of each other—quarks make atoms, atoms make molecules, molecules make cells—we are able to predict that there will be ever smaller versions and ever larger manifestations of the same principle. We can predict an infinite regress where we discover that quarks are made of an even more fundamental particle that is also made from an even more fundamental particle, and so on forever. Going the other way, we can predict that this same building process will be used forever to make ever larger versions—whatever was made below will continue to create more of itself as the flow of evolution continues. The world is ever more complex going forward, and ever simpler as we glance behind—exponentially so. The Uncertainty Principle is a correspondence, a fundamental law of nature, another *way* the world was crafted.

The mathematician Benoit Mandelbrot provided a geometric picture of "as above so below." He called these repeating patterns "fractals." When I say that our existence is based on a fundamental on-off frequency, I am postulating a fractal—a pattern that repeats. Think of fractals as building blocks. In the beginning there was one block, but this doubled into two blocks. The doubling is exponential, so from two blocks we get 4, then 8, 16, 32, 64, 128, and so on, forever expanding in size and complexity.[1] In other words, movement comes in quanta, in discrete chunks, and it appears to follows musical octaves[2]. The Uncertainty Principle is a fundamental fractal—all of existence as we know it must follow this law.

Thomas Young: The Original Mind-hurting "Logic"

The famous Thomas Young double slit experiment[3] contains a conundrum that seems impossible. Shoot a series of single photons from a photon-gun at a surface that has two vertical slits. On the screen behind the slits we expect to see two dots (or two lines), one for photons that went through the left slit and another for photons that went through the right slit. Unfortunately, this doesn't happen. Instead, we see interference patterns[4] that can only be possible if *a wave* passed through both the slits.

A single photon starts off as a particle, travels forward as a wave, passes through both slits as a wave, and then hits the screen as a particle again. This is a classic quantum conundrum that has driven many good minds to the pub to drink and forget.

However, I suggest that this confusion is in our minds, especially in our sense of vision, rather than in any peculiarity of objective reality. Here's what I think is happening (my theory)[5]: "Particle" really means "exact location," or no-movement. Thus, at the photon gun, we measure the "particle" and we observe that there is "something" at this location that is not moving.

Then we pull the trigger and shoot a stream of photons toward the double slit.

Now the photons (the particles) are moving. But they no longer are defined by location—they no longer *have* a location. We call this "energy without location" a "wave" because "wave" actually just means "movement." If we watch the wave travel from photon gun to screen, we are measuring movement. This is Heisenberg's Uncertainty Principle: we can measure movement (waves) or we can measure location (particles), but not both at the same time.

When the photons arrive at the screen, they are no longer moving. Once again they have a location—they have stopped moving. We measure them, and sure enough—because we are measuring location, no-movement—we again see "particles."

In my opinion, we are fooled by our use of the terms "particle" and "wave;" these seem like objective entities that can be perceived egocentrically. However, they are not entities; they are processes that must be perceived allocentrically. Wave means movement. Particle means no movement.

Metaphors and Mentality

In the beginning, according to the Bible, God created man using mud, water, and air.[6] Like a master potter, God made human beings using the same technology that craftsmen used to make bowls. That is how early men saw their own creation, and this made sense for a clay-driven culture.

Later in history, early scientists in the Western world thought human beings were made from hydraulic fluids that moved "medieval spirits." According to this version of God's handiwork, human beings were crafted using mysterious vapors and animated spirits.

In the industrial age we proposed a machine metaphor to explain how our bodies work. Human beings were now made from gears and levers; we were bio-machines with moveable parts manufactured in different regions of heaven.

Next came the age of biology; the microscope ushered in the notion that human beings were made of cells, tissues, and organs—we believed, and many still do, that God made us from blood vessels, electrical nerve impulses, and lymph.

Our current era is called the computer or information age, and we have a new batch of metaphors that explain the human mind as networked-processors, data-banks, and memory storage units. We speak of algorithms, nested programming, and quantum computing as if brains and bodies were designed by God's IBM research center, on a hilltop just south of Heaven.

Consciousness: A New Slant on an Old Conundrum

This version of how God created mankind has been swallowed hook-line-and-sinker by many "modern" scientists. However, looking back through history, we are naïve to think that we have arrived at the ultimate metaphor that describes how minds work.

I confess that I use the computer metaphor often in my own writing, especially from a quantum perspective. This is a modern way to give analogies, as long as we keep it clear that we are using a metaphor.

On the horizon looms virtual reality (VR), with avatars, holodecks, and embodied communication. This VR revolution in technology will allow us to create imaginary worlds that we will visit anytime from anywhere. Scientists are busy finding ways for the virtual experience to be ever more realistic. A time will come when virtual worlds are as real as "ordinary" reality, which will get boring in short order. I would guess that human beings in the future will use virtual reality metaphors to explain how minds work. This has already begun, of course.

A well-accepted understanding in cognitive science is that we perceive a "good enough" representation of objectivity. However, in this text, I have gone a step beyond representation to say that brains *create* the world on-the-fly, moment-to-moment. Notice that this is how virtual reality works—the computer generates a world as needed.

Notice that when we dream, the brain can create any kind of world: it can mess with time, with space, with colors, movement, shape, emotions, and so on. In other words, the brain has the raw ingredients to manufacture realities because that is what it does moment-to-moment. This is a key idea, especially when we look at our concepts of eternity and infinity. The brain can manufacture time, on-the-fly, so it can make endless time. Likewise, the brain can manufacture endless space, on-the-fly, so it can conceive of infinity. The brain, using this metaphor, is a virtual reality machine.

To recap what I have postulated throughout this book, a major contention of dual-process theory is that the *brain manufactures* a world out of raw sensory information. The egocentric mind *manufactures time*—it creates a time-matrix that is projected as reality. However, the egocentric (time-based) mind is, ironically, blind to space; it cannot manufacture space—that is not its evolutionary mandate. *Therefore, the egocentric mind cannot perceive or comprehend its paradoxical spatial twin.* Likewise, the allocentric mind *manufactures space*—it creates a spatial-matrix. The allocentric (space-based) mind is, however, blind to time; it cannot manufacture time—that is not its job. *Therefore, the allocentric mind cannot perceive its perplexing temporal twin.* Two minds exist in the same head but they cannot perceive each other—this mind-blindness is a fundamental characteristic of the human condition.

To successfully navigate through an environment requires exact spatial-temporal coordination, so nature found a way to combine *space*-perception (the self, the allocentric mind) with *time*-perception (the ego, the egocentric mind). A summary of these two navigational abstractions is that egocentric processing takes place "one-thing-at-a-time," from the perspective of an ego (a definition of time perception), while for allocentric processing, "everything-is-perceived-at-once" from the perspective of a self (a definition of space perception).

There is no proof, of course, that infinity and eternity exist outside human minds. We are simply creatures who manufacture experiences, moment-by-moment. This virtual reality perspective, I contend, is just the most recent metaphor for dialoging about our cognition. I agree with the philosopher Ludwig Wittgenstein:

> The solution to the riddle of life—in space and time—lies outside space and time. ~ *Tractatus Logico-Philosophicus, Ludwig Wittgenstein, 1921.*

What Wittgenstein means by this statement is that we can only know the world with the senses we are born with. But these senses are limited; they give us only one way to understand a universe that is much more complex than we can imagine. A spider living on a tree in a rain forest does not know about a universe beyond the branches and leaves where it spends a lifetime. We are like that spider; we know only the space and time we are born into.

To suggest that we can comprehend all of existence because we know about space and about time is very naive. Beyond our physical senses, something exists that is not space and not time.

In the next section, I will compare two fundamental concepts in science, entropy and syntropy. To introduce this line of thinking, it is necessary to deconstruct our illusory reality to get an idea where we might have come from. This is just some fun food-for-thought before we dip into heavy biology in Chapter Eight.

Riding the Horse Until It Drops Dead

Let's deconstruct reality and see where it goes. Earlier I asked you to imagine yourself standing beside a tree. I asked that you take a photo of the two of you—tree and human together. Now I ask you to look at the picture again. There you are standing beneath the canopy of your favorite tree. Look carefully at the two of you, majestic tree and wondrous human standing proudly together. There is something in the picture that you need to notice. Yes, the tree is your cousin, you share a unique space-time in history, and you have some common genes. Together you are the miracle and tragedy of existence. But there is more.

Instead of playing your developmental life-cycle forward, let's play the video backwards through evolution. Behind you and your tree a spectacular sunrise is taking place. You are bathed in pink pastels, enveloped in compassion as a new day dawns. Now begin to reflect on how you got to this wonderful moment in history. Especially reflect on your ancestors—who were the people who had to exist for you to be here now; when did they live? Ask yourself: How did I get to be this miracle? Let's follow your genealogy back into history.

You had two parents. Two mysterious forces merged together and created you. Your parents had sex and the sex made you. But wait a minute;

both those parents were created by two other parents. You couldn't be this miracle unless you had four grandparents plus two parents. Six people had to exist before you got to exist. Six other miracles had to happen before your miracle got to take center stage in the book of life. What a nice feeling— except it opens up an endless regression, an eternity.

You had two parents, four grandparents, eight great grandparents, 16 great-great grandparents, 32, 64, 128, 256, 512 (look familiar?) and well, go ahead and do the math. Exponential doubling is going on here, and there is no endpoint reached as we flow backward in perspective. You don't go back to a big bang or to Adam and Eve. You go back to the Big Kahuna, to eternity.

At the time of Christ, which is not so long ago in evolutionary terms, you could have had 1,200,000,000,000,000,000,000,000 great grandparents[7]— as long as no relatives had sex. But of course, they did. Exponential doubling slows down when cousins marry cousins, or when sisters and brothers make babies—or (back in the caveman days) fathers with daughters, or mothers with sons. Indeed, marrying first, second, and third cousins is the rule in prehistory—you had sex with whomever you could when villages were far apart. So the doubling slows way down—it is not exponential.

Yet, no matter how it slows down the number of "relations" you have in your ancestry gets larger—insanely larger—because there is never an end point at which you can say, "Well, that relative had no parents." Go back a few million years and you find that you are related to every human creature that ever walked the earth. And they had parents, those early proto-humans, and you are related to all of them as well; they all had to exist for you to exist. Keep marching backward through evolution. The proto-humans had ever more primitive apelike parents and they had even more primitive parents—all of them are your relatives, every single one of them. By the time of the Cambrian Explosion over 600 million years ago, the number of your mammalian ancestors exceeds comprehension. As long as sex has existed, where two creatures combine to make a third, the regression has existed.

Consciousness: A New Slant on an Old Conundrum

At some point, billions of years ago, sex—two equals one more—was invented. Prior to sex, cells just split into two. One cell became two cells. Yet something older still gave birth to something new; the process of creation changed styles, but the regression did not end—always there is a cause and before that another cause, and so on and on.

The doubling rolled on past the Cambrian all the way back to the beginning of life-soup, 4 billion years ago. But it didn't stop there either; it just kept doubling: single cells to molecules, to atoms, to quarks in the "lifeless" sea. *Therefore, everything that exists at this moment is absolutely dependent on the past being exactly as it was.*

Take another look at the award-winning photo of you and your ideal tree. Not only is the tree your cousin, but so are all the animals and plants in the picture. So are all the human beings who ever existed and who exist now, and who will exist in the future—all of them are your relatives. All the single cells that ever lived are your cousins. So are all the molecules that made up all the cells of existence related to you. Every one of those molecules had to exist in the past for you to exist today. Logically then, the earth itself is your "cousin." Look at the photograph. The ground that supports you is your relative.

But the eternal doubling doesn't stop there, does it? Every atom that ever existed had to have the "parents" it did; everything had to be as it was for you to stand in the morning glow. But the doubling goes back still. Every quark, every vibrating string, every multiverse, every cosmic brane had to be exactly as it was for you to be exactly as you are. The universe is our cousin. All the stars and the Field, the Void, all the black holes, dark energy, little green aliens from Sagittarius, all these "guys" have to be invited to the family reunion. Look at the picture again. The sky all around is your cousin. Everything in the snapshot is related. Everything was born from something previous.

This is called a "vicious regression," or "infinite regress," but it is not like philosophical regressions that ask, for example, what God created God?

And what God made the next God, and so on. The difference is that our existence is much more tangible than philosophical speculations. This is a phenomenological conundrum, not fallacious logic. It is pretty clear you had parents and they had parents, and everything had parents—it's hard to deny the reality of this causal regression.

We are the whole of eternity and infinity, all the questions and all the answers. We are one big "happy" family built out of each other, creating each other. But, wait, the doubling of our relationships goes back further and further, unending. I don't know about your wonderful mind, but my three pound sponge is at a loss to explain this. I'm going to the Red Eye Café for another double mocha.

Let's give this problem to the mathematicians. Maybe they can figure the odds of our existence. Show them the video of you standing with your arm around your favorite tree, waving at the camera. So, here, math-guy, help me out:

Every time 200 million sperm race for the ovum—on any given conception day—only one of the sperm will be the groom that fertilizes the egg. That's one chance out of every 200 million for every offspring born in the eternal regression. And what are the odds that your parents would find each other in the mass of humanity, indeed, that all your grandfathers would find the right grandmothers to ask out for dinner—what are the odds they would cross paths, have sex, and create a life? It looks like the odds are against you being here. You are a mathematical impossibility with a passion for double mochas. And I get it why God might love you—as they say—because you are the end result of eternity; and besides, if the cause and effect regression continues backwards through time, then *you* plus everything else *is* God (at the least, God is your cousin).

When your parents did have sex, an egg cell from your mother encountered a little fish (a little sea squirt) called a sperm cell which came from your father. This was the beginning of your "forward evolution." The single fertilized cell that was now "you" had no neurons, no sensory organs, and no muscles. It was a sack of instructions, a bag of information. In the beginning you were invisible, just a potential to evolve into a very specific form. All that you are is the result of this cell doubling itself continually, cloning itself for about 85 years before doubling stops, cells age and die, and then the body surrenders back into nature.

It seems sad that after all of eternity and infinity, after all that incomprehensible series of manifestations, that the end result would be our disappearance from the story—we are suddenly, one fateful day, completely removed from the picture. We drop back into the canvas, into the scene; we become the gestalt. My ego is shaking its head in disgust and despair.

Good thing we aren't just egos. There is more to this story.

A Head-on Collison

Quantum physicists, as they pondered the big bang and the mathematics that came out of inflationary theory and multiple dimensions, saw clearly—using the mathematics of the second law of thermodynamics—that entropy[8] was increasing. American journalist and author Lynne McTaggart gives this layman's explanation:

> According to the second law of thermodynamics, all physical processes in the universe can flow only from a state of greater to lesser energy. We throw a stone in the river and the ripples it makes eventually stop. A cup of hot coffee left standing can only grow cold. Things inevitably fall apart; everything travels in a single direction, from order to disorder. ~ *The Intention Experiment, Lynne McTaggart*[9], *2006*.

The big bang was like a cosmic rock tossed into a calm sea; the biggest impact was at the point when the rock first landed in the ocean—powerful waves rippled outward. However, over time, the waves had only one thing they could do: get ever smaller and less powerful. At the moment of the big bang, according to science, everything had ideal order. After the bang, however, "order" immediately began to unravel. There has been an evolution of increasing disorder since the big bang happened; everything is coming apart, becoming more random, and getting colder. "Dying" is built into the equation—no getting around the math. Eventually even the universe is scheduled to grow cold and drop over dead. But wait a minute.

If our lovely human form, and the majestic form of our beloved tree, came out of ever more creation, from endless exponential doubling—two combining, giving birth to another life, stretching back forever—then our very existence is evidence for syntropy (negative entropy), the opposite of entropy. We are proof of ever more life emerging out of the background. We are part of an evolution that is the opposite of entropy. We emerged out of the "quantum soup," and our evolution has been an exponentially unrelenting *life-building* process.

Notice that allocentricity—as a background—not only gives birth to forms, but also pulls forms back into itself; the background never completely releases the forms that manifest from it. Objects that manifest from the background (call the background "heaven" for this discussion) eventually and inevitably dissolve back into the background—they are pulled into "hell" (from their perspective); they drop into the hell of nonexistence. For example, human beings are "objects" that have manifested, and as we age, as we are dying, we experience the hell of having to return back into that which had manifested us. But when new objects (babies, for example), bubble out of the background field and become manifest, they go from the allocentric background into the egocentric light of existence. The overarching concept of the *figure and the ground* that gave birth to dual-process theory, to allocentric and egocentric minds, is a universal process that permeates all of existence. Entropy and syntropy are just another slant on this basic perspective that life and death exist in parallel.

Our greatest poets, philosophers, and holy men knew about this struggle between death and birth, entropy versus syntropy. Take William Blake's book *The Marriage of Heaven and Hell*. Blake was trying to understand our fundamental duality. He saw in the universe a fierce determination that was "hell bent" on death and destruction. Something at the core of existence rips to shreds anything that tries to manifest; it is a fundamental "tearing down" force. Physicists called this force "entropy." Blake and everyone else in the Christian world called it "Hell." However, balancing these hellish forces of destruction was another force, something that created rather than destroyed, something that repaired and constructed. Life itself was a generator of more life, and that new life innovated, built, repaired, planned, and followed-through. Life got in the face of death. On reflection, it is plain to see what the goal of life is. The goal of life is to make life eternal, to create Heaven—or more heavens.

The very notion of "grace" in Christian theology looks very much like the force that pushes forms out of allocentric consciousness, and then pulls those forms back into the background. Grace is an unending well of compassion, unending energy for potential goodness, bubbling out from the dark background to create and perpetuate life. We can see that the background is paradoxically both Heaven *and* Hell. It slowly "destroys" all that it creates, yet it *does* "create."

To give two more examples, Rudolf Steiner and Frederick Nietzsche, both writing in the late 1800s, wrote about this battle between the forces of destruction and the forces of construction. Steiner set Lucifer against Ahriman, while Nietzsche contrasted Dionysus with Apollo. Both philosophers saw that life did not create passive entities. Instead, life created fierce creatures obsessed with survival. In other words, life co-creates and perpetuates ever more life. And you, standing beside your beloved tree, are the sum total of all there is at the moment. You are the crest of the creative wave. You are the best-hand-dealt. You are the current great hope— you and all your co-existing fellow relatives. Mr. Roger's Neighborhood is called to battle against hellish entropy. Humans are mandated by eternity to create, and to stay alive long enough to keep the life-force active.

I want to make an outrageous suggestion: You know all that dark energy/ matter we can't find? Maybe that is cumulative syntropy coming from all the life-forms in the universe. I know, totally nuts, but it did pop into my mind as I walked into the grocery store this morning to buy cat food, the pink sunrise at my back. Maybe the mathematics of syntropy equals the mathematics of entropy. Maybe—here's another bit of flaky speculation—the universe only appears like a flat disk. Maybe, instead, the universe is really a sphere because dark matter exists above and below our visible, material universe. This kind of brilliance is routinely shared over morning coffee at the Red Eye Café.

Or maybe, because life has such a long history of getting ever more sophisticated, ever more conscious, ever more capable of working collectively, ever more able to communicate at faster and faster speeds, and ever more capable of building and rebuilding, well, maybe syntropy is winning over entropy in the big Super Bowl of existence. Maybe God has the ball on the two yard line and is close to scoring in the fourth quarter of a tied ballgame—the Devil is nervously pacing the sidelines.

Notice that we are able to create because we purposefully move—sex is certainly a kind of purposeful movement. Relative movement slowly gave birth to time, space, space-time, creativity, double mochas, and "meaning." Trees don't write novels because trees don't purposefully navigate. *Movement* is the primal verb for us. Our existence, or so it appears from this logic, is modeled on something that has been going on forever—we are just the latest construction. Background entropy is churning, gobbling up all that dares to manifest; yet out of the devilish turmoil, life-forms still emerge, manifest, and repair what entropy has rudely trashed. The duality that is found in our minds, our bodies, our perceptions, and our consciousness has a universal mirror: as above, so below.

Quantum Science and Duality

This law whereby two opposing states cannot coexist without one undermining the other is the key premise in the Buddhist argument for the transformability of consciousness. ~ *The Universe in a Single Atom, by His Holiness, the Dalai Lama, 2005.*

In terms of quantum theory, what the Buddhists mean by "the transformability of consciousness" in the quote above is the same as that which is expressed by the equations of the Heisenberg Uncertainty Principle. There is something fundamental about our universe: we can have things one way, or another, but not both at once. In a more spiritual translation, the Buddhists mean that we can have a compassionate mind that includes all sentient beings under its benevolence, or we can have a non-compassionate mind that is concerned with its own welfare above all else. We cannot have both at the same instant. Furthermore, we can transform a mind through meditation and discipline from a frenetic egocentric entity into a serene, joyful, empathetic entity. Creation is syntropy at work. Destructive egocentricity is entropy at work.

Buddhists essentially figured out 2500 years ago that two mutually exclusive forces were at the heart of consciousness. In one sense, the Western world is still figuring this out, because the leap has yet to be made that we have two minds. We can use one of our styles of consciousness, or we can use the other, but not both at the same time. The recent cooperation between Buddhist scholars and Western neuroscientists holds great promise for the eventual understanding of mind and consciousness.

In 1979, the physicist David Bohm and the 14th Dalai Lama, Tenzin Gyatso, met for the first time; they were to become lifelong friends. Their discussions over many years set the stage for the blending of Buddhist philosophy with quantum theory. Prior to meeting the Dalai Lama, Bohm had developed friendships with Albert Einstein and Jiddu Krishnamurti. Therefore, David Bohm's mind was a blend of the Eastern and the Western; he was a bridge between two worlds. The Dalai Lama, before he met Bohm, had friendships with remarkable people like Thomas Merton, Karl Popper, and Huston Smith. He had also met and dialogued with Chairman Mao Zedong, Premier Chou En-lai in China, and he knew—and had the blessings of—the heads of state in England and in India. The 14th Dalai Lama is a world treasure, a man who understands Eastern and Western perspectives, a man with a mission to make the world a kinder place through dialogue, intention, and practice.

In 1980, Bohm wrote a book called *Wholeness and the Implicate Order*. Many have tried to explain what Bohm defined in the book as the "explicate order" and the "implicate order." I hesitated, like others must have done, to read Bohm's book because quantum math is well beyond my finger-counting brilliance. I also assumed that Bohm's complex mind would weave abstractions into a meshwork of nested abstractions that would send me to the Red Eye for a double mocha. But when I actually read Bohm's book, I was delighted. Here was a kindred spirit.

Bohm understood the concepts that I call allocentric and egocentric, but he used the language of quantum theory rather than dual-process theory to explain what he had discovered. Bohm saw that one brain "fragmented" the world (the egocentric mind) and it had gone too far. It needed to be balanced by what he called "wholeness" (the allocentric mind). Bohm felt that we had built a world—cultures, governments, philosophies, and scientific disciplines—with fragmented minds. Without a sense of wholeness, we had crafted a sick, hostile, fragmented world. This world, so evident all around, was not healthy for individual humans or for the collective whole:

> It is instructive to consider that the word "health" in English is based on an Anglo-Saxon word "hale" meaning "whole:" that is, to be healthy is to be whole . . . Likewise, the English "holy" is based on the same root as "whole." All of this indicates that man has sensed always that wholeness or integrity is an absolute necessity to make life worth living. Yet, over the ages, he has generally lived in fragmentation . . . Surely the question of why this has come about requires careful attention and serious consideration. ~ *Wholeness and the Implicate Order, David Bohm, 1980.*

David Bohm and his friend Albert Einstein shared common misgivings about quantum theory. They both felt intuitively that something was wrong either with the equations or with the conclusions that were being drawn from the calculations—important ingredients seemed to be missing. Bohm eventually came up with an explanation that made sense to him. He concluded that order can be manifest (explicit) or hidden (implicit).

Bohn's implicate order has many names, including the zero point field, the unified field, the superstring field, the Higg's boson field, or just "the field." Hindus call it the Akashic field. From an allocentric, evolutionary, and navigational perspective, "the field" gave rise to a unique kind of perception— a constant background awareness. This became our allocentric mind.

Bohm had taken a long journey, as a brilliant physicist and as a student of Eastern philosophy. In the final analysis, he concluded that there was a world hidden from view; from this hidden world emerged objective manifestations. This is exactly, I suggest, how allocentric and egocentric consciousness works within our minds.

I am fascinated by the correspondence between the peculiarities of the quantum world—as described in the quote below—and what I call the allocentric or background mind. At the level of the ultra-small, according to quantum theory, the world of objects disappears and we are left with a field phenomenon, an amorphous background that is pure potential—Bohm's implicate order. Our allocentric minds seem to have been modeled after this phenomenon:

> My discussion of quantum mechanics has stressed the three most fundamental and disturbing aspects of quantum mechanics: (1) its acausal nature, (2) the participatory nature of the universe or its lack of full objectivity, and (3) nonlocality or interconnectedness. Although these findings run contrary to our commonsense ideas (or projections), they are precisely what is to be expected from the view of emptiness, from the notion that the highest truth of objects is the interdependence and relatedness, their lack of an independent or objective existence. All this, of course, is in stark contrast to the old materialism enshrined in Newtonian physics and our unreflective

view of the world. ~ *Synchronicity, Science, and Soul-Making, Victor Mansfield, 1995.*

What Victor Mansfield is describing in the quote above is the bewilderment that the egocentric mind faces when it tries to comprehend the nature of its allocentric twin. That twin is acausal, lacks objectivity, and is non-local—all these concepts are "impossible" from an egocentric perspective.

The background field, the implicate order, gives rise to manifestations that appear to be caused. The egocentric mind exists in this causal reality. However, when objects disappear into the "background," this is an apparent acausal event. The egocentric mind deals with objects and an environment that is separate from the egocentric body. The allocentric mind, however, "participates" in nature because it contains no separate ego—it is one-with-nature. There is no ego located in the scene for the allocentric mind—there is no center, no locations—there is only nonlocality. From an overhead perspective, from the perspective of the allocentric mind, everything can be seen to be interconnected in both time scales and spatial perspectives; there is only interdependence and relatedness. Quantum strangeness seems to be the same as allocentric strangeness.

Another bit of speculation arose as I revisited Kurt Gödel's two incompleteness theorems.[12] Gödel used the logic of the Liar's Paradox based on this phrase: "This sentence is false." If the sentence is true then it is false; if it is false then it is true—thus the conundrum. Gödel used a similar phrase: "I am not provable." He made this sentence into a mathematical formula that bewildered the mathematics community; he showed that mathematics—set theory in particular—would forever have to deal with blind spots and paradoxes; there was no chance for a grand unified mathematical theory of all sets.

From a dual-process theory perspective, Gödel found the dividing line between the background (which is not supported by proof), and the foreground, which is composed of manifest objects, sets, that are subject to proof. Gödel simply found that there is a mutually exclusive duality in the universe and in the mind—we can solve for one or the other, but not for both at once—when one is true, the other one is always false.

Mathematics has a symbol for zero and a symbol for infinity. Does it, I wonder, have a symbol that acknowledges the fundamental oscillation of our universe as manifest in our two minds? I am too mathematically naive to know, so this is a sincere question. "True and false" oscillate. Rhythmic fluctuation is the nature of our reality. Perhaps taking oscillation into account for the Gödel equation would help with the confusion? The same would apply when we look at the mathematics of the Heisenberg's Uncertainty Principle.

Allocentric Space

Space is not empty even though, to our vision system, it appears to be obstacle-free. Physical space, the kind we walk through every day, is filled with sounds, smells, light waves, echoes, and pressure (tactile) frequencies. These various sensory inputs reverberate and blend to make each space unique. Everyday space is also filled with air molecules, dust, dead skin, numerous tiny "living" creatures—like bacteria, viruses, mites, and fairy flies—animal dander, and homeless fleas. No two spaces are ever the same. The vacuum of space is also not empty; although it does not contain matter—there are no lawn chairs or bowling balls in outer space—however, it does contain hydrogen and helium plasma, electromagnetic radiation, cosmic rays, space debris, neutrinos, magnetic fields, astronauts, and more dust (dust is a universal problem for cosmic cleaning ladies). This is the standard accepted understanding of space.

Technology has also given us the concept of "smart spaces;" we can deliberately put information into physical locations. Therefore, we must add "intelligence" to our list of what space can contain. If a sentient creature occupies a space, we can also say that degrees of consciousness are contained in that space.

Our bodies emit frequencies that mingle with spaces; for example, our hearts emit a powerful toric field, and our brains emit various, much more subtle frequencies. When we are in a space, we broadcast our frequencies into that space. We must also be open to the possibility that space—earth space and outer space—contains dark matter and dark energy. These cannot be detected by our sensory systems, and yet mathematically we can infer their presence. Subtle dark matter and subtle dark energy may contain information, consciousness, and intelligence—although this falls in the flaky speculation category, according to the egocentric mind.

Allocentric meditation—sensing the surround, becoming the background—is a way to "ask questions" of space. It is also where answers come to us from embedded spatial intelligence—so say the mystics and creative types. When we are in our egocentric mind, we are deaf-blind to the messages available in space. We cannot hear the messages, and we cannot envision the "messenger" that holds and manifests spatial intelligence.

However, the above look at space does not capture the background that I call the allocentric mind. The allocentric mind is a perspective, a frame of reference, as well as a potential substrate for the manifestation of matter. Another way to look at this is to consider the allocentric mind as scaffolding, a framework that holds a holographic projection. Let's give this holographic space some further thought.

Holographic Minds

From different perspectives, neurosurgeon Karl Pribram and physicist David Bohm arrived at the conclusion that "the mind" was the result of a quantum process that operated like a hologram, or *was* a hologram. How the mind could be holographic was not clear and the evidence was not totally convincing when they first introduced this hypothesis. However, the notion that a holographic mind resides inside a universal hologram, a universal mind, is now no longer outlandish.

Pribram and Bohm arrived at the holographic hypothesis separately, Pribram because he was trying to comprehend how memory could be stored in a distributed fashion throughout the brain, as research suggested, and Bohm because he was frustrated with quantum physics—a holographic universe seemed to answer questions that quantum science could not. The two men worked together for a while as they explored the implications of this line of thinking. Their thoughts became known as "holonomic brain theory."

In his book *The Holographic Paradigm*, author Ken Wilber provides this introduction to holograms and holographic brains:

. . . Pribram's studies in brain memory and functioning led him to the conclusion that the brain operates, in many ways, like a holo-gram. A hologram is a special type of optical storage system that can best be explained by an example: if you take a holographic photo of, say, a horse, and cut out one section of it, e.g., the horse's head, and then enlarge that section to the original size, you will get, not a big head, but a picture of the whole horse. In other words, each in-dividual part of the picture contains the whole picture in condensed form. The part is in the whole and the whole is in each part—a type of unity-in-diversity and diversity-in-unity. The key point is simply that the part has access to the whole.

Thus, if the brain does function like a hologram, it might have ac-cess to a larger whole, a field domain or "holistic frequency realm" that transcends spatial and temporal boundaries. And this domain, reasoned Pribram, might very likely be the same domain of tran-scendental unity-in-diversity described (and experienced) by the world's great mystics and sages. ~ *The Holographic Paradigm, Ken Wilber, 1982.*

Holograms create totally life-like miniature three-dimensional illusory images. You can walk all around these images and view them from many angles. If you chop a hologram into parts, you will perceive smaller versions of the original image in the fragments: still three dimensional, yet ever smaller as the fragments are divided further. The images are eerily real—no matter what size. If "mind" is the brain-plus the body-plus the environment, then holonomic theory suggests that we are part of the whole and we contain the whole. The implications are mysterious and powerful.

Karl Pribram

Karl Pribram was perplexed by the neurology of memory. As he searched for the locations of specific memories, he failed to find any anatomical centers; memory didn't seem to be located anywhere in the brain. His best guess was that memory was somehow distributed, but how could that be? The hologram seemed a perfectly plausible explanation if he could figure out how it might work. Some scientists, who were still not convinced that memory was distributed, wondered what would happen to memory if an animal brain was dissected—would memory be destroyed or preserved?

In his book *Shufflebrain* (1981), biologist Paul Pietsch, describes experiments using salamanders. Pietsch was initially a critic of the distributed holographic model of memory, and he set out to prove that memory was local, not distributed. He removed salamander brains and did a series of drastic surgeries to show how the salamander would be unable to remember specific tasks after the brain was altered. In Michael Talbot's book *The Holographic Universe* (1991), he says of Pietsch's work, "In a series of over 700 operations he sliced, flipped, shuffled, subtracted, and even minced the brains of his hapless subjects, but always when he replaced what was left of their brains, their behavior returned to normal." In *Shufflebrain*, Pietsch himself wrote:

> Memory often survives massive brain damage, even the removal of an
> entire cerebral hemisphere. In the 1920s the celebrated psychologist

Consciousness: A New Slant on an Old Conundrum

Karl Lashley, with whom Pribram once worked, demonstrated that the engram, or memory trace, cannot be isolated in any specific compartment of a rat's brain. Certain optical holograms invented in the early 1960s, the most common today, exhibit just what Lashley had alleged of memory: A piece cut from such a hologram—any piece—will reconstruct the entire image. For as unlikely as this may seem, the message exists, whole, at every point in the medium. ~ *Shufflebrain, Paul Pietsch, 1981.*

Using our current understanding of brains and behavior, we might say that what Paul Pietsch discovered was that the brain is not the mind. The mind is distributed throughout the body, so memory for running a maze, for example, is embedded in the proprioceptive system and in the muscles of the joints. Salamanders evidently don't need their brains to run a maze or to remember where food is hidden—the body remembers.

In a TV interview (2010), on a program called "Thinking Allowed" hosted by Dr. Jeffrey Mishlove, Karl Pribram talked about the interface of quantum physics and neurology. When I watched this interview with Karl Pribram, I was amazed how well it fit with my notion of allocentric and egocentric navigation systems. A quantum mind makes much more sense than a biologically-based mind, where our thinking is stuck at the level of neurons, synapses, and biochemistry. This is useful but not a deep enough view of "reality" to enable us to make further breakthroughs. Here is a summary of Pribram's half-hour discussion:

- The mind is not a noun; there is no such thing as "the mind," or "consciousness." There is only *a process* going on that allows for sensation and behavior. We get into trouble when we talk about the mind as if it was a singular "object." The brain is an object, but the mind is not.

- One of the main principles of holonomic brain theory is that there is a relationship between what we call the mind (egocentric attention) and a hidden process (allocentric awareness). According to Pribram, David Bohm's approach to quantum theory postulates

that besides an "explicit order"—everyday egocentric reality—
there is something hidden that he calls the "implicate order."
This hidden "universe" is a blur of potential and probability. Our
physical world emerges, manifests, from this quantum soup of
possibility.

- We have an evolving system of mathematics that explains brain neurology.[13] This set of mathematical equations is the same as, or very
nearly identical to, the mathematical systems that describe quantum
theory. The implication is that the brain uses quantum processes to
contribute to what we call "the mind."

- Therefore, what the holonomic brain theory asserts is that the brain
"obeys the same rules" as quantum mechanics. This also means the
mind is subject to quantum weirdness, such as non-locality and paradoxes like the inability mathematically to calculate for location at the
same time we solve for velocity [Heisenberg's Uncertainty Principle].

- If the brain follows the rules of quantum mathematics, then we have
to take a fresh look at religion, spirituality, and mysticism. In the
years ahead, science will find explanations for spirituality, and then
we will enter a new age where science and religion are siblings.

- There is no causality in this mathematical theory. Nothing causes
anything else. "Things" just emerge. Causation is only apparent. It
is as if the senses were projection systems, secreting things like visual
images and sound patterns.

- The external senses are like camera lenses. They focus on patterns:
sound patterns, visual patterns, etc. If you slowly take away the focused pattern by defocusing the senses, you get a blur that dissolves
the pattern back into potential . . . even though the "pattern" is still
inside the blur.

- Creativity occurs when we allow ourselves "to get into" the wave
form, which is part of the implicate order. When we meditate, for
example, dropping the egocentric perspective and entering into the
allocentric mind, we unleash creative power.

I find myself in agreement with the views of Pribram as expressed in the summary above. The logic that I followed to arrive at a dual-process theory of cognition is reinforced by his studies with memory.

Suppose that we live inside virtual reality, as the mathematics of quantum theory is now suggesting. Suppose we accept our holographic fate. Walk around during your day and pretend that nothing is as it seems. Consider that it is all a projection from the allocentric mind, busy creating a background for us to play within. Your eyes are projecting the world in front of your face. Your ears are projecting sounds to match the visualization. Strangest of all your body is just a projection as well. There is no "you," just as there is no "other" and no "stuff;" there is no "matter."

This exercise—pretending that the holographic theory is correct—totally alters our sense of who we are. At first, for me anyway, the feeling is one of existential dread, but then, strangely, it becomes very comforting. A greater reality is beyond space and time and is apparently, at-first-glance, unavailable to everyday beta consciousness. Spiritual thinkers might say that the allocentric mind has a power that goes beyond space and time, so that it is possible to probe beyond physical reality.

The notion of a holographic universe and of a mind that is modeled on this universal design—as above, so below—is a paradigm seemingly beyond even quantum perspective. It is a new scale, a new zone of perception, a "Virtual Reality" that provides a new way to construct thoughts, concepts, and theories.

Notes

(1) **Moore's Law.** Doubling is exponential, so from two blocks we get 4, then 8, 16, 32, 64, 128, and so on forever, expanding in size and complexity. Exponential doubling is a given in our universe—it is a fundamental law.

(2) "Movement" comes in quanta, in discrete chunks, and it appears to follows musical octaves. Many people have discovered this correspondence. Harmonics show up in the orbits of the electron around the atom, in the formation of planets, in the energy levels of the chakras, in brain wave frequencies, and so on.

(3) The Thomas Young double slit experiment. During Thomas Young's lifetime there was a debate between those who believed that light traveled as a series of particles (photons) and those, like Thomas Young, who believed that light was a wave phenomenon. Young believed that the results of his experiment proved that light must be a wave.

(4) Interference patterns. Light waves interfere with each other. Sometimes waves of light double in size, and sometimes they eliminate each other. This interaction of light waves can be witnessed experimentally. Streams of particles do not interfere with each other and, therefore, offer a different pattern when observed.

(5) Here's what I think is happening. I cannot bring myself to assert with total confidence that my logic and my interpretations are entirely correct. The path I took as I tried to solve various puzzles led me to these suppositions. Different minds, with different life experiences, may be able to offer greater clarity. Quantum theory made it clear that we could not ignore the researcher conducting experiments, nor could we ignore the spatial and temporal circumstances of each unique moment. In other words, we cannot ignore our own senses. Our sensory system is dual, and it is constrained by the Uncertainty Principle. Sometimes I boil in my own mental stew as I try to figure out whether mathematics is explaining the objective world or is explaining how we perceive.

(6) God created man using mud, water, and air. I am fascinated with the wisdom tradition and with spirituality, especially as it is connected with our dual cognition. There are two generic ways to be religious because we have two minds—and lots of individual complexity, of course. This is discussed in more detail in my book *The Confusion Caused by Being Your Own*

Twin. I could put words like "God" within quotes, but there is no need—just realize that I am staying neutral.

Neuroscience plainly reveals that no two brains are the same, or ever will be the same—the complexity of the brain and body rules out duplication. This means that you are unique, and so is everyone you encounter. You will have to decide for yourself about religion.

(7) 1,200,000,000,000,000,000,000,000 great grandparents. This comes from Steven Pinker's essay "Strangled by Roots, The Genealogy Craze in America," written in 2007.

(8) Entropy and Syntropy: It is easier to think of this distinction using the differentiation of background and foreground. When things are returning to the background, that is the process of entropy. When things are manifesting from the background, then that is the process called syntropy. Entropy means that clearly evolved entities are returning to a state of equilibrium, to an unmanifest state of probability and potential. Syntropy means that the undifferentiated background is giving birth to perceivable entities. The in-breath is entropy; the out-breath is syntropy. Life is one long exhalation, an out breath.

Perhaps the undifferentiated background state—the ever-present origin of Jean Gebser—sends forth manifestations that are probes. These probes, the manifestations, gather information and then return that information to the background. In this way, the background develops a memory of all patterns. This would be the Akashic Field of the wisdom tradition. The claim of mystics and psychics is that these Akashic records are available to review.

Thinking of entropy as chaos or disorder is not correct. Entropy is undifferentiated potential; it is a process in synchrony with creation. What we experience in life informs the whole of creation.

(9) Lynne McTaggart is the author of three books that influenced my philosophy: *The Field: The Quest for the Secret Force of the Universe* (2003); *The Intention Experiment: Using Your Thoughts to Change Your Life and* *the World* (2007); and *The Bond: Connecting through the Space Between Us* (2011).

(10) "Flaky speculation" is a self-effacing, tongue-in-cheek phrase that I use when confessing my lack of in-depth knowledge of a subject. I am not a neuroscientist, Buddhist, or quantum physicist. I am a special education teacher with a doctorate in optometry. I am part of the wave of fellow human beings who are exploring the evolution of consciousness.

(11) Heisenberg's Uncertainty Principle. I recommend the book *Uncertainty: Einstein, Heisenberg, Bohr, and the Struggle for the Soul of Science* (2007), by David Lindley. This well written book tells the story behind Heisenberg's thinking as he tried to make sense of the quantum world. It is interesting that the journal article in which the word "uncertainty" first appeared was called "On the *Perceptual* Content of Quantum Theoretical Kinematics and Mechanics." Apparently, the early quantum physicists understood intuitively that what they were finding mathematically was tied to the human senses, especially vision.

(12) Kurt Gödel's famous two incompleteness theorems: What fascinates me about Gödel's theorems is that they seem to be another case of "as above, so below." The theorems are statements about mathematics, but they seem to have relevance in other analytical systems. The first incompleteness theorem states that no single algorithm can prove all truths about natural numbers. There are always statements about natural numbers that are true, yet they are unprovable within the system. The second incompleteness theorem is an extension of the first: a mathematical system cannot demonstrate its own consistency.

(13) We have an evolving system of mathematics that explains brain neurology: There is actually a discipline called mathematical neuroscience and a professional journal called *The Journal of Mathematical Neuroscience*.

Eight

Do not feed both sides of yourself equally.
The spirit and the body carry different loads
and require different attentions.

Too often we put saddle bags on Jesus,
and let the donkey run loose in the pasture.

Do not make the body do what the spirit does best,
and don't put a big load on the spirit
that the body could carry easily.

~ "Different Loads," *A Year with Rumi*, 2006.

Hard-to-Figure Cookies

"Hello, Surge, what's happening?"

"Everything at once. How's the book coming along?"

"I'm about done. I have to present the hard evidence in court, however. I am a little nervous about that. There's too much circumstantial evidence mixed with theory. Got any advice?"

"Of course. Get enough sleep, eat less, and visit your aging mother more often. You could also bribe the jury. That often works. Are you hungry? We have just the right meal for your dilemma."

"I could eat a cookie or two."

"We make a delicious Hard-to-Figure Cookie, a staff favorite. Our Hard-to-Figure Cookies are made with the finest Circumstantial Wheat. We blend the mysterious flour with Theoretical Vegan Sugars extracted from Sweet Conversations. We harvest most of our Sweet Conversations from the Berkeley School of Physics Cafeteria—home of the Quantum Oreo."

"Yeah, that's what I need. Give me a half dozen of those Difficult-to-Digest Cookies and a quart of cold Vegan milk for dunking."

"Would you like something to go with the cookies?"

"Like what?"

"Well, we have a large selection of after-dessert meals, made especially for seekers like yourself and the reader."

"Okay. I'm game. What have you got?"

"I suggest, given your nervousness about the courtroom, that you have our Meal-of-Chance. It's a roll-of-the-dice concoction that comes in a Dr. Who invisible mug. It's a Time Lord drink, so timing is everything. You have to drink it at just the right sip-speed or it relocates. But tell me what your plan is. What will be your opening remarks to the jury?"

"You want me to role play?"

"That's a great idea. Why not practice before you do heart surgery. Nervous patients love doctors who have practiced. Go ahead. Take a deep breath and pretend that your presentation is a matter of life and death. Because, of course, it is. Go ahead."

"Okay, I can do this."

"We know you can."

"Ladies and gentlemen of the jury . . ."

"Sorry, I have to stop you there. It's best to have just empathetic ladies on the jury. If there are any analytical males on the jury, they will disappear into their mind after your opening sentence—and never hear another word you say. Just have women on the jury. So, go ahead. Talk to the ladies."

"Ladies of the jury, what I present to you today is evidence that males forget to feed the kids because they are busy trying to solve the problems of the universe using their over-developed egocentric minds."

"Excuse me. I have to stop you there. The jury has heard all that ego and allo business for seven chapters. They get it. And stop beating up your brothers. Stand up for masculinity just as you stand for the rights of your sisters. And don't repeat. Okay, go ahead."

"Are you going to interrupt me after every sentence, Surge? Because it's annoying."

"I just want you to get it right. But go ahead. I'm listening."

"Ladies and spectators, seekers and prisoners, I come today to show you the hard evidence, the in-your-face proof that duality is natural to nature. Why are there just two eyes in human beings? Why not have eight eyes like spiders or 100 eyes like sea scallops? How do your two eyes project a single world in front of your face using just light as the raw material? Why are you equipped to project and walk about within a holodeck of your own creation?"

"Hey, that's good, Dutch. I like that beginning. It reminds me of some lines from a Rumi poem:

"Sunlight looks a little different
on this wall than it does on that wall
and a lot different on this other one."

319

"Thank you, Surge. Maybe I'll use that, just as soon as I figure out how it is relevant to what I am trying to say."

"Rumi is telling us that manifestations of light vary, but the background canvas does not."

"Okay. I get that. It's the same message he keeps repeating. I'm getting a little Rumi-weary, Surge. Just saying."

"That's like saying you are too tired to be alive. Leave your depression on the couch, Dutch, and walk away."

"Can we get back to the topic at hand, Surge?"

"Sure. By the way, you do know that everyone has a theory about consciousness, right? And it is only going to get more complex as more accountants and plumber's assistants write their own books about consciousness. I read your grandmother's theory of consciousness, by the way, and it is off-the-wall hilarious."

"Leave my grandma out of this."

"Just saying she had a theory of consciousness."

"Well stop it. I realize that the reason I can add my theory of consciousness to the common pot is because so many others are self-publishing whatever their tiny minds have coughed up. This is my hairball, and I am proud of it."

"You know that even your Uncle Bill has published two books on consciousness, and one was a comic book—that was the better one, by the way, because it had pictures. Uncle Bill and the Bowling Team—that's their pen name—just published their second book: *How Bowling Can Save your Marriage: A Ten Step Guide, Plus*—as an added bonus—*Consciousness Solved!*"

"You are very funny, Surge."

"What comes next?"

"I then tell the stunned reader that quantum physics crafted the retina of the eye."

"They already know that."

"No they don't!"

"Yeah, they do. The only readers who will pick up a book about consciousness are people who have their own books about consciousness—or are working on one. They want to see how far off the mark your hairball is."

"This is not a fun lunch, Surge. Get me a mug of that Time Lord brew and bring me my crazed cookies. I need some silence before I write the next chapter."

"Of course. I will bring you some silent tea and quiet crumpets. No charge. Good luck with the new chapter; it is very important."

The Hard-Evidence

I have a certain knowing.
Now I want sight.

~ "What I See in Your Eyes," *A Year with Rumi*, 2006.

Now we arrive at the heart of the matter. It is easy to see our dual world in religion, history, literature, and philosophy; easy even to postulate a biological substrate for the duality, as many have done using hemispheric brain specialization, and as I do using purposeful movement (navigation). But all that is not enough—we need hard, in-your-face evidence that we have two minds. Indeed, this evidence is staring us right in the face and has been in plain-sight for decades.

Let me emphasize at the beginning of this important discussion that my theme in this book is based on a theory, a set of logical observations that led me to the conclusion that navigation is the evolutionary creator of our mental duality. Consequently, we cannot leave theory and logic in the shadows as we look at the hard evidence. I am assembling a gestalt from the many notes that are scattered about my mind. This cognitive gestalt is composed of knowledge coming from anatomy, physiology, logic, and theory.

I first saw the hardwired evidence while studying optometry in the 1970s. I saw it there but, like most others, I did not realize the significance of what I observed. The hard evidence begins in the retina, the delicate photoreceptive neural tissue that lines the back of the eye. In that tissue there are millions of photoreceptor cells of two types: there are over 6 million cone cells clustered together in a pinprick-sized area of the retina called the macula; and about 125 million rod cells are spread throughout the rest of the retina in the non-macular periphery. My postulate is that these two cell types are the beginning of a trail that leads all the way to dual-processing—to our twin minds.

Thinking back, I wonder why none of my professors ever asked us to consider why there were only *two types* of photoreceptor cells. Why not just one type, or if two is good, why not ten or twenty types? Why only two? Furthermore, I wonder why no one saw the possible connection between Heisenberg's Uncertainty Principle and the design and function of the two types of retinal photoreceptors.

In the 1970s, we talked about refraction, reflection, and absorption of light waves—we drew arrows to show how light moved and was curved by lenses. The eye was understood as a biological computer then, and it was the job of the two photoreceptors to turn light waves into biochemical and electrical discharges that magically created what we call vision. What we missed then—and are still missing—is that the retina is a quantum computer. As I thought about this supposition, that the retina was a quantum processor, it occurred to me that there are two types of photoreceptor cells because quantum energy is a duality, and Heisenberg's Uncertainty Principle makes it necessary to create two mutually exclusive processing systems.

It is theoretically and logically plausible that a quantum retina would have two separate ways to decode quantum patterns. The Heisenberg

Consciousness: A New Slant on an Old Conundrum

Uncertainty Principle, a basic mainstay of quantum theory, says that it is impossible to measure for momentum the same time as we measure for location. I would rephrase that important observation to this: it is impossible to measure movement the same time as you measure no-movement. I believe this is the explanation for the retina's duality: cone cells measure objects that are stationary, frozen in their location, while rod cells measure momentum and relative movement. Another way to say this is that rods are the beginning of the mind that creates space and, therefore, creates a stable background medium for navigation, while cones are the beginning of the egocentric mind which creates time and assigns relevance and meaning to the invariant "objects" in our reality. This becomes more plausible as we trace the nerve fibers from the separate cell types through the brain. Hold that thought.

In the 1970s we were making another mistake, an assumption that wasn't true, but it seemed correct at the time. Aristotle had told us over two thousand years ago—and we repeated the great man's wisdom so often it got hardwired—that there were just five human senses. Indeed, in his treatise *De Anima*, Aristotle spent an entire chapter arguing that there could *only be* five senses. However, this observation is false. It has thrown us off-course repeatedly. It might be more helpful to begin with a contrary statement: *There is no such thing as the five senses.*

The empiricists, especially Locke, Berkeley, and Hume reinforced Aristotle's misperception. The empiricists spoke of the senses as totally separate from each other; this kept the mythology going for a couple more centuries. Finally, in the 1960's, J. J. Gibson and his wife E. J. Gibson challenged this popular mythology. The senses, according to the Gibson team, *never* work as isolated entities; they *always* work coherently as one unit:

Gibson (1966) proposed that the sensory modalities are *complex perceptual systems*. They do not correspond to the popular notion of the "five senses." Single stimuli are rarely, if ever, sensed as such by the organism. ~ *Space and Sense, Susanna Millar, 2008.*

The Hard-Evidence

The so-called five senses are really just the most obvious *sensory portals*—they evolved from the body's skin, on the boundary of our form. If we see ourselves—for a moment—as a huge single-cell creature, then the external senses formed on our cell membrane, on our skin. Eyes, ears, nose, mouth, and tongue—each external sensory portal—has a distinct and obvious shape that stands out in contrast to the skin surface. The skin-membrane itself, with all its sensory divisions (for pressure, heat, etc.), is the most obvious of the external sense organs. Therefore, we can forgive Aristotle for the over-simplification; he saw what was obvious to any casual observer. But we won't get any further with our thinking if we don't transcend this limited belief that we have five isolated senses.

The body behaves as a whole. Psychologists speak about embodiment, which is, from one perspective, an acknowledgement that we act as a single unit. Dissecting out sensory systems is a convenient way to communicate, to do research, and to share information, but it gives a false impression that some entity called "vision," or "hearing," exists independently from whole behaviors. Obviously, the body acts *all-at-once,* instantaneously combining all motor, sensory, and mental processing. This is an over-simplification, a useful generalization, but it holds in most cases.

Therefore, as we look at the vision system and find evidence for duality in the architecture, we must conclude that this same dual-neural design *must* be present in the other sensory systems, as well. What we discover about one sense is duplicated by the other senses. For example, if I reach down to turn the volume up on my car radio, what my eyes see, what my ears hear, and what my sense of touch tells me must all coincide. Sensory input must be precisely integrated for coordinated actions to be efficient and accurate.

Furthermore, my contention is that there are *only two ways* to pay attention, either allocentrically to movement, or egocentrically to non-movement. Each of these opposing processing systems is *served by all the sensory channels working synchronously.* For example, if vision is attending egocentrically, then at the same time, so are the auditory, tactual, kinesthetic, proprioceptive, and vestibular systems. If there is an inherent oscillation discovered in visual processing, an alternation between egocentric and allocentric frames of reference, for example, then we would expect to find this very same oscillation in all the other sensory systems. Indeed, this is exactly what we observe.

Back to my statement and proposition that the retina is a quantum computer and that it, therefore, processes for location (no-movement) in one stream and for relative motion in a second stream. Adding now the understanding that the senses always process coherently—either allocentrically or egocentrically—we can conclude that the whole system is quantum. Furthermore, each of the sensory systems works either to solve for location (no-motion) or to solve for momentum (motion). Each of the sensory portals—vision, hearing, touch, proprioception, olfaction, and so on—employs quantum-based mechanisms. Quantum biology is beginning the exploration of this proposition, and the evidence is starting to trickle in.

As I pointed out earlier, from an anatomical perspective, the brain is clearly divided into two hemispheres, a right side and a left side—not unlike our entire physical construction. This anatomical fact led many to conclude that our duality might be found in this obvious and quite observable fact. However, research did not entirely support this theory because the brain is a vastly interconnected network that does not lend itself to absolute modularity. We had to look elsewhere for the origin of our duality.

If we look at the brain not from above but from the side, there is another obvious anatomical divide. The Sylvian fissure is a long and deep cut that divides much of the brain into upper and lower divisions. Very generally and incompletely, the lower half contains the temporal and occipital lobes, while the upper half contains the parietal and the frontal lobes.[1] Physiological evidence supports this top-bottom anatomical divide. In 1982, a now-classic research report from the National Institutes of Mental Health, based on experiments with monkeys, seemed to show that the top brain was being used for allocentric processing (for spatial/flow processing, enabling orientation while navigating), while the bottom half was being used for egocentric processing (for no-movement, shape analysis):

The results of these operations were dramatic: The animals that had a portion of the bottom brain removed no longer could do the shape task—and could not be taught to perform it again—but they could still perform the location task well. The animals that had a portion of their top brain removed had exactly the opposite problem: They could no longer do the location task, and could not relearn how to perform it—but they could still do the shape task well.

Many later studies, including those that relied on using neuroimaging to monitor activity in the human brain while people performed tasks analogous to the ones the monkeys had performed, have led to the same conclusion. ~ *Top Brain, Bottom Brain, Stephen Kosslyn and G. Wayne Miller, 2013.*

Somehow our split brains, whether divided into left and right brains or top and bottom brains, must factor into the overall division between allocentric and egocentric processing. I believe there will be a grand unification of anatomy and physiology as the years unfold. What is clear is that duality is fundamental and cannot be dismissed.

There is another mistake that we make about the brain, which has confused our thinking for decades. Somewhere along the way, we started referring to brain lobes as the "visual cortex" or the "auditory cortex," and so on. This is at best a partial reflection of the truth, but it obscures the real functions of brain regions.

For example, researchers discovered that non-visual activities show up on MRI scans in the so-called visual cortex. When a blind individual reads braille, or actively echolocates, the occipital (visual) cortex will "light-up" as if a visual image was being processed. *This means that tactual images and auditory images both activate the so-called visual cortex.* The logical conclusion is that the visual cortex is a region that processes for spatial images (patterns), some of which are visual. It is no longer valid to refer to the occipital cortex as the visual cortex.

Researchers have also rewired the brains of fetal animals. They have sent nerves from the retina to the auditory cortex and nerves from the ears to the visual cortex. The postulate was that the animals would see sound and hear images. What actually happens is *nothing*. The brain still sees images coming from the retina and hears sound coming from the ears. This further shows that labeling the temporal cortex as auditory and the occipital cortex as visual is wrong. My contention is that the brain is processing for *space*, or it is processing for *time*, using whatever sensory input it receives.

I will begin the discussion of duality by discussing the dual architecture that flows from the retinas; this is my area of expertise—I spent 50 years studying how the visual portal works. I will then discuss the other sensory systems and—knowing that they must overlap—I will show the same duality in all the other sensory systems. I will show how each has an allocentric component as well as a separate but synchronized (alternating) egocentric component.

This is, as I have said throughout the book, my own theory of dual-processing. I have lots of near-allies, experts who see the duality and hunt for origins, but at this point I must stand alone. No one has conclusive evidence to validate the overall theory that I present here. I am confident, however, that researchers will slowly find the evidence in their respective fields and that the evidence will eventually catch up with the theory. There is simply too much circumstantial, logical, and hardwired support for dual-process theory.

Vision

It is clear, from a dual-processing perspective, that there are two anatomical systems for processing light waves, starting at the level of the retina—one for allocentric processing and the other for egocentric processing. I suggest that these correspond to the spatial and temporal aspects of quantum waves. For example, from two specialized retinal cells, the rods and the cones, there arise two kinds of neural tracts that traverse separately through brain lobes—the two systems are plainly processed separately and differently. That there are two processing streams seems undeniable. From the simple process of assessing spatial characteristics, an entire brain mechanism evolved for analyzing and using space. Likewise, from the assessment of the frequency of a quantum wave there evolved a complex mechanism for analyzing and using time. Below is a summary of the complexity that arises from retinal cell processing through various brain regions, to the highest processing centers in the brain. I will look first at the rod cells and then compare their processing with the cone cells.

The retinal *rod cells*, which outnumber the cones 20:1, are the initial *spatial processing* units. They are the beginning of vision's contribution to the allocentric mind. Rod cells are highly motion sensitive; they perceive a flat world, in black and white, with a maximum acuity of 20/200. Called "peripheral vision," this brain mechanism monitors the whole visual scene. It is a spatial perception system that answers questions—at high levels in brain processing—like: Where are things in relationship to each other? What is the relative motion of objects within a scene? What is the spatial arrangement, the relationship of invariant objects that constitute a scene—what is moving and what is not moving? What pathways exist through the scene? This allocentric system creates a gestalt. It is tied into a higher order collection of neural brain maps. Most nerve fibers that originate in this rod system travel to the occipital lobe, then go to the parietal lobe, frontal lobe, and end, for final processing, at the pre-frontals—and probably at the hippocampus, as well.

A key element in allocentric processing is called "flow." All *flow* means is that the allocentric mind is aware of peripheral wave-like sensations occurring during the act of moving. For example, if you are driving down the expressway, you can use your peripheral (allocentric) visual system to perceive the flow of the environment as it passes by on either side. The trees flow past like a river in your peripheral visual fields. If you suddenly, perceptually "lock-on" to the car in front of you, then your awareness of the optical flow disappears. If you again focus on the peripheral flow, the awareness of the car in front fades. Optical flow must have a multi-sensory correlate—the senses are either all contributing to allocentric awareness or they are contributing to egocentric attention, but not both at once. We would expect to find, therefore, the equivalent of optical flow in all the other senses—auditory flow combines with tactual flow, and so on. Indeed, this is the case; we find synchronous flow in all sensory systems.

In contrast to the rod cells, the retinal *cone cells* are the initial *temporal processing* units. They are the beginning of vision's contribution to the egocentric mind. The egocentric visual system originates in the fovea of the retina. Cone cells "see" in color, in depth, and with 20/20 acuity. Cone cells are the front-end of a pattern recognition system that locks onto objects-of-regard and analyzes what is being perceived. Egocentric processing then compares images with what is stored in memory for long-term recognition; this is—at the highest levels of processing—a search for meaning, relevance, and knowledge. The cone cells are the beginning of the "central vision system," which answers questions like: What is this? Who is this? What meaning does this moment, this object, this person, have for me? Nerves in the retina, which leave this cone-based vision system, project fibers initially to the occipital lobe and temporal lobe; from these two regions, neural tracts pass through the frontal lobes and culminate in the prefrontal lobes and hippocampus.

Early researchers traced optical fibers, both from the cone cells and from the rod cells, directly to the occipital lobe. These early scientists missed other neural tracts that left the retina but went to other locations besides the occipital lobe. As research moved forward over the years, for example, it became increasingly obvious that fibers leaving the retina went also to the suprachiamatic nucleus in the hypothalamus to regulate body clocks. Retinal nerves also were found to travel to the pretectum to control pupil size and thus control the amount of light flooding the eye. Additionally, retinal nerve fibers were traced to the superior colliculus, which has the critical function of coordinating the movement of the head and eyes and, therefore, whole-body alignment and posture.

The superior colliculus is very important because it is more or less where vision was processed in creatures before the neocortex evolved. It is part of a subconscious, allocentric vision system. Rich networks of afferent and efferent nerve fibers connect the superior colliculus with cortical and subcortical regions of the brain. Nerve fibers leaving the superior colliculus join either a temporal (ventral) or a parietal (dorsal) processing stream. Significantly, the pineal gland, the so-called third eye, is embedded within the superior colliculus and appears to be a primary organ in subconscious visual processing.

James T. Fulton, in his online book *Processes in Biological Vision, The Electrolytic Theory of the Visual Process* (http://neuronresearch.net/vision) challenges many of the biological assumptions underlying how the vision system works in human beings. At the onset, he says, as I am saying, that the retina is a quantum processor, not just a biochemical processor. Quantum effects, Fulton says, are occurring at the level of the retina and throughout the brain. Relevant to the dual-processing hypothesis, Fulton argues that there is a system for processing "coarse imagery" (allocentric) and a system for processing "precision imagery" (egocentric).

Researchers also discovered that the occipital cortex lights up on MRIs whenever blind individuals read braille, echolocate, or explore surfaces with their fingers. This suggests that the occipital lobe is a center for scene analysis—for the first stages of image assembly. The occipital lobe is look-ing for patterns, *coming from any and all of the senses* that contribute to the creation of invariant forms and invariant flow.

The understanding that the vision system had an inherent duality be-gan with the two-stream hypothesis defined in a paper written in 1992 by Canadian neuroscientists David Milner and Melvyn Goodale at the Brain and Mind Institute at the University of Western Ontario. Several books were published relevant to their theory of dual-visual processing, includ-ing *The Visual Brain in Action* (1996) and *Sight Unseen: An Exploration of Conscious and Unconscious Vision* (2004, revised 2013). In *Sight Unseen*, Goodale and Milner wrote:

. . . we need vision for two quite different but complementary reasons. On the one hand, we need vision to give us detailed knowledge about the world beyond ourselves—knowledge that allows us to recognize things from minute-to-minute and day to day [the egocentric mind]. On the other hand, we also need vision to guide our actions in that world at the very moment they occur [the allocentric mind]. These are two quite different job descriptions, and nature seems to have given us two quite different visual systems to carry them out. One system, the one that allows us to recognize objects and build up a database about the world, is the one we are familiar with, the one that gives us our conscious visual experience. The other, much less studied and understood, provides the visual control we need in order to move about and interact with objects. This system does not have to be conscious, but it does have to be quick and accurate. ~ *Sight Unseen: An Exploration of Conscious and Unconscious Vision David Milner and Melvyn Goodale, (revised 2013).*

This statement by Milner and Goodale supports dual-process theory. There is one processing system that "guides action" and allows a person "to move about and interact with objects." This is allocentric processing—it starts at the level of the retina with the rod cells. The other system gives us "detailed knowledge about the world." This is the egocentric system and it also begins in the retina, with the cone cells.

Before I move on to show how the other senses have both allocentric and egocentric functions, I must pause to look at the findings and conclusions of the Milner and Goodale research study. There is serious confusion that arises when we try to marry dual-process theory to their excellent work.

Major Confusion: a Terminology Nightmare

There is major confusion that arises when the Goodale and Milner model described above is compared to what I am presenting here, so I need to give this well-respected theory some attention. First, Goodale and Milner *do* support the proposition that we have two vision systems that have two distinct jobs to do; we are in agreement from the beginning. The two divisions they define also fit nicely with what I am suggesting: There is a relatively slow and conscious visual system that stops and explores, searching for meaning and relevance; and there is a second, faster and mostly unconscious visual system that is responsible for orientation and mobility, for flowing around solid objects to reach a destination.

Goodale and Milner focused on the two visual processing streams that seemed to originate, as they saw it, in the occipital cortex, and then flowed through two distinct pathways toward the front of the brain. One of the streams went over the top of the head, so they called it the dorsal stream. The second stream went along either side of the head, through the temporal lobes, so it was called the ventral stream. They did not suggest, as I do, that these streams end at the prefrontal lobes and hippocampus. Their hypothesis stated that the streams, as far as they could tell at the time, terminated before they went to higher processing regions. Later research did, indeed, affirm that the streams went beyond the temporal and parietal lobes:

> . . . neural tracts leading from the occipital lobe do not stop in the parietal or temporal lobes, but rather continue on to the top and bottom parts of the frontal lobe, respectively. ~ *Top Brain, Bottom Brain, Stephen Kosslyn and G. Wayne Miller, 2013.*

The 1992 paper that explained the two-stream visual perspective was also not generalized to include the whole sensory system. The authors did not state that visual duality was a subset arising from two separate brain architectures, as I suggest. As they speculated why there might be two separate processing streams, they decided that the function of the ventral stream was to recognize and then to identify people and things; it extracted meaning and held a database of knowledge about the world. They also saw the ventral stream as a way for vision to store long-term memories of invariant patterns—like faces, words on a page, and objects. In contrast to the ventral stream, the dorsal stream was for visual manipulation, for visually-guided action; it was an unconscious visual motor system. The dorsal stream was used so that animals could flow through the environment in a purposeful way using vision. Short-term memory, as used by the dorsal stream, was needed to allow for fast navigation—it is important to quickly forget what is being passed as we move about. So far, so good—my dual process theory matches perfectly with Goodale and Milner's perspective. From here, it gets complicated.

The most confusing difference between the 1992 theory and my perspective is that the authors completely reversed allocentric and egocentric—what I call allocentric, they called egocentric, and vice versa! From my perspective, after building a dual-process theory around a specific definition for allocentric and egocentric, this is a nightmare, especially since many researchers after Goodale and Milner simply repeated their initial definitions. I can see how the confusion came to be—the authors are perfectly justified in their perspective—but clarification is definitely in order.

Goodale and Milner based their definition around *active* processing versus *passive* processing. However, my perspective is based on the two ways we pay attention. Here is Goodale and Milner's reasoning: Movement—processed in the dorsal stream—is an *active process*. In the ventral stream, however, "meaning extraction" is a *passive cognitive process*. Goodale and Milner reasoned that navigation, being an active process, involved knowing where the body was in relationship to the environment during movement. They could see that the body moved in relationship to objects, so the system responsible for navigation they called egocentric. However, when the focus switched to the perspective of cognitive processing—which is a passive mental process—then they called this allocentric.

My distinction between allocentric and egocentric—the complete opposite of the Goodale/Milner distinction—is between *two ways to pay attention*, and is *not* about passive or active movement. I am talking about attention, they are talking about movement. Here is my perspective again. When a person stops to explore, they do so from the perspective of egocentricity, as if they are the center looking outward. When actually flowing through the environment, however, processing is not egocentric; there is no search for meaning or relevance. Allocentric awareness is based on sensory flow.

It is easy to get confused as we review the literature about human navigation, especially when we try to comprehend and differentiate these conflicting uses of "allocentric" and "egocentric." Almost the entire field has adopted Goodale and Milner's differentiation of egocentric and allocentric. My perspective centers on how we pay attention, so it flies in the face of the literature up until now. I am saying that the dorsal system—the "where is it" guided-action system—creates a background, a canvas (a set of cognitive maps) from which the ventral system can manifest figures.

Obviously, there has been much research and rethinking since 1992. The authors themselves have adjusted their theory—allowed it to evolve— and the idea of total separation between the two processing streams has been repeatedly questioned because plainly we are one organism; the two streams must somehow work in harmony. The original hypothesis that two visual processing streams exist has not been discredited, however, even though neuroscience has evolved considerably since the idea was first presented a quarter century ago.

That dual-process theory is basically sound is given support when we consider what the other senses are doing during any activity. Vision, as we would expect, is not the only sense with this two-stream neurological distinction. All senses contribute to either an allocentric or an egocentric processing stream.

Hearing

After I got my doctorate in optometry and went on to graduate school at Western Michigan University to study blind rehabilitation, I reviewed the literature on echolocation that my friend and colleague Daniel Kish[2] had collected for his master's thesis on human perception. I saw clearly that the bat—a master at echolocation and a focus of Daniel's research—has two auditory systems that are interconnected but anatomically distinct. One system is linked to egocentric processing and is synchronized with vision to answer the "what is it" question. Likewise, the allocentric auditory system of the bat is coherent with the allocentric visual system and answers the "where is it" question. Therefore, in the bat we find a corollary between vision and hearing, just as we do in human beings.

James T. Fulton wrote a second online book (besides the text on vision mentioned above) called *Processes in Biological Hearing*. As with vision, Fulton states that hearing is a result of quantum processing and not just biochemical processing. He also states (I am using my terminology) that there is an allocentric and an egocentric dichotomy within hearing; he calls these neural processing streams a *communications tract* and a *source location tract*.

> It is important to recognize the primary purpose of the hearing system. It is to extract information describing the external surroundings from the acoustic stimuli received . . . Information is extracted from the stimuli via two distinctly different functional tracts, the communications tract and the source location tract. Each of these tracts involves a variety of subsidiary tracts . . . ~ *Processes in Biological Hearing, James T. Fulton, available online at http://neuron-research.net/hearing/*

Hearing is used primarily for egocentric communication, which is why Fulton labels one tract the communications tract. The individual ego is either receiving or broadcasting information from an egocentric perspective. However, when we navigate, when we flow through the environment, hearing can also assume an allocentric role, as when a blind individual uses passive echolocation to gather spatial information about a location; that is why Fulton identifies a source location tract—hearing contributes to our awareness of position in space.

Creatures like bats and dolphins generate clicks that cause environmental echoes. The reflected echoes that return to their ears and their brains create spatial images. Animals that use active echolocation generally have poor or no eyesight—so their navigational abilities are not coming from vision. Therefore, there must be an auditory imaging system that rivals visual processing and that has navigation as its primary mandate. There is reason to believe that the images made possible by sound reflections are as good as images created by the reflection of light. In other words, in animals that evolved sophisticated echolocation ability, ears can perceive images as distinctly as eyes can perceive images—we have the observation of bats and dolphins navigating with great speed and precision using active echolocation to avoid obstacles or to catch prey.

Bats have a set of low frequency clicks that provide them with a sense of the surround; this is a system that monitors movement/flow in a stable background. However, when movement is detected, high frequency clicks are able to pinpoint the exact location of a fast-flying insect. The bat homes in at great speed and picks the insect out of the air using its egocentric attention system. However, it spends most of its time monitoring the surround, staying within allocentric awareness. When it detects movement within the background, it suppresses allocentric processing and maximizes egocentric attention.

Bats emit such a powerful burst of sonar energy during egocentric targeting that their ears would be damaged unless they had a compensatory protective mechanism. Therefore, bats can occlude sound the same way that closing an eyelid shuts off seeing. Bats protect their ears just before each high sonic emission. After the blast is away, the bat's "auditory lids" open back up to receive returning echoes. This extremely rapid and precise alternation is very typical in nature—it is another example of inhibition alternating with excitation.

People who are blind also use echolocation very efficiently in the same way as do bats, although with much less acuity. Hearing uses a very effective 360-degree allocentric awareness mechanism that monitors a steady-state set of auditory background frequencies. This allocentric system sits and waits for a change in the environment. For example, when a person who is blind hears another person speak, or hears a sound source, attention becomes egocentric. As I mentioned earlier, one of my best friends and a professional colleague is Daniel Kish. Daniel is both an orientation and mobility specialist and a developmental psychologist specializing in perception. Being blind himself, Daniel is living proof of the allocentric/egocentric duality. Daniel monitors the surround with a monk-like enlightenment, with acute allocentric awareness. Then, when he wants to examine objects or locate the exact position of people in a room, for example, he emits occasional clicks from his mouth, activating his egocentric attention system.

Blind individuals use two kinds of echolocation to extract information from their environment: passive echolocation and active echolocation. This division is exactly a corollary of peripheral allocentric processing and central egocentric processing. Peripheral vision is a scene analysis system, a way to take in the gestalt using parallel processing. Likewise, passive echolocation uses ambient sound waves to give a blind individual a gestalt; it is also a scene analysis system that uses parallel processing. Echolocators will also occasionally emit a lower-pitched click to get a more diffuse read of a total scene. However, a blind individual who wants details about a particular space will emit a sharper click from the mouth aimed toward an object-of-regard. The returning echo-pattern provides a snapshot of a foreground figure against a less clear background.

Brain areas for *visual* processing, such as the color-processing cortex, the shape-processing cortex, and the motion-processing cortex, have corresponding and overlapping responses to echolocated sound patterns. Here is a report from the Durham University team in England that found this evidence:

Using fMRI, we found that while listening to echolocation sounds as compared with control sounds, participants showed significant increases of brain activity in BA17 [occipital cortex for early visual processing] . . . A recent fMRI study by Wallmeier and colleagues has since confirmed the involvement of BA17 in echolocation in the blind. My colleagues and I also have found that echo motion (i.e., motion perceived via echolocation) activates brain areas that might coincide with the brain's visual motion area MT+ and that the shape of echolocated surfaces might activate the lateral occipital complex (LOC), a brain area thought to be involved in the visual processing of shape. We also have found that both blind and sighted people show activation in the posterior parietal cortex during echolocation of path direction for walking, and the location of this activation might overlap with areas involved in the processing of vision for motor action.

In sum, although there are only a few studies to date about neural substrates of natural echolocation, it is increasingly evident that traditional "visual" brain areas are involved during echolocation in blind echolocation experts and that this activation appears to be feature specific. ~ *"Using Sound to Get Around; Discoveries in Human Echolocation." Observer, 28 (10), Lore Thaler, 2015.*

This overlap between visual and auditory processing makes perfect sense if all the senses contribute to either allocentric or egocentric attention; we would *expect to find* this overlap of sensory processing. All the visual centers are also auditory centers and tactual centers; they are *sensory centers* for specific tasks like movement perception or shape perception.

Auditory flow also contributes to the allocentric sensation of relative movement. As images flow across the peripheral retina, the optical experience of image-flow is matched exactly by the flow patterns of sound.

Just as light enables scene analysis, so too does sound provide for scene analysis. In other words, both light and sound have overlapping allocentric systems. Auditory processing is not just an egocentric process that localizes sound in space. In his extensive review of auditory scene analysis, McGill Professor Albert Bregman identifies two kinds of auditory processing: sequential integration and simultaneous integration. These correspond with our egocentric and allocentric minds: sounds in a sequence are processed by the egocentric system; sounds that are grouped in gestalts for scene analysis are processed by the allocentric system.

The environment contains auditory invariants. For example, the sounds we find in bathrooms differ from the sounds we encounter outside in our yards. Furthermore, the sounds that repeat in a specific bathroom that is familiar, and the sounds that occur in a familiar backyard, stand out from generic patterns—from universal bathroom or backyard sounds. Therefore, spaces have sound signatures. These overlap or overlay onto the visual scene, reinforcing our perception that sound identifies specific spaces.

An important distinction Bregman makes is between primitive seg-regation [allocentric processing] and schema-based segregation [ego-centric processing]. Primitive segregation is a bottom-up process whereby streams are parsed according to the correlations of acoustical cues. By contrast, scheme-based segregation is a top-down process that arises from experiential and cognitive factors. Schema-based streaming is characterized by voluntary or effortful listening—an ac-tive "hearing-out" for a given pattern. ~ *"Review of Albert S. Bregman's book Auditory Scene Analysis: The Perceptual Organization of Sound," Journal: Psychology of Music, David Huron, 1991.*

In other words, Bregman discovered through his research that we could di-vide hearing into two distinct streams, one that is egocentric—voluntary, slow, top-down processing—and one that is allocentric—non-voluntary, fast, bottom-up processing. This is exactly what we would predict using dual-process theory.

Vision and Hearing

Vision and hearing are the two primary external sensory systems that evolved to probe the world beyond our immediate bodies; they are distance receptors.

But nature evolved two ways to explore and experience the world beyond the body. Vision specializes in one of these ways and hearing in the other. When the attributes of our two minds are laid out, what immediately becomes apparent is that the description of allocentric processing looks a lot like a description of vision, while the description of egocentric processing looks a lot like a description of hearing. This does not mean that vision is purely allocentric and hearing purely egocentric, but it does suggest that the two major sensory modalities are unequally weighed in their contributions to our two minds.[3]

Vision is indeed the primary sense for perceiving spatial *layouts* and invariant spatial *arrangements* (like faces). It essentially defines (invents) space. Space is the background, the scene, the stage wherein our lives are lived. Allocentric *whole-body* perception probably gave rise to the visual system. Therefore, vision seems to be the primary sensory contributor to the allocentric mind and is the main, but not exclusive, factor in allocentric consciousness.

Hearing is the primary sense for perceiving sequentially; one sound follows another. This is the way we crafted language and music. Hearing is temporal processing, and it can be said to have invented time. The allocentric background for hearing is relative silence. From the silence, sound patterns manifest. Close your eyes and listen to sounds emerge from the silent surround. All sound comes to us as if we are the center, the egocentric heart of existence. Therefore, hearing appears to be the primary, but not the only, contributor to the egocentric mind and is the main factor in egocentric consciousness.

Despite their disproportionate contributions, I want to strongly emphasize that both hearing and vision perceive using allocentric awareness as well as egocentric attention, as described above. The two minds never perceive in isolation; they can never be dissected out from the behaviors of the embodied organism.

Touch

There are also two kinds of touch. Egocentric touch is what we are most aware of using. When we examine objects with our hands—especially with our hands—we use hand-and-eye and hand-and-ear coordination. Suppose in your hands you hold a radio. What you see, hear, and feel must be coherent. The source of the input must appear, for each sense, to be located at the same position in space. All sensory inputs must combine within the architecture of the brain and body to give the illusion of "object." This egocentric system seeks to gain meaning and relevance from the moment, from the object, from the event, and from others.

However, allocentric touch is different. Allocentric touch is a whole-body phenomenon that is related to flow, to navigation; it is not concerned with extracting meaning from the surround. Allocentric input from the tactile sense is a summation of all the various skin receptors firing synchronously and in patterns. Not surprisingly, given the billions of years that membranes have evolved, the skin contains a variety of very complex systems. In a general sense, the skin has cells called mechanoreceptors that either fire fast or fire slow. This corresponds with the very fast allocentric system—System One in dual-process theory—and the relatively slower egocentric system—System Two in dual-process theory.

There are four principle types of mechanoreceptors in hairless skin.[4] Tactile corpuscles called Meissner Corpuscles respond to light pressure, and they adapt to textural changes. Bulbous corpuscles called Ruffini Endings detect deep pressure. Nerve endings using Merkel Discs detect sustained pressure. Lamellar corpuscles called Pacinian Corpuscles detect rapid vibrations. The skin, like the eyes and ears, detects frequencies. Each of these somatic cell types specializes in different kinds of frequency analysis.

Fast responding cell systems in the skin contribute to allocentric flow. These are the quick responders that work in harmony with the peripheral retina and passive echolocation in the auditory system. Slower responding somatic systems serve egocentric attention and the exploration of meaning. Given the complexity of the skin receptor systems, this is a very big generalization, but it does underscore my major point that all sensory portals have two ways to pay attention, and every sense fires coherently with all the other senses.

Experiments with blind individuals suggest that there might be such a thing as "eyeless sight"—also called dermal-optical perception. Research suggests that running the fingertips over a colored surface, for example, might provide enough clues to differentiate colors. This appears to be statistically valid although not consistently reliable—there is a skill involved that requires training. This is a form of egocentric touch.[5]

As the body flows through space, there is an allocentric tactual flow that occurs synchronously with all the other senses, but especially with optical, auditory, and vestibular flow. Also, there is a whole-body tactual sensing that gathers ambient information for the allocentric mind. For example, the sun on a sunbather or the wind on the body, as you lay on the beach on a summer day, are whole-body sensations. This allocentric system can also provide synchronous "where is it" information, since the sun and wind are directional.

Blind author Jacques Lusseyran[6] would often explain his ability to sense the world using the metaphor of pressure. After becoming blind, he said that he noticed that everything had a subtle *signature pressure*. We know that the ear drums are sensitive to pressure differentials, and we know that blind individuals use hearing to sense objects, especially as they near the head; early research called this facial vision. Logically, air flow, air pressure, and frequency differentials also contribute to a whole-body ambient scanning of the surround. Our membrane of skin, our boundary, our interface with the atmosphere is much more sophisticated than we at first surmised. Evolution has produced delicate and accurate ways to immediately "read" the world. We probably read atmospheric pressure changes very expertly as well as reading air flow differentials with minute specificity; these factors partly explain Lusseyran's use of pressure to comprehend the surround.

What is clear and relevant to our discussion is that the sense of touch is dual and that all aspects of touch can be reduced to either of its two roles: allocentric touching and egocentric touching.

Smell

In 2015, psychology researchers at U.C. Berkeley found that smell was used as a navigational tool. In other words, they found the allocentric component of the olfactory sense in the human being. Their findings show that not only is smell used to identify a substance using egocentric processing, smell can also be used to decode the gestalt using allocentric processing. Using a combination of ambient odors, a *smell signature* is available for any given space. This is a background olfactory steady state against which distinct changes can be detected.

Thinking only egocentrically, earlier researchers and the lay population assumed that humans used smell simply to identify entities like popcorn, leather, and gas fumes. It had not been verified that smell also contributed to allocentric processing—the mapping of space as an aid to purposeful movement through a domain. This changed when Professor Lucia Jacobs published an article in the June 17, 2015, issue of the *Journal of the Public Library of Science* that showed the allocentric component of smell.

> You may not realize it," Professor Jacobs said, "but you're identifying each space by the smells of the space, and that may be a big part of how you map that space . . . The human olfactory system is a very underestimated system; very understudied. ~ *"Olfactory Orientation and Navigation in Humans," Journal of the Public Library of Science, Lucia Jacobs, with Jennifer Arter, Amy Cook, and Frank J. Sulloway, 2015.*

This U.C. Berkeley research study opens a new avenue for future research on how olfaction contributes to navigation among different species, and within different domains. This research also reopens all studies of navigation that did not control for odor perception.

We would expect to find that there are two systems within the olfactory neural network, just as for all the other sensory systems. We do know that the olfactory pathways go to the thalamocortical system, just as do all other sensory pathways. The thalamocortical region of the brain is a primary area for sensory integration. This is where olfactory input would divide into allocentric and egocentric neural pathways. The olfactory tracts also connect to the limbic system, the hippocampus, and the frontal lobes. In other words, olfaction connects at the primary levels of the brain with both the allocentric and egocentric processing networks.

When we deliberately smell something, like a flower, we are using our egocentric mind to determine meaning: what kind of flower is this? How shall I categorize this flower? We do this with the in-breath. *This suggests that breathing in is part of the egocentric processing system. When we breathe out, we switch to allocentric processing.* The philosopher Goethe saw a relationship that he called "the great game of life." This game involved what he called "expire and expand." In other words, expiration, breathing out, caused our consciousness to expand, while breathing in caused our consciousness to narrow. He could see, even in the 1700s, that there were two minds; one mind narrowed our focus, while the second mind widened our perspective. Smell is intimately associated with the in and out breath.

Quantum biologists are rapidly adding to our understanding of the dual nature of olfaction. They find that the nose "listens" to odors; it uses the mechanisms of quantum mechanics to contribute to both of our minds.

When creatures like the ant follow scent trails, they are using an allocentric flow of information. Recall that there is optical flow on the retina as well as auditory and haptic (tactual) flow. Smell flow should come as no surprise then, nor should we be amazed to discover that optical flow, auditory flow, and smell flow combine naturally in any embodied creature. Insects follow trails, pathways, and routes. This is also what humans do; we move on roads, trails, and pathways using our spatial cognitive maps. Human beings may be using subtle smell trails that are not conscious, but which combine with overall flow perception.

Humans easily cede the olfactory Olympics to other creatures. We know that other mammals, even insects, have superior olfactory capabilities compared to human beings. Perhaps, however, this is not so much superiority as it is a matter of employment—human beings don't need the sense of smell because vision is dominant, so we don't rely on olfaction even when we have the capacity. Because I was a teacher of blind children for over 30 years, and because many of my friends are blind, I am always on the lookout for ways that human beings compensate for the loss of sight. I know that Helen Keller, for example, developed her olfactory skills to a remarkable degree—she could smell her friends coming, knew them by their characteristic odors, and she knew rooms by their olfactory signature. Keller suggests that we have the capacity to use smell for navigation, but don't need it (or develop it) unless we have damaged vision and hearing.

In 2016, at the Max Planck Institute, research examining the olfaction of mice revealed that these rodents can smell oxygen levels in the environment. In human beings, we have a complex dual olfactory capability. This ability probably includes whatever we discover in "lower" animals, like mice. I would not be surprised to discover that humans also can detect levels of oxygen through olfaction. The nose has been overlooked as an avenue for navigation—this is especially important for those of us who help blind and visually impaired individuals navigate.

The Hidden Senses

I said at the beginning of this section that the popular notion that we have five senses is a myth. We can divide the senses for research purposes and for communication, but when we do that, and decide there are only five, we invariably leave out at least four of the most fundamental sensory systems: the vestibular, the proprioceptive, the kinesthetic, and the photo-sensitive. These four whole-body senses have been called the *hidden senses*. The so-called five senses—because they are on our body surface—are just the most obvious sensory portals, easily seen by the naked eye.

Of course, the hidden senses are also coherent; they cannot be disembodied—they do not exist in isolation. Internal and external senses fire synchronously. The body always acts as a single embodied system. Therefore, there is really no such thing as the "vestibular sense" or the "proprioceptive sense" because they are a unified team. Furthermore, like the external senses, the inner senses respond either allocentrically or egocentrically; they work in unison to map the background, or they map foreground "objects," but never both at the same time.

Vision, hearing, and smell are distance and directional sensory systems. Touch and taste are also a direct way to probe the world. These five external sensory portals are located on the skin surface, on the boundary of our bodies, and although processing takes place in the brain and nervous system, our awareness is such that we perceive the sensation at our skin boundary. For example, we perceive as if vision was at the location of the eyes—projected outward—even though visual processing is a total-body phenomenon. The image on the retina is actually flat and inverted—the brain flips the image and gives it 3-D representation. We also hear at the level of our ears, yet auditory processing, like vision, is a total-body process.

As I said above, there are four sensory portals that are internal, not on the skin surface. These are whole-body systems that have been hidden from awareness—from egocentric attention and study—for at least two thousand years.

Proprioception is sometimes used as an umbrella term for all the hidden senses. At other times, in a narrow sense, it has a specific role. As a specific entity, a sensory system, proprioception has been defined as our perception of position. We know, for example, where our body parts are and what our posture is from moment-to-moment. We also know where our body is in space relative to the scene we are in. Proprioception tells us where our body is in space without needing visual or auditory feedback. Proprioception allows us to walk without looking at our feet. In the discussion of Zoltan Torey's ideas in Chapter Three, the importance of proprioception as the source of consciousness was discussed. This is a critical understanding. Proprioception may well be the reason for what we call consciousness.

Kinesthesis is the sensation of moving the whole body. Vestibular sensations monitor balance and adjustments to gravity. The photo-sensitive system adjusts to the constant fluctuation in the quantity and quality of light. Obviously, these whole-body processes work together, but they are mixed up in the literature, often overlapping, and often dependent on working definitions used by various authors.

Together, the internal senses monitor whole-body movement, posture, balance, and diurnal and seasonal variations in light. In this regard, each of the hidden systems has both a conscious egocentric and an unconscious allocentric component.

A Few Spicy Hors d'oeuvres

"Hey, Surge."

"We're closed."

"Yeah, I know that. But this is important. I need some seasoned help."

"I have a few spicy hors d'oeuvres left over from a fraternity party, but that's it. Come back at happy hour. We don't usually open in the middle of a chapter."

"I know that Surge, but I have a recurrent fear that I might fumble on the two-yard line in the championship game. My hands are trembling."

"Why, at your advanced age, are you still on the playing field? Why are you still trying to score? Hand the ball off to a rookie and take up coaching."

"I'm carrying a ball of anxiety, Surge. I'm close to the end zone, the goal line is at my feet, but suddenly I have severe doubts."

"Doubts are good, Dutch."

"I need your help, Surge. I told the world that consciousness was a proprioceptive phenomenon. What if I am wrong and send great minds off on a wild goose chase? I need some appetizing insights to still my angst."

"You didn't tell the world anything; you told yourself. Stop fretting about illusory readers. There's no enemy over the horizon. The person reading this book was meant to find the thoughts and ponder them. There are only a few readers at this level; very few human brings care about this line of musing about musing. The right people will come and absorb the ideas. They will carry forward the thoughts that are relevant to their lives. Just accept that. And remember that life is not a football game—there is no need to score points."

"Okay, whatever. Give me some spicy appetizers, Surge. I need to wake up."

"We have a few crumbs left of the old reliable *As Above, So Below* Layer Cake. How about some Complementarity Cookies to quell your unnecessary qualms?"

"Been there, done that. I need to taste something new and refreshing."

"Okay then, how about One-or-the-Other-But-Not-Both-at-Once Fruit Cake?—sometimes called Crazy-as-a-Fruitcake Cake?"

"That's stale news, Surge. It's bland and not motivating. What else have you got?"

"How about Exponentially Doubling Blue Dolphin Crackers? We inject tiny dolphins up your nostrils."

"Sounds truly insane, Surge. You see, what troubles me about these old hors d'oeuvres is that they don't complement each other. It's a Hodge Podge dish of soul-like foods that don't go well together. Or am I missing something? Shouldn't these wisdom dishes go together to make a perfect meal? And what about Natural Selection? Shouldn't evolution be on the menu?"

"Nature always takes the easy way out, Dutch. Yes, we do serve Consistently Simple-Solution Saltines; they go well with Monkfish Caviar. However, Natural Selection stops caring about you after the reproductive years have ended. Nature likes to make babies, but then doesn't give a rip after the babies grow up."

"Am I wrong, Surge? Is this latest message from the gods just the most recent joke? Is proprioception a key puzzle piece, or is it a diversion?"

"Take a deep breath, Dutch. Call time out. Go to the sidelines and talk it over with the internal staff."

"That's why I'm here, Surge. I came for your advice. I don't want to make a fool of myself at the family reunion."

"It's too late for that, Dutch. The damage is done."

"Well then, help me save face, Surge."

"Which one?"

"Both of them."

"Sit down on the bench, Dutch. We need to talk. Be quiet for a few minutes."

"I'm sitting."

"Listen to me. You are right about proprioception; just remember that this is Zoltan Torey's insight, not yours. Indeed, proprioception did invent the two voices in your head. But the first voice—it came first in evolution—is a village idiot. It sings old college fight songs and catchy commercials. It is often prejudiced and loud-mouthed; it keeps repeating the same vapid nonsense over and over—it rants and raves. That voice is a moron, Dutch. Too many people think this voice is who they are. By the way, do you suppose there might be levels of consciousness? Levels of *who we think we are?*"

"I think that must be true, Surge."

"Yes, it is true. There *are* levels of consciousness. Ironically, you have to be at a high level of consciousness to realize *that there are* levels of consciousness. There are people who believe they *are* the monkey mind; they stay at that low level of consciousness because they don't know there are other layers— so they don't become seekers. There are other people who are sure that level two *rational consciousness* is who they are. And so on it continues as we rise through higher and higher proprioceptive levels. The most basic level is that imbecile in your head. He is a ball of loquacious anxiety."

"Okay, I get that. There is a voice that is not really me and I shouldn't listen to it so much."

"That's a good beginning; *voice one* is not you. Whatever you turn out to be, don't identify with that crazy voice. Proprioception is a system of layered watching—memory stacked on memory.[7] People with level one proprioception hear others speak, and they repeat in their minds what they hear. They join with groups who have similar monkey minds, similar mental loops. The ego that they develop is defined by the expectations and values of these like-minded groups. They are conformists who follow their internal dialogue, as if they actually were that stupid voice. That is the ball of anxiety guy who came in the door a few minutes ago all concerned about winning or losing. You didn't arrive here with a high level of awareness. You were a *level one* drone when you walked into the restaurant. So tell me, Dutch, what did proprioception invent after the first primitive internal voice? What happened when proprioception evolved the next layer of memory-of-memory? What came after the Mad Max voice in the head?"

"I don't know."

"Oh yes, you do, Dutch. Let me give you a hint. Who do you consult when times get hard? Who calls you out when your first mind wonders and reacts prematurely? Who makes you rethink and reconsider? Who calls you Dutch?"

"Oh my God! You aren't real, Surge? You are just a voice in my head!"

"I am more than that, but you aren't ready to handle anything further. I am your logical voice, Dutch, your alter ego. Just accept that for now. But more importantly, what layer of proprioception comes after *voice one* (monkey mind) and *voice two* (rational mind)? Consider the alter ego, the voice of Surge, the mind that speaks to you now. This is the beginning of rational *thought*—what we call *thinking*. It overrides the monkey mind when it can. It does battle with the monkey mind. It is a form of executive functioning in which thoughts are ordered, controlled, and carefully expressed. This is the more solid ego, and it becomes who most human beings think they are—the so-called personality. This proprioceptive level of awareness believes in the objective world, that everything has a cause and effect. There is a maximum separation between this ego and the world of objects and other sentient creatures. This is a science mind, a philosopher's mind, a rational mind, the main mind of the modern world of collective consciousness. It is also easy to get stuck at this professorial stage, as if it was reality—which it is not."

"But, it's a good mind, right Surge? I mean, you. You're a good guy, right?"

"I have been the only *good mind* in town for centuries, although most of the planet is still a slave to the first monkey-like mind, the ranting lunatic fellow. But now another layer of memory-of-memory has evolved beyond the monkey mind and beyond the rational mind. A third level of proprioception has evolved. A witness has developed. This witnessing consciousness is a memory of the second mind. It can witness and record the alter ego as it reasons, as it manipulates patterns. Therefore, a third level of internal mind has evolved, but only in a few people. These more evolved individuals have a higher level of consciousness; they are not a slave to the earlier proprioceptive layers, to the monkey mind and the rational mind. Individuals at this high level of awareness do not identify with any kind of voice in the head—they have transcended the level of internal dialogue. They fall silent as they watch and witness."

The Hard-Evidence

"Okay, Surge. This is what I understand so far. Proprioception starts as muscle memory, a system for remembering movement patterns. So the level one voice in the human head is just a memory of the actual sensation of proprioception. The second voice is a memory of the first voice. Therefore, there must be a memory of the second voice also. There must be another layer to proprioception beyond the second voice, a witnessing mind."

"There you go. Yes, there is a witness to internal dialogue. It does not speak—it just notices."

"Okay, I get it. So what is this witness level of consciousness?"

"Individuals who have transcended their internal voices, those with memories of memories, become fascinated as they learn to witness themselves trying to make sense of themselves. They suspend their rational analytical voice and become more holistic—they begin a long journey to understand the allocentric mind. The obsession with reason and explanation decreases. *Experience itself becomes the new attraction.* The journey of discovery becomes more interesting than reaching a goal or creating a product. Human beings who reach this witnessing level of consciousness can watch as words pour out of their mouths or run in thought-loops. This witness is a new entity. It is the dawn of self-consciousness. It is a cut above mental garbage and mental reasoning. Then what do you supposed happened next? What new proprioceptive layer emerged? What is the memory-of-the-memory of the witness?"

"I give up."

"Exactly! Very good! I didn't think you knew."

"There you go again, Surge, messing with my innocence."

"Individuals who develop a monitor, a witness to the witness, fall into a deeper silence. They can watch experience unfold moment to moment. The seeds of enlightenment can only sprout after an individual reaches this level of sophistication. Very few human beings have reached this level of awareness. Human beings are confused because they routinely operate on a low level of proprioception."

"So, let me review again, because this is hard: We have a sensory portal called proprioception; its job is to know the location of body parts and to record when the body moves. Therefore, a memory system developed for remembering useful behaviors. This memory for behaviors evolved and became more sophisticated, enabling ever more complex behaviors as the eons rolled forward. The specific behaviors that enabled speech somehow became internalized so that a voice began to speak inside the skull. This voice scared the crap out of our ancestors. This first voice, this level one proprioception just repeated vocalizations—those that were self-generated, and later, those that others spoke. That first voice, the so-called Monkey Mind, evolved. Brain maps combined to make super-brain maps that were proprioceptive memories-of-memories. This eventually enabled a second *rational* voice. That thinking voice evolved and then enabled a witness, an executive function. Rationality and ego were born. Then the process of proprioception evolved more sophisticated super-brain maps that joined with other super-duper brain maps that eventually enabled a silent monitor to watch the witness. How am I doing, Surge?"

"Close enough, for now."

"Okay, so what is the next level of proprioceptive multi-memory brain mapping? Is there an observer of the monitor of the watcher, a watcher-watcher?"

"You aren't ready to find out. You don't have the tools to cope with the next level of evolution. However, I can tell you two things. First, the voice in your head doesn't age, so humans say that they don't feel older inside. The body grows old, but the voice in the head remains fresh. Humans feel like they should live forever because their voices feel eternal. Second, as you know, the human nervous system manufactures space and time, so human beings evolved the concepts of eternity and of infinity. Likewise, proprioception manufactures consciousness. This means that human beings have a concept of infinite regress—for example, watchers of watchers of watchers. As with space and time, consciousness may be simply a human experience. What is beyond space, time, and consciousness, is beyond human capacity to comprehend. Do you understand this?

"Sure, Surge. Whatever."

"Human beings spend most of their time at level-one consciousness, at a primitive level of proprioception, at the mercy of the monkey mind. They are slaves to word-circles, propaganda, memes, word-soup, and word-insanity. If you want to spiral up to a higher level of mental development, talk with the Dalai Lama."

"Maybe later."

"Get back to writing, Dutch. Stop running with the ball. This is a book, not a sport."

Proprioception in a Narrow Sense

When I wrote about the contributions of psychologist Zoltan Torey in Chapter Three, I discussed proprioception as a global (whole-body) sensory system. I don't need to repeat that lengthy discussion here, except to reiterate my theory that proprioception is responsible for the evolution of dual consciousness. In other words,—both allocentric and egocentric processing is enabled by proprioception.

Proprioception can also be thought of in a narrow sense, rather than as an umbrella term for all the internal senses. In this narrow regard, proprioception, like all the senses, has a dual nature. The discussion below is from this narrow definition for proprioception.

Our skin provides us with a boundary. Inside the skin is "us." Outside the skin is "other." Sensors in muscles allow us to know where our limbs are in space relative to the rest of the body and to register when our body parts move. Maps in the brain have location cells that know where every square inch of "us" is located relative to the whole body, and relative to the environment. The sensation of *having a body*, of embodiment, comes from proprioception. Also the mental feeling of wholeness, of wholes nested within wholes, arises from our sense of proprioceptive oneness.

In a book called *How the Body Shapes the Mind* (2005) author Shaun Gallagher divides proprioception into two categories that are clearly egocentric and allocentric–although Gallagher did not use that terminology. *Body image*, as defined by Gallagher, results from egocentric processing, whereas *body schema* results from allocentric processing. Explaining egocentric proprioception, Gallagher writes:

> With respect to attention, this has the status of a general law. Attention is always structured in terms of a figure on ground in which some figural aspect is maintained in attentional focus. In the case of the consciousness of one's own body, the limit of the perceptual field would be the whole body, and attentional focus would necessarily be less than that. Thus, my body appears in consciousness with certain parts emphasized or singled out. ~ *How the Body Shapes the Mind, Shaun Gallagher, 2005.*

Body image, therefore, is the result of paying attention to the parts of the body rather than to the whole body at once. When we see our body as a collection of features—eyes, hair, teeth, legs, and so on—and especially when we attach judgment, we perceive our body image.

Regarding allocentric proprioceptive processing, Gallagher says:

> The body schema, on the other hand, functions in a more integrated and holistic way. A slight change in posture involves a global adjustment across a large number of muscle systems . . . The holistic nature of the body schema can also be seen in the fact that various proprioceptive inputs originating in different parts of the body do

not function in an isolated manner, but add together, in a non-linear fashion, to modulate postural control. ~ *How the Body Shapes the Mind, Shaun Gallagher, 2005.*

Gallagher also points out that the two major processing streams of the nervous system, *perceiving for knowledge*, which is egocentric, and *perceiving for experience*, which is allocentric, are reflected in proprioceptive awareness:

> I suggested that when the body appears in consciousness, it normally appears as clearly differentiated from its environment. In experimental situations, body-image boundaries, for example, tend to be clearly defined. When I am immersed in experience, however, the limits of the body and environment are obscured. ~ *How the Body Shapes the Mind, Shaun Gallagher, 2005.*

Furthermore, Gallagher proposes that we cannot understand the development of consciousness, or cognition, or any part of human evolution unless we perceive ourselves as a body and not just as a brain. As I stated above, the term that philosophers and psychologists use to encompass this concept is called *embodiment*.

As I stated earlier, navigation is a fundamental behavior of all animals. To figure out how we navigate, and how the evolution of wayfinding contributed to attention, perception, and consciousness, we had to understand the concept of embodied cognition. In other words, our *behaviors* are the sum total of *all body activates* every moment—no sense stands alone, no muscular movements stand alone, and no mental processes exist in isolation. Both Gallagher and I agree about the fundamental concept of embodiment, except that he does not discuss his work from a navigational perspective.

Proprioception, along with vision—which evolved later in evolution—is the chief contributor to the allocentric mind. Its egocentric role is minor in comparison to its foundational role as the parent of allocentricity. Gallagher understands that proprioception is not a system that uses the ego's frame of reference:

. . . we can ask about the distance and direction of a perceived object in terms of how far away it is, and in what direction. But these spatial parameters are meaningful only in relation to a frame of reference that has an origin. This does not apply to proprioception. Proprioceptive awareness does not organize the differential spatial order of the body around an origin . . . Thus, whereas perception organizes spatial distributions around an egocentric frame of reference that is implicitly indexed to the perceiving body, somatic proprioception reflects the contours of my body, but not from a perspective of another perceiver. Proprioception operates within a non-relative, non-perspectival, intra-corporeal spatial framework that is different from both egocentric and allocentric frameworks. ~ *How the Body Shapes the Mind, Shaun Gallagher, 2005.*

Gallagher's use of allocentric and egocentric as frames of spatial reference is correct; indeed, that is how these terms are narrowly defined. He correctly sees that proprioception is primarily allocentric. However, his observation that "proprioception operates within a non-relative, non-perspectival, intra-corporeal spatial framework that is different from both egocentric and allocentric frameworks" is confusing from the perspective of dual-process theory. *In my view, proprioception can extend beyond the body and, in its allocentric form, is the foundation for the allocentric mind.* I believe what Gallagher is emphasizing in the above quote is that proprioception (in a broad sense, encompassing all the internal senses) is not egocentric. It is a whole-body "perceptual" system. Gallagher is drawing a distinction between "perception"—which he sees as an egocentric process—in contrast to "proprioception in a broad sense," which has no egocentric perspective.

The important idea for this discussion is that proprioception, like hearing, vision, smell, and touch, is a dual phenomenon. Gallagher's discussion supports dual-process theory.

Brain Maps Beyond the Body

As an orientation and mobility specialist working with blind children, I was delighted to discover a book called *The Body has a Mind of its Own* (2007) by Sandra and Matthew Blakeslee. One paragraph from that book leaped out at me immediately:

> Your brain faithfully maps the space beyond your body when you enter it using tools. Take hold of a long stick and tap it on the ground. As far as your brain is concerned, your hand now extends to the tip of that stick. Its length has been incorporated into your personal space. If you were blind, you could feel your way down the street using that stick. ~ *The Body has a Mind of its Own, Sandra Blakeslee and Matthew Blakeslee, 2007.*

So when a blind person uses a cane, the brain incorporates that cane into the body schema—the cane becomes part of the person's allocentric proprioceptive system. Therefore, allocentric proprioception maps not only our bodies, but also our tools and our immediate environment as part of the mind! This is a huge leap of understanding. Awareness "pushes itself outward;" we end up mapping not only our body but the objects *and spaces* surrounding us—*as if they were an extension of the body.* Consequently, the brain's body maps also include our tools and our domain, as if the tools and environment were an integral part of us. *This means that consciousness is hardwired to extend beyond the body.* It also means that *the mind extends into and includes the environment*—as I emphasized earlier from a philosophical perspective.

As a teacher working with children who struggled with navigational disabilities, I taught my students to be aware of what I called "perceptual zones." I told them that they could "move" their awareness from their bodies outward into space.[8] We would start with *personal space*, something easy to understand and not far removed from the physical body. Personal space belongs to the individual—no one may enter this space unless invited. Personal space extends around the body to various distances depending on individual personalities and cultural norms. For example, North Americans tend to have more extensive zones of personal space—we don't like to be crowded or bumped. However, in other cultures, personal space shrinks, and people have no problem squishing against each other in public spaces. The boundaries of the perceptual zones of space vary from individual to individual and from culture to culture, but there is definitely a beginning and an end to a spatial zone, an "edge" that can be "felt."

After my students grasped the concept of personal space, I then divided space beyond the personal aura into three other perceptual zones: communication space, landmark space, and beacon space. Beyond personal space is a perceptual zone inside of which we communicate with others—face-to-face interaction. When we communicate, we can sense a comfort zone. People are either "too close for comfort," invading our personal space; or they are too distant—so we say: "Come closer so we can talk."

Beyond communication space is a zone of landmarks. Landmarks are sensory entities used for navigation. Landmarks have a message for the navigator depending on the destination. The same landmark can have different messages: for example, "turn left at the landmark if you are going to the grocery store," or "turn right at the same landmark if you are going to the gas station."

"Pushing" proprioceptive awareness outward into space is an allocentric process. However, actually pinpointing a landmark is egocentric, as is knowing the meaning and the use of the landmark. Being able to discourse about the landmark is also egocentric. There is a mental oscillation between the two awareness systems: you sense the world egocentrically—with your body as the center of everything—or you sense your relative position within a scene (from an overhead perspective). It is no wonder that children with severe damage to the body and brain have navigational disabilities, given the complexity of this dual-awareness system.

Beyond the landmark perceptual zone is a region of space that I call *beacon space*. The distance receptors, vision and hearing, evolved to monitor this perceptual zone, especially the visual system. Blind kids are left with just hearing to monitor distant sounds like traffic flowing on a nearby street, or tactual clues like the position of the sun or the direction of a stiff breeze. The further out we push our sense of space, the more we must rely on allocentric awareness. *The closer in we get perceptually, especially in our modern world of technology, the more we get "stuck" inside our egocentric minds.* Indeed, egocentric awareness has gotten so over-developed from excessive practice that our allocentric "depth-perception," our mapping of external space as if it was part of our body, has deteriorated—dying allocentric neurons are piling up from lack of stimulation.

A useful meditation is to sit in stillness and practice pushing awareness further and further beyond the body. The mind will build body maps to include ever-greater regions of space, especially with practice. When monks meditate, they push their awareness outward to levels well beyond what we common folk can comprehend. The ability to push awareness outward, to build ever wider spherically-extended zones of perception requires a gradual increase of mental frequency beyond beta-wave frequencies into gamma-wave frequencies.

When I taught kids to perceive zones of perception, I was concentrating on a flat-land terrain. We are earth-based creatures and we have body maps that reflect this circumstance. However, it is also important to perceive zones of perception from a universal rather than from an earth-bound perspective. Allocentric spatial awareness spreads out from our physical body as a sphere. We can get ever wider perspectives as we "journey outward spherically." A commonly understood way to do this is to perceive using a bird's-eye-view of our terrain. As we push our "imagination" further and further from our egocentrically-bound bodies, we gain ever more insight into the spiritual world that is made available through allocentric consciousness. This is a good meditation technique: sit still while mentally projecting further and further above the body—map ever-wider spaces from a bird's-eye perspective. This is just one more way to shut down the egocentric mind and exercise the allocentric mind.

Consciousness: A New Slant on an Old Conundrum

Notice that when monitoring allocentric perceptual space, the awareness and monitoring of other allocentric spaces is blocked. In other words, we can only be aware of one perceptual spatial zone at a time. For example, if we are monitoring personal space, we cannot at the same time be consciously aware of deeper perceptual zones like beacon space. Switching to another level of spatial awareness requires an inhibition of all other possible perceptual depths. What is important to see, in this regard, is that being in an egocentric zone inhibits social interaction and concern for others—we don't move into communication space—we remain behind a psychic wall when using the egocentric mind. Empathy requires a "moving beyond" the body into more expansive mind frames, specifically into the zone of spatial communication. To connect with another sentient creature requires our allocentric mind to expand and include the other. We map others as part of our own body space—they are included as part of our mind.

Recall the development stages you went through as you got older. The spaces you were comfortable in, the spaces that "held" you, expanded as you grew. When you were an infant, for example, you found comfort in the arms of people who loved you. This was an emotional *holding space*. As your body and mind matured, you became comfortable within a crib, an emotionally secure holding space that was somewhat removed from the loving arms of adults. Later, as you developed, you became master of your bedroom, and then all the rooms in your house became your domain. Later still, you were let out of the house to experience more challenging environments; you slowly found comfort within ever larger spheres of perception. You eventually crossed streets, followed pathways, and visited friends in the neighborhood. Preteens roam ever farther from the family compound, and young adults use transportation systems that allow them to push perception even farther from home base. As adults, we push our perceptual zones to an extent that includes whole countries and continents. From there, humans are pushing into the universe, into the depths of the mind, and into virtual realties. Remember that brains and minds evolved to allow for navigation. It is no wonder that we have such a thirst, such an insatiable and driven passion to explore farther and farther from home base.

Proprioception probably deserves a book of its own because in its role as monitor of movement and position, it is the biological substrate for both egocentric and allocentric consciousness. Proprioception is the reason that allocentric *awareness* and egocentric *attention* evolved in the first place, which then gave rise, over eons of evolution, to our two forms of consciousness. Here is how Shawn Gallagher explains it:

> Movement and the registration of that movement in a developing proprioceptive system (that is, a system that registers its own self-movement) contributes to the self-organizing development of neuronal structures responsible not only for motor action, but for the way we come to be conscious ourselves, to communicate with others, and to live in the surrounding world. ~ *How the Body Shapes the Mind, Shaun Gallagher. 2005.*

Shaun Gallagher, along with Zoltan Torey, stated years ago that proprioception was the key to comprehending consciousness. My own "revelation" evolved from the work of these two remarkable men.

Kinesthesis

The philosopher Edmund Husserl agreed with Aristotle that movement was the mother of all cognition; human beings in all their complexity are the offspring of the primal process of motion. Movement is more basic than atoms, or quarks, or the material of creation.

Very early in evolution, at the cellular level, living creatures became unconsciously "aware" of self-movement; they could discern when they were relatively still and when they were moving relative to their surroundings. This primal awareness of moving, discerning no-movement from movement, and sensing the flow of movement, is what we call kinesthesia.

There is a fine line between kinesthesia and proprioception. These two terms seem to overlap in the literature or sometimes to be two names for the same process; one or the other is used as the all-encompassing perspective to explain the internal sensory system. I decided to use proprioception as the umbrella term but could just as easily have selected kinesthesia. They are both whole-body sensorimotor systems that are reliant on sensory fibers inside muscles to measure relative motion. Proprioception is more a sense of position and balance, while kinesthesia is about movement and flow. Because of embodiment, no isolated behaviors can exist—neurons that enable proprioception and kinesthesia always fire coherently.

The *kinesthetic egocentric system* is the awareness that the organism as a whole is still, not moving relative to the surround. *Allocentric kinesthetic awareness* is a discernment that the whole organism is moving relative to a stable surround.

A single cell living in the ocean can find the best location where there is warmth, food, and shelter simply by riding the currents in nutrient-rich seas. However, when wayfinding arose very early in cellular evolution, self-moving organisms had to develop an internal coordinated memory system. A way *to know* the internal state of the whole body had to evolve in order to determine temperature, hydration, acidity, hunger, energy levels, and so on. This internal memory-mapping system resulted in an organism's "awareness" of front and back, top and bottom.

Eventually, there evolved a bias for moving forward. Although it was possible to move backwards or sideways, nature preferentially evolved a system in which forward movement dominated. This primitive internal mapping system became the substrate for our internal sensory systems, for proprioception (in the narrow sense), kinesthesis, and the vestibular sense. This entire internal sensory matrix evolved solely so that organisms could move with a purpose. Rather than riding the currents and trusting in the benevolence of drifting, organisms learned to self-move toward things that had meaning for their survival.

Therefore, there is a primary, internal, evolutionarily ancient, proprioceptive-consciousness. Kinesthesis is a prime member of this proprioceptive team.

Proprioception, in a broad sense (with kinesthesis, the vestibular system, and photo-sensitive processing), is the mother of all the variations in consciousness that we struggle to understand today. This primal consciousness has an inherent duality that is observable in all the sensory systems in human beings.

The Vestibular System

The vestibular system also has an egocentric component and an allocentric component, exactly as we would suspect. The vestibular system affords balance and posture in relationship to gravity. It is closely tied to vision and to eye movements especially. There are at least two ways that the vestibular system contributes to the sensory mix. When the eyes are focused on something—using foveal vision, the egocentric visual system—the head and body are held relatively still so the eyes can localize, isolate, or track an object in space. This relatively frozen posture is the perceptual stance used to extract meaningful information. For this reason, vestibular egocentricity is called an active process—we are probing the surround, examining the world using our senses to gain knowledge about the environment.

In contrast, the allocentric vestibular system measures velocity, as well as stopping and starting (in close association with kinesthetic processing). Synchronous with the processing of flow in vision and hearing is a vestibular flow and a kinesthetic flow, all part of the allocentric system. Even though the body is in motion, this is called a passive process because there is no effort to extract meaning from the environment; it is totally a navigational process, allowing us to get from one place to another.

The vestibular system has two kinds of hair cells, one for allocentric processing and the other for egocentric processing. Type one hair cells are newer in evolution and they are the neural substrate for egocentric vestibular processing. Type two hair cells are older in evolution and are the substrate for the vestibular allocentric system. There is relativity to these two systems, just as we saw for vision. When the whole body is moving through space, the allocentric vestibular system is active, while the egocentric vestibular system is inhibited. Contrary to this, when the body is not flowing through space, the egocentric system is active and allocentric processing is inhibited. The online article referenced below supports the dual-process theory in regard to the vestibular system:

> Two dynamic, integrated functional systems are currently iden-
> tified for maintaining balance: one orienting the body to evoke
> anti-gravity support, the other providing perception-action cou-
> pling . . . we could say that the former has more to do with
> "attention," while the latter [has to do with] "intention"—being
> able to stay attentive while we carry out our goals. ~ *Online PDF:*
> *"Graviception, F.M. Alexander's Science of Poise," Alexander Studies,*
> *Glenna Batson, 2015.*

The above quote makes an interesting use of the word "intention." Egocentric processing depends upon a vestibular system that maintains stable balance so that attention can be localized in space. However, during allocentric processing, the vestibular system uses a method of process-ing that occurs when the individual is moving. I equate *awareness* with allocentric consciousness, using dual-process theory. *Intention*, from my perspective, is an egocentric, executive function. The allocentric system cannot "attend" or "intend." In this case, the author, in the above quote, is perhaps using the word "intention" as a synonym for "awareness." At any rate, she is aware of our two vestibular processing systems, and her perspective supports dual-process theory.

The Photo-Sensitive System

The photo-receptive sensory system is rarely considered during the discus-sion of the hidden senses. Yet ambient light energy has been a constant since the dawn of life. Light is a consistent and reliable variable that affects organisms in a whole-body manner:

One of the mechanisms for sensing light without eyes appears to be a substance called melanopsin, first discovered in 1998 in the skin of frogs. It allows mammals to detect light beyond and separate from the retina's rod and cone cells. This photopigment reveals a previously unknown and primitive visual photoreceptive system. ~ *Beyond Biocentrism, Robert Lanza, 2016.*

A simple cell, early in evolution, floating in a medium of salt water, can only exist if it is bathed in light of a specific quantity and quality. This light ebbs and flows with the movement of the earth around the sun, the spinning of the earth on its axis, and the cycles of the moon. Seasons change—there is rhythm and a pattern to the ebb and flow of light that bathes single cells. These cells need light as a power source and for temperature balance.

These rhythmic cycles have been going on for 3 billion years. When single cells combined to make multi-cellular creatures, each cell brought its own photo-sensitive and photo-reactive capacities to the marriage. Nature's creations are solar-powered, driven by fluctuations in the quality and quantity of light. The Sun is the ultimate source of energy for almost all cells. For example, algae and plant cells use light to harness the energy needed to sustain growth, metabolism, and reproduction.

It is no wonder, then, that our total-body photo-sensitivity co-evolved with all the other sensory-motor systems in the body. This invariant process is often overlooked when we talk about the senses. Animals "know" that the sun is above and that light shines at different angles, with different polarity, and with varying quality as a day progresses. Light comes from above for all creatures, and they have evolved with this invariant knowledge. For example, fish in the ocean orient to light coming from above, and migratory birds calibrate their magnetic compass based on polarized light patterns at sunset and sunrise.

Light is also the chief energy source for powering movement. It is the fuel supply for purposeful movement and has been since the dawn of cellular life. This "fueling of cells" is also a whole-body, molecular phenomenon. It is a reliable invariant during the lifetime of all creatures. Light takes its place, as a steady influence on life, alongside gravity, atmospheric pressure, temperature, and the chemical composition of any environment.

The photo-sensitive system is linked to allocentric awareness and egocentric attention. For example, the entire skin surface is part of all-at-once allocentric processing. Light is ambient and diurnal—it has a varying intensity and quality (wavelength variation). Our internal body clocks set all our hormonal levels and regulate body temperature. These clocks are set by the quantity and intensity of light. Light in the morning helps us wake up, while dimmer light at night cues us to go to sleep. Therefore, light is critical for our health and wellbeing.

Egocentrically, light highlights whatever we are observing. Direct light from the sun—the sun's position in the sky—or the position and intensity of the moon—helps animals navigate. It also triggers the cone cells of the retina so that pattern recognition is amplified. This enables object perception, and maximizes one-thing-at-a time attention.

Two Nervous Systems

We divide our nervous system into dualities: peripheral nervous system versus central nervous system; cortical nervous system versus non-cortical nervous system; somatic nervous system versus autonomic nervous system; and sympathetic nervous system versus parasympathetic nervous system. We speak of spatial processing and temporal processing, but we don't add a third or fourth variety of cognitive processing. This binary arrangement is exactly what we would expect to find if dual-process theory is correct. We would expect to find two divisions, one allocentric system for all-at-once processing, and one egocentric system for processing one-thing-at-a-time. In other words, given dual-process theory, we expect to find a nervous system that has two divisions: a system for attending to the background (allocentric), and a system for attending to what manifests from that background (egocentric).

Whatever contributes to a steady, restful state, whatever allows the body to restore itself, whatever we do that is experiential, and whatever contributes to the background is allocentric. Also, when the body is moving, flowing through space, it is defined as allocentric processing because the system is not gathering information. On the other hand, whatever gets the body ready for action, whatever freezes the body in a posture that gets it ready for extracting meaning from the environment, and whatever manifests forms from an amorphous background is egocentric.

Each of the two nervous systems is supported by a specific kind of memory. Using dual-process theory, we would expect this dual-memory arrangement. In a book chapter called "Imaging, Embodiment, Phenomenology, and Transformation," authors Francisco Varela (philosopher and neuroscientist) and Natalie Depraz (French philosopher) discuss a memory system for space and a memory system for objects—in my terminology they are finding an allocentric spatial memory system as well as an egocentric memory system for form perception:

> There seems to be a differentiated participation of object and spatial memory and imagery. In fact, it has been shown that at least both types of memories can be differentiated as to the regions they mobilize. The first circuitry is frontalized and seems to be active when the image is dynamic (i.e. spatial transformations of the image). The second is more ventrally located in the middle frontal gyrus and is better related to figurative working memory. This distinction also holds for mental images. - *"Imaging, Embodiment, Phenomenology, and Transformation," Buddhism and Science, Breaking New Ground, Francisco Varela and Natalie Depraz, Ed. by B. Alan Wallace, 2003.*

Varela and Depraz also say that there are two kinds of neural activity: propagated nerve impulses that send signals along neural tracts, and slow potentials that do not propagate—these slow potentials form holographic-like patterns. This is probably how sensory "images" are actually formed. Notice that this is more evidence for dual-process theory: neural tracts handle egocentric processing, while non-propagating neurons handle allocentric processing.

The distinction between the somatic nervous system and the visceral nervous system is also important in our discussion about the duality of the senses. This division is very ancient in evolution. Dr. Paola Bertucci, Associate Professor of History of Medicine, and biologist Dr. Detlev Arendt have studied this historic distinction between somatic visceral nervous systems:

> The vertebrate nervous system is deeply divided into 'somatic' and 'visceral' subsystems that respond to external and internal stimuli, respectively. Molecular characterization of neurons in different groups of mollusks by Nomaksteinsky and colleagues, published in *BMC Biology*, reveals that the viscero-somatic duality is evolutionarily ancient, predating Bilateria . . . Taken together, this is strong evidence that the somatic-visceral duality predates the bilaterian divergence. ~ *"Somatic and visceral nervous systems - an ancient duality," BMC Biology, Paola Bertucci and Detlev Arendt, 2013.*

The above quote is another example of how nature crafted duality using the building blocks of the universe, with their quantum dual nature, to create living creatures. We have a visceral, internal mind which is concerned with flow, with purposeful movement, and with unconscious navigation. We have a second mind that is concerned with extracting, exploring, and remembering that which has meaning for an organism. This is the familiar allocentric/egocentric conundrum.

Before bilaterality evolved in the Cambrian Age, before the concept of "animal" evolved, before neurons, heads, and brains were invented by nature, there existed a fundamental biological contrast. The visceral system is automatic, reactive, and very fast. This is allocentric. On the other hand, the somatic system came from the evolution of egocentric processing.

The Hard-Evidence

There is another perspective, another brain organizational schema, which illustrates our overall inherent dual-processing capacity. The fronto-insular cortex (FIC) has been described as the center for the *self*, a seat of consciousness, and the location of self-identity. This region of the brain (one in each hemisphere), working within a network with other brain regions, is involved in switching between allocentric processing and egocentric processing. In the terminology of neuroscience, the egocentric processing system is called the central-executive network (CEN) and the allocentric system is called the default-mode network (DMN). The quote below is from the Proceedings of the National Academy of Sciences of the United States of America:

> Cognitively demanding tasks that evoke activation in the brain's central-executive network (CEN) have been consistently shown to evoke decreased activation (deactivation) in the default-mode network (DMN). The neural mechanisms underlying this switch between activation and deactivation of large-scale brain networks remain completely unknown. Here, we use functional magnetic resonance imaging (fMRI) to investigate the mechanisms underlying switching of brain networks in three different experiments. We first examined this switching process in an auditory event segmentation task. We observed significant activation of the CEN and deactivation of the DMN, along with activation of a third network comprising the right fronto-insular cortex (rFIC) and anterior cingulate cortex (ACC), when participants perceived salient auditory event boundaries. Using chronometric techniques and Granger causality analysis, we show that the rFIC-ACC network, and the rFIC, in particular, plays a critical and causal role in switching between the CEN and the DMN. We replicated this causal connectivity pattern in two additional experiments: (i) a visual attention "oddball" task and (ii) a task-free resting state. These results indicate that the rFIC is likely to play a major role in switching between distinct brain networks across task paradigms and stimulus modalities. Our findings have important implications for a unified view of network

mechanisms underlying both exogenous and endogenous cognitive control. ~ *"A critical role for the right fronto-insular cortex in switching between central-executive and default-mode networks," Proceedings of the National Academy of Sciences of the United States of America, Devarajan Sridharan, Daniel J. Levitin, and Vinod Menon, edited by Marcus E. Raichle, 2008.*

What this quote shows is that the allocentric and egocentric networks are mutually exclusive—as I have argued throughout this book. When one system is active, the other is inhibited. The quote also acknowledges that these two processing systems are widespread across the brain—they represent two vast, generalized networks. Exogenous control refers to allocentric awareness, while endogenous control refers to egocentric attention.

Hardly a day or week goes by for me without my finding more anatomical and physiological evidence for our duality. This morning while reading a book called *The Man Who Wasn't There* by Anil Ananthaswamy (2015), I found this statement about the frontoparietal network in the human brain:

... the frontoparietal network associated with consciousness is actually two different networks. Activity in one correlates with awareness of the external [world] . . . The other correlates with awareness of the internal [world] and is potentially related to aspects of the self . . .

Studies in healthy patients showed that these two dimensions of awareness are inversely correlated: if you are paying attention to the external world, then activity in the network associated with external awareness goes up while the regions associated with internal awareness dampen down. And vice versa. ~ *The Man Who Wasn't There, Anil Ananthaswamy, 2015.*

This is clearly the allocentric and egocentric networks at work. Evidence for dual-process theory will continue to be complied until a threshold is reached and the awareness that we have two minds finally sinks in.

The New Brain: One Mysterious King to Rule Them All

The cortex of the human brain is about the size of a cloth table napkin. It is a convoluted covering for the brain—"cortex" means "bark" in Latin. The cortex is composed of a delicate layer of neurons—only six wafer-thin layers thick. Therefore, most of the human brain is not cortical—almost the whole brain is a subcortical structure. This can be misleading since most references to brain functioning refer to the cortex as if it was the whole of the brain.

When we talk about brain lobes and streams-of-processing, we are primarily referring to this thin layer of tissue. This presupposes that the rest of the brain is not that important relative to the cortex. There is also a concerted focus on the neurons of the brain with little discussion about the significance of the many varieties of support cells that share space in the cortex. Keep these caveats in mind as we explore the physiology and function of our "highest" level of brain development, the neocortex.

In an overly generalized way, we can think of allocentric processing as taking place in the bulk of the brain below the cortex, while egocentric processing is mostly cortical. This is not strictly correct, since the cortex processes both allocentrically and egocentrically. However, the bulk of automatic processing is unconscious and tends to be primarily concerned with the management of the whole body at once. The ability to coordinate the whole body at once is very old in evolution and, consequently, most whole-body processing is on autopilot—like our ability to move with a purpose to do routine tasks like walking or eating.

An earlier brain in evolution ended at the thalamus, the bulb-like area at the head of the brainstem. The thalamus was a pretty sophisticated brain back in the day, and it is still very important, especially because it is richly networked with the cortex. Remember that *wayfinding* in primitive organisms required no cortex and no thalamus. *Navigation*, however, does require both an intact thalamus and a fully evolved cortex. With this caution in place, we can turn to the human cortex, which is responsible for differentiating human beings from the rest of the animal kingdom.

In the book *On Intelligence* (2004), author Jeff Hawkins described the anatomy and physiology of the human cortex—sometimes referred to as the neocortex.[9] One of the amazing conclusions discussed in the book is that the anatomy of the cortex is the same at almost all locations in the brain. Apparently, the same anatomical and physiological algorithm is used for processing information coming from all of the senses. In other words, a common neurological architecture evolved for comprehending the world. It is as if evolution discovered a cellular architecture that was so efficient that it was used repeatedly throughout the cortex.

Hawkins concluded that the neocortex is anatomically designed the same throughout the cortex because it has the same job throughout the cortex: processing invariant patterns. The cortex—whether visual cortex, somatic cortex, or auditory cortex—creates patterns, makes partial patterns whole, generates motor patterns, perceives patterns nested inside of other patterns, and remembers patterns. Patterns are invariants that enable us to move through the environment, and to make decisions about the relevance of what we perceive.

Hawkins' observation that the same anatomical algorithm is used to process all sensory information—at the level of the cortex—supports my position that the senses are all doing the same thing: either gathering information (patterns) egocentrically, or gathering information allocentrically. After some reflection, it becomes clear that there are *two kinds of overall patterns*: *spatial patterns* based on map-like "arrangements" and *temporal patterns* based on sequences. The dual-process theory put forth here asserts that one mind, the allocentric, evolved to comprehend spatial arrangements, while the other egocentric mind evolved to comprehend sequential events. Therefore, the cortex processes both allocentric and egocentric *patterns*.

One observation made in *On Intelligence* turns out to be especially important for my argument that our two minds evolved so we could navigate.

Hawkins realized that a brain structure called the hippocampus kept coming up in the research literature. He realized that the hippocampus was not a minor brain organ left over from the evolution of the mammalian brain, but rather it was perhaps one of the most important processing regions of the entire brain:

> One of my colleagues at the Redmond Neuroscience Institute, Bruno Olshausen, pointed out that the connections between the hippocampus and the neocortex suggest that the hippocampus is at the top region of the neocortex, not a separate structure. In this view, the hippocampus occupies the peak of the neocortical pyramid . . . ~ *On Intelligence, Jeff Hawkins with Sandra Blakeslee, 2004.*

If the brain evolved so that creatures could navigate, then the hippocampus is a "top-level manager" in charge of the brain's navigational control system. The hippocampus, therefore, may be the "end of the road" for the dual processing streams: the allocentric dorsal (where is it) processing stream, and the egocentric ventral (what is it) processing stream. However, the hippocampus never operates as an isolated modular organ. It is an important component of the patterned-firing of cell assemblies as proposed by neuroscientists Susan Greenfield and Gerald Edelman.

Consider that an experience is always tied to a location; experience and spatial position are intimately linked. Therefore, it makes sense for the hippocampus to be both a center for experiential (episodic) memory and for space perception—location and experience cannot be separated out from each other. Discharging cell assemblies for episodic memory, as well as for spatial awareness, invariably include hippocampal recruitment.

In 2014, the Nobel Prize for physiology went to three experts on the hippocampus: English professor John O'Keefe and the Norwegian husband-wife team of Edvard and May-Britt Moser. These pioneers found a cellular basis for navigation. Without a normally functioning hippocampus, we would have no memory for specific spaces. We would not be able to navigate from one place to another. The press reported, somewhat incorrectly, that the hippocampus was the GPS system of the brain. The part that the popular press missed is that the hippocampus is implicated in the evolution of consciousness. What the research experts uncovered was the cellular basis for comprehensive, high-level processing of brain maps that determine where things are in relationship to each other and to the self.

Professor O' Keefe discovered *place cells*. Wherever we stand within an environment, there is a cell assembly that is designated to denote that location—for example, if we put an X on the floor of a room and stand there, the brain will assign a cellular network to that location—call the network "Y." This Y cell assembly will be silent anywhere we walk about in the room, but when we go back to the X and stand again upon it, the Y cells will fire.

Researchers also found what they called *head direction cells*. These only fire when the head is pointed in a specific direction. *Head direction cells* define (create) the concept of "straight ahead." The animal body is designed in such a way that the organism can predominantly move forward. The importance of *head-direction cells* is that they are a neural substrate for what is called "dead reckoning." Here's an example of dead reckoning:

Imagine a football field. The player with the ball can head toward the goal in multiple ways, zigzagging around other players but always heading in a general direction using a goal, an end zone, as a beacon—a distant landmark keeps calling "come this way." This is dead reckoning, using a distant beacon to maintain a course. *Head-direction cells* are closely tied to the vestibular system and to control centers in various locations in the brain, including the thalamus and basal ganglia. The hippocampus is just another control center for the head-directional neural network.

Head direction cells were discovered in 1984 by Dr. James Ranck, in the presubiculum, a neural structure near the hippocampus. Jeffrey Taube, a postdoctoral fellow working with Ranck, made *head direction cells* a lifelong focus of his research. Taube, Ranck, and Professor Robert Muller wrote two papers in the *Journal of Neuroscience* in 1990 summarizing their findings. These important papers became the foundation for all future explorations of the head direction neural-net. Dr. Taube devoted his entire career to the study of head direction cells, and he is responsible for many of the most important discoveries in this research arena.

Head direction cells go back deep into evolutionary history. They are found, for example, in rodents, non-human primates, and bats. Primitive head direction cellular-nets are also found in insects. It appears that head direction cells are a key component that accompanied the evolution of the head. Therefore, any creature with a head-like appendage could be expected to have evolved head direction neural nets and to have a sense of moving in a constant direction.

Norwegian scientists Edvard and May-Britt Moser added to the list of hippocampal anatomy when they discovered *grid cells* in the hippocampus, a kind of nautical chart composed of longitude and latitude lines. This cellular-chart allows for route travel, based on domain maps. Using a sports analogy again, if a football player is injured and has to go to the locker room, dead reckoning is not of much value. There is a shortest possible, least time-consuming, and exact route leading from the field to the locker room. This route is embedded within a total map of the area. *Grid cells* are associated with mapping spatial domains.

Place cells become active in the brain when we recognize familiar places, while grid cells provide us with an absolute reference system, so we can determine exactly where we are on a map. The online science network called *Phys Org* discussed a robotic navigation system designed to mimic hippocampal functioning:

> The human brain uses grid cells, which provide a virtual reference frame for spatial awareness to handle this type of relative navigation. Each time we move through and pass one of the virtual grid points that the brain has set up, the respective grid cell becomes active, and we know our relative movement in relation to those coordinates. By using both place and grid cells for navigation, humans and animals [and robots] are able to accurately move through the environment.

~ *"A robot computer algorithm that copies the navigation functionality of humans and animals," Phys Org, 2015.*

Notice, in the above quote, that the description of the "grid cell neural network" parallels our understanding of the allocentric attention system. The description of the "place cell neural network" parallels our understanding of the egocentric attention system.

Another important discovery, by John O'Keefe and Neil Burgess, was *border cells*. *Border cells* (also called *boundary cells*) use information from solid objects to discern how far we are from different surfaces. One of my first blind students surprised me when he walked down a hallway exactly in the middle. When I asked him how he did this, he said that he balanced the sound from each wall and this guided his navigation. If he got too close to one wall relative to the other, the sound increased on the near side and faded on the far side. At a high level of processing, in the hippocampus, *border cells* were firing as my student balanced the sound between the two walls.

So we have *place cells*, *grid cells*, *head-direction cells*, and *border cells* located in or near the hippocampus. Obviously, all these cells work in concert to create space and to enable navigation. Also, these cellular systems are not isolated modular areas; they are part of whole-brain networks.

If the hippocampus is damaged, as often happens with strokes or diseases of aging, the result is some degree of navigational disability. Unfortunately, quite often, the hippocampus is severely damaged by blood loss and, consequently, the ability to navigate is lost completely. The hippocampus can also be affected by developmental or congenital anomalies; the most dramatic and telling of which is DTD—developmental topographic disorder.

The press that followed the Nobel Prize awards to these remarkable scientists stated that this explained why people with Alzheimer's disease lost their ability to navigate. What the press did not say was that the hippocampus is unfortunately not served by a rich network of blood vessels; it has only a single vein and a single artery feeding the brain tissue. When there is any kind of ischemic insult to the brain, the hippocampus is extremely vulnerable. This is why navigational impairments often accompany strokes and diseases of aging.

Brain processing is often explained as a river: tributaries flow from the end organs—the retinas of the eyes, for example—toward the occipital cortex at the back of the head. From there, sensory streams gather together and an ever greater river of information flows forward, through the parietal, temporal, and frontal lobes of the brain, to the prefrontal cortex behind the forehead, and then, according to Jeff Hawkins, processing is probably "consolidated" at the hippocampus. As the river swells, the brain does ever more complex higher-order processing. Starting with fragments of data, the cortex assembles patterns. The egocentric processing stream and the allocentric processing stream split off early, making two parallel rivers that flow to a common processing ocean in the prefrontal cortex and hippocampus.

I have oversimplified and speculated as I crafted the above relationship between hippocampal research findings and the allocentric and egocentric processing streams. The hippocampus is a high-level integration system and not the whole answer to navigation. It is not the brain's GPS system, only a kind of mastermind overseeing both subcortical and high-level processing that leads to sophisticated awareness of location. What is obvious from the anatomical and physiological studies of the neocortex is that there are two processing streams, allocentric and egocentric. These two neurological networks cooperate, but they remain mutually exclusive.

J. J. Gibson

Professor James Jerome Gibson[10] observed that human beings are always moving—we are animated creatures, not statues. Therefore, according to Gibson, the research lab is not the right place to figure out how we work. We are land-based, earth-bound creatures who move within a repetitive, invariant environment. We are created to survive in this land-locked existence. Gibson felt that human beings could only be understood in non-clinical settings. Isolated lab experiments were always suspect for Gibson:

> Gibson's' concern with the characteristics of the information responsible for perception led him to emphasize the fact that real life perception involves not a stationary observer fixating on a small light in a laboratory, but, rather, an active observer who is constantly moving his or her eyes, head and body relative to the environment. ~ *"The Ecology of J. J. Gibson's Perception," Leonardo, E. Bruce Goldstein, 1981.*

When we observe human beings in the real world, we quickly observe that they react effortlessly and immediately to the environment. There seems to be a fast and direct "perceptual" system that operates independently from the rational mind. In his first book *The Perception of the Visual World* (1950), Gibson explained that we come into life equipped to handle the environment of our species. We can move, make do, and get by, through directly perceiving and then adjusting to our reliable surroundings. There is no need for complex cognitive processing. Gibson challenged the notion that "perception" required high-end processing:

> James Gibson put forward a radical approach to perception that was largely ignored at the time. The dominant approach until 30 years ago was that the central function of visual perception is to allow us to identify or recognize objects in the world around us. This often involves extensive cognitive processing, including relating information extracted from the visual environment to our stored knowledge of objects. Gibson argued that this approach is of limited relevance to visual perception in the real world. Vision developed during evolution to allow our ancestors to respond rapidly to the environment. ~ *Cognitive Psychology, A Student's Handbook, Michael Eysenck, 2013.*

It is plain to me that what Gibson "discovered" was the allocentric background mind at work. He cannot be clearly understood, in my opinion, unless we realize that the word "perception" has two meanings. The egocentric system does, indeed, need to heavily process information to do its job, which is to extract meaning from the objects in the environment. Analysis and long-term memory storage are required for egocentric processing. However, the allocentric system is a whole-body, all-at-once, deeply hereditary mechanism that reacts quickly to change in the surroundings. Gibson was an enigma to the scientists of his time because most of them had no concept of our dual nature; they saw only the egocentric system and believed this to be the essence of the brain and mind. What Gibson did, in my opinion, was to unearth evidence of allocentric processing, of a second mind.

Because of his unique perspective, Gibson showed us why navigation is automatic, reliable, and immediate. Navigation is primarily governed by allocentric processing, and this ancient whole-body system comes endowed with a hard-wired knowledge of the environment. Allocentric "perception" allows for *immediate knowing* because of our genetic heritage; learning is not needed. Furthermore, understanding and knowing how to use the important invariants in our environment is hard-wired. I remember watching a documentary on wild turkeys and being amazed that they came out of the egg knowing what was good to eat and what wasn't. They avoided snakes, but ate grasshoppers. No parent was around to teach the young birds; they were born with a sense of meaning. So are we, according to Gibson. His word for describing this innate knowing was "affordances:"

> Thus, according to him, perception of an object involves not only perception of the visual characteristics of that object, but also involves perception of what the object affords. And this perception of the object's affordance, like the perception of the object's visual characteristics, occurs directly. ~ *"The Ecology of J. J. Gibson's Perception," Leonardo, E. Bruce Goldstein, 1981.*

I think what Gibson is getting at is that the allocentric mind has a different kind of perception. Actually, the word perception is best left as an egocentric concept. The sister to this in the allocentric mind Gibson called "affordances."

Perception requires processing and categorization. Affordances are immediate kinds of knowing. There is no need to heavily process at the allocentric level—just a need to quickly and appropriately respond in a way that suits the survival of the organism.

In his final book *The Ecological Approach to Visual Perception* (1979), Gibson discussed these affordances of the environment. Affordances are opportunities for action, choices the organism could make depending on circumstances. Affordances don't cause anything; they just make certain behaviors possible and optional.

In his second book *The Senses Considered as Perceptual Systems* (1966), Gibson discussed the idea that invariants in the environment were the origin of affordances. Gibson's main area of research was vision, so his theory of invariance was optically based. He said that invariants were properties of the optical array that remain constant even though other aspects vary. Gibson defined an invariant as "non-change that persists during change." That sounds a lot like the definition of the allocentric "background."

Gibson also studied what was happening at the level of the retina. How, he wondered, could we maintain a reliable, stable visual scene when the image on the retina was constantly changing? He reasoned that there must be invariant patterns that hold true no matter what images are flying around the retina:

> . . . Gibson notes that although an observer's movement may cause the image on the retina to be in constant flux, there is information on the retina that remains constant. ~ *"The Ecology of J. J. Gibson's Perception," Leonardo, E. Bruce Goldstein, 1981.*

In other words, there are invariant patterns that remain stable on the retina no matter how we move our body, head, or eyes. There is an invariant background from which objects manifest. Essentially, what Gibson found was that our allocentric mind could build a steady state background.

One of Gibson's most important contributions was the awareness of optical flow on the retina. Think about driving your car down the expressway. The direction of travel is a point of stillness ahead on the horizon. However, if you engage your peripheral vision while driving, you will perceive that the scenery is flowing—moving quite fast on the peripheral retinas. This contrast between a distant still-point and the optical flow that results as we move is an environmental invariant. Optic flow is always present when we move forward; it is part of what defines a steady background state.

Knowing that we either attend egocentrically or allocentrically, and knowing that the senses must always fire synchronously, we can conclude with Gibson that "peripheral flow" is an invariant *that all the senses contribute to*. When we look at experimental findings we do, indeed, find perceptual flow in the auditory system, the vestibular system, the proprioceptive system, and the visual, olfactory, and tactual systems.

Gibson was a pioneer who only became revered after his death. Ecological psychology,[11] the profession that his ideas spawned, can be seen as the science of allocentric perception. In light of dual-process theory, ecological psychology and environmental psychology[12] become central disciplines in the effort to comprehend human consciousness.

Summary Observations about Attention

I based this entire book on the duality of attention: allocentric awareness and egocentric attention. Of course, we are embodied creatures and attention and awareness cannot be dissected out from perception, intention, memory, and so on. With that caveat in mind, we can cautiously make some summary observations.

Professor William James, the Father of American Psychology, strongly believed, as I do, that attention was at the root of consciousness:

Consciousness: A New Slant on an Old Conundrum

Millions of items of the outward order are present to my senses which never properly enter into my experience. Why? Because they have no interest for me. My experience is what I agree to attend to. Only those items which I notice shape my mind—without selective interest, experience is an utter chaos ~ *Principles of Psychology, William James, 1950.*

Here is one of William James' most famous statements about attention:

The faculty of voluntarily bringing back a wandering attention, over and over again, is the very root of judgment, character, and will. No one is compos sui [master of the self] if he have it not. An education which should improve this faculty would be the education par excellence. ~ *William James: Writings 1878-1899, Ed. Gerald E. Myers, 1992.*

William James also believed that we have two ways to pay attention; these are analogous, in my opinion, with egocentric and allocentric processing:

William James (1890) distinguished between "active" and "passive" modes of attention. Attention is active when controlled in a top-down way by the individual's goals or expectations [egocentric attention]. In contrast, [allocentric awareness] is passive when controlled in a bottom-up way by external stimuli. ~ *Cognitive Psychology: A Student Handbook, Michael Eysenck, 2013.*

Dr. Evan Thompson, professor of philosophy at the University of British Columbia, and a noted authority on Buddhist philosophy, agrees with James:

If what we experience is what we attend to, and what we attend to is reality for us, then what we ignore or fail to notice will have no reality for us, even though it may affect us in all sorts of ways. ~ *Waking, Dreaming, Being, Evan Thompson, 2014.*

Attention is a fundamental activity that sets the stage for consciousness and cognition. The idea that consciousness evolved from a process like attention has historical precedence. Here is a quote, written in 1976, by Professor Joseph Neisser at Grinnell College in Iowa:

A better conception of consciousness, which has been suggested many times in the history of psychology, would recognize it as an aspect of activity rather than as an independently definable mechanism. - *Cognition and Reality, Joseph Neisser, 1976.*

Consciousness is indeed an aspect of activity; it is the result of two ways to pay attention. Attention's primary place in this architecture of processing is to structure our internal world so that the thoughts, emotions, or motivations that are most relevant to our goals will get preferential processing. Almost all higher-order cognitive functions, such as memory and language depend on attention. Without attention all other cognitive functions would be impaired or dysfunctional.

Blind individuals, especially very astute people, observe what happens to their attention after they become blind. One of the most remarkable blind individuals I studied, as an orientation and mobility specialist, was Jacques Lusseyran—the first Knight for the Blind in my book by that title. Lusseyran writes as if most human beings don't even have a concept of the term "attention:"

Because of my blindness, I had developed a new faculty. Strictly speaking, all men have it, but almost all forget to use it. That faculty is attention. In order to live without eyes it is necessary to be very attentive, to remain hour after hour in a state of wakefulness, of receptiveness, and activity. Indeed, attention is not simply a virtue of intelligence or the result of education, and something one can easily do without. It is a state of being. It is a state without which we shall never be able to perfect ourselves. In its truest sense it is the listening post of the universe. - *Against the Pollution of the I; Selected Writings of Jacques Lusseyran, 1999.*

Here, Lusseyran sounds like a Buddhist monk explaining how to reach nirvana. Most of the time, our minds freewheel, flying from one point of attention to the next. What expert meditators know is that we will not "awaken" unless we control attention. In a chapter of the book

Buddhism and Science called "Imaging: Embodiment, Phenomenology, and Transformation," authors Francisco Varela and Natalie Depraz state the following:

> The very basis of the training of mind is, first and foremost, grounded on cultivating the stability of attention. ~ *"Imaging, Embodiment, Phenomenology, and Transformation," Buddhism and Science, Breaking New Ground, Francisco Varela and Natalie Depraz, Ed. by B. Alan Wallace, 2003.*

In Dr. Wayne Wu's book called *Attention*, he states:

> . . . attention is not just pervasive but plays a fundamental role in all . . . aspects of mind. Accordingly, philosophical accounts . . . will not be adequate unless they come to grips with attention. ~ *Attention, Wayne Wu, 2014.*

The mystic G. I. Gurdjieff put it this way:

> . . . mobilization of attention is the first step toward the possibility of self-remembering [waking up, being mindful]. Without a different [kind of] attention, we are obliged to be automatic. With an attention that is voluntarily directed, we go toward consciousness.

> . . . for transformation to take place, there must be a total attention, that is, an attention coming from all the parts of me. In order for a certain blending to occur, my thinking, my feeling, and my sensation must be together. ~ *The Reality of Being; the Fourth Way of Gurdjieff, Jeanne de Salzmann, 2011.*

Alain Berthoz in his book *The Brain's Sense of Movement* (2000), defines perception as *active exploration*. Perception, he tells us, is not a passive process; it is an engagement with an environment—an active search for that which has meaning, toward the good and away from the bad. From his perspective, he has included attention within the concept of perception saying in effect that they are inseparable, one process, not two. He is talking about egocentric processing, which is, indeed, a search for meaning and relevance.

However, in this context, he is ignoring the allocentric mind, which is passive and does not search for meaning. When speaking of the egocentric mind, I agree that perception, attention, and memory are embodied and synchronous. I would add that allocentrically, awareness is also an all-body receptivity and is an immediate reaction system—like egocentric processing, allocentric processing is embodied and synchronous.

Dr. Berthoz also defines perception as *simulated action*, not just the registering, identification, and early processing of sensory input. Here, he has included motor planning in his definition of perception, saying in effect that as a whole organism, we cannot separate out the response from the sensation. Therefore, using Dr. Berthoz's viewpoint, perception sets the stage for purposeful movement. We are able to make intelligent choices about how to move based on what the moment reveals. Motor planning is both egocentric and allocentric. Hardwired motor responses are more the responsibility of allocentric processing. If consideration, strategy, and future scenarios come into play, then the process is egocentric.

At the biological level, attention is a complex set of neural activities incorporating different brain maps. Like so many other complex concepts, when we train our microscope on the subject of attention it quickly breaks down into many subcategories. The different stages or varieties of attention have different sets of neural nets dedicated to that particular function. There is a neural network for the following kinds of attention: selective attention, alerting attention, orienting attention, open-awareness attention, and embedded attention—to name just a few. These categories, according to dual-process theory, fall into either the allocentric or egocentric processing streams.

In the previous chapter, I pointed out that researchers had used the same terminology (egocentric and allocentric) that I use, but in opposite ways from my understanding. A similar confusion arises when researchers use the words awareness and attention. Here, for example, is the understanding of professor of Psychology and Neuroscience at Princeton University Dr. Michael Graziano:

For decades scientists used the terms "awareness" and "attention" more or less interchangeably, as though both referred to what happens when your mind takes hold of something . . . We now know that we need a better theory of what they are and how they relate to each other.

One such theory is the Attention Schema Theory (AST), first proposed by my lab in 2011. In that theory, attention and awareness have a precise relationship to each other. Attention is a data-handling trick. It's the brain's way of focusing resources on some signals, boosting them and processing them at the expense of other signals. It's a mechanistic process. Awareness is different. It's more like the brain's explicit knowledge about what it's doing. The brain doesn't have information about the microscopic details of attention, the neurons and the electrochemical signals, but it can give you a general account. It can say, "Yeah, I've got hold of that dot. I'm processing it. I have a kind of mental possession of it." Awareness is the brain's schematic description of attention.

In AST, attention is a constant process like a factory stamping out parts, and awareness is a constantly updated account of what the factory is doing, for quality control purposes. If you want to control something carefully, monitor it. ~ *"Can Your Brain See Things That You Don't?" The Atlantic, Michael Graziano, 2016.*

Dr. Graziano's Attention Schema Theory is close to what I am proposing in this book. I can see that navigation gave rise to one system, the egocentric, that "focuses resources on some signals, boosting them and processing them at the expense of other signals," and a second system, the allocentric, that is holistic. However, disagreeing somewhat with AST theory, the allocentric system deals with implicit memory, while the egocentric handles explicit memory.

Attention and awareness came before language and socialization. In the final lecture of John Long's The Great Courses series on Robotics, Dr.

Long makes a simple statement that has great import: "Attention is communication." Two of the most compelling reasons given for the evolution of consciousness are that human beings are social and that human beings have language. From the perspective of dual-process theory, what Dr. Long has expressed in the simple phrase *attention is communication* is that the evolution of attention resulted first in our becoming conscious, and then, as consciousness evolved, it led to culture and also to language.

Relationships are based on how much egocentric attention we afford each other. The depth and quality of a relationship depend on the depth and quality of our attention. Our deepest relationships are based on trust and compassion, which arise from the allocentric mind. Without "paying attention," without "sharing awareness," nothing social happens and language is not addressed to anyone.

Egocentric attention and allocentric awareness are the foundational abilities that eventually led to rich cultures and complex languages. The act of egocentric attention is the act of manifestation. We manifest the world as we gaze in certain directions, and we are "blind" to what we don't pay attention to. In a way, *we manifest each other* with attention. We also manifest our own bodies and personalities as we pay attention to our needs and desires.

Notes

(1) **Brain anatomy and physiology:** I am making the assumption that the reader has a fundamental understanding of the brain and the body.

(2) **Daniel Kish's master's thesis:** This treatise on echolocation is available online at: Kish, Daniel (1995), "Evaluation of an Echo-Mobility Program for Young Blind People," [Master's thesis], San Bernardino, CA: Department of Psychology, California State University. http://www.martin-naef.ch/index.php?menuid=39&reporeid=66

(3) The two major sensory modalities are unequally weighed in their contributions to the two minds. This seems to explain, in part, what we are doing when we meditate. Vision is primarily silent and holistic, while hearing is verbal and divided into temporal events. As we meditate, we need to know which mind we are trying to cultivate. When we meditate and try to be more allocentric, we are trying to be more aware of allocentric flow. When we are stuck in our egos, we are stuck in auditory mode—dominated by the internal voice or the need to communicate verbally. Knowing the different roles played by hearing and vision also sheds light on personality differences. Some of us are more visually oriented, while others are drawn to auditory processing.

(4) The complexity of skin. Reference in the text is to skin without hair follicles. However, skin *with* hair follicles is also highly complex. The evolution of hair cells goes all the way back to the cilia found in bacteria. These follicles are concerned with flow, with initiating movement, with stopping, and with maneuvering. As such, they are part of the allocentric processing system. When we refer to touch as one of the five senses, we are being naïve. The membrane that holds us together has 3 billion years of evolutionary magic—the sense of touch is a multiverse of complex sensory subsystems.

(5) Dermo-optical perception (DOP) has many names, including eyeless sight, finger vision, and cutaneous perception. It is often associated with parapsychology and deemed to be of dubious scientific merit. Like remote viewing or other extra-sensory abilities, DOP has statistical validity but not consistent reliability. Using the fingertips *to see* also depends on individual abilities, since some people seem to be more reliably accurate than others.

(6) Jacques Lusseyran is a remarkable French author who was blinded at the age of seven. I wrote a chapter about Lusseyran in my book *Knights for the Blind in the Battle Against Darkness* because his insights about vision and blindness are so unusual and powerful. As a teenager, he was a resistance fighter in Paris and was eventually arrested, tortured, and sent to the Buchenwald concentration camp. He was one of only a handful of survivors who were liberated when the American Army drove the Germans

out of Buchenwald. Lusseyran is important in the history of blind rehabilitation because he was a master at using passive echolocation to perceive his
environment. He probably also had a form of synesthesia that perceptually
turned sound images into visual images. Lusseyran claimed to be able to still
see after his blindness using what he called his inner vision.

(7) **Memory-of-memory.** The theories and research of Professor Allan
Paivio in the 1960s concluded that memory was the key to consciousness—
rather than attention. The truth is that memory and attention cannot be
separated out as distinct systems. However, as I thought more deeply about
proprioception, I came to see that layers of memory were indeed a key to
consciousness. Proprioception records sequences of muscular activity. These
sequences enable purposeful movements, behaviors.

(8) **I told my students that they could move their awareness** from
their bodies outward into space. However, my students were often developmentally delayed. They were capable of experiencing concrete space—most
were at Piaget's concrete operational stage of cognitive development—but
they were not necessarily able to grasp concepts and abstractions. I created
a community-based education curriculum for these students so that they
could learn through experience.

(9) **Differentiate cortex from neocortex.** Of course the details are always more complex than the generalization. In humans, almost all cerebral
cortex is neocortex, so in everyday language cortex and neocortex are synonyms. The difference is that the neocortex has 6 horizontal layers of differing neuronal structure, while the rest of the cortex (the allocortex) has just
three layers.

The hippocampus, part of the allocortex, has three layers and is much
older in evolution than the neocortex. In a way, the hippocampus is the "beginning" of the neocortex, as if the neocortex grew out of the hippocampus
as the trunk of the neocortical tree.

(10) **Professor James Jerome Gibson** (1904 to 1979) is one of the
most important psychologists in modern times to study visual perception.

Using my terminology, Gibson found the allocentric importance of the visual system. He founded the field of ecological psychology, which holds that the [allocentric] mind directly perceives environmental stimuli without the need for cognitive oversight or additional processing. Vision, according to Gibson, is more complex and immediate than the egocentric processing that dominates our understanding of the world. Not only did Gibson argue that an organism must be studied in a natural setting, where the environment is filled with three-dimensional invariants, he also insisted that we had to account for an organism's constant movement—movement was the foundation for behavior. Incessant activity is fundamental and cannot be ignored when we do experiments. Gibson was concerned with information processing, with meaning, rather than with raw sensations. He held that the invariants in an environment came to the human sensory system in the form of patterns that were then useful for survival.

Gibson's wife **Eleanor Jack Gibson** (1910 to 2002) became famous and honored in her own right. She was an expert in the development of reading in children. She also did pioneering research—the visual cliff experiments, for example—on perceptual development in young children. With her husband James, she helped to define and develop ecological theory. Eleanor Gibson received the National Medal of Science award in 1992. This is the highest scientific honor in the United States.

(11) Ecological psychology and environmental psychology. These are similar fields of study that were founded at about the same time in history—mid 1900s. Ecological psychology mainly evolved from the research and insights of J. J. Gibson and his team. However, similar work was being carried out at the same time at the University of Kansas by Roger G. Barker, Herb Wright, and their associates. Barker's team called their approach "environmental psychology." Both schools did research in real world settings rather than in the artificial circumstances of the laboratory. Barker's observations suggested that the environment directly affected behavior. The mind was altered by the circumstances afforded by the surroundings.

Nine

The eye is meant to see things;
the soul is here for its own joy.

~ "SOMEONE DIGGING IN THE GROUND,"
A YEAR WITH RUMI, 2006.

The Soul is Here for its Own Joy

"Time to celebrate, Surge. I am writing my final chapter. A grand dinner fit for a king is in order."

"Okay, well, let's see what we have for King Dutch at his last meal in the Hotel California."

"Something Rumi would approve of."

"Of course. You want a Sufi meal, something mysterious and magical."

"Whatever. Just make sure it tastes heavenly and is fit for royalty."

"Ah, yes, we have just the thing for the likes of you."

"Serve it on a silver platter, please."

"Just like Rumi?"

"No! Not like Rumi. Like King Tut. Don't be serving me brown rice in a wooden bowl. There weren't a lot of classy restaurants in Rumi's village."

"Like I said, we have just the meal for you and your kind."

"What do you mean *my kind*?

"People with your kind of disability."

"I don't have a disability!"

"Sure you do. You are deaf, blind, and emotionally impaired. Aren't you the guy who invented allocentric and egocentric disabilities? We should build a monument to your flaws, King Dutch."

"My heart was in the right place. I tried to make a contribution."

"Here's what I suggest for dinner. Sauté your own words in buttery compliments, then serve them on a bed of minced but well-meaning abstractions. Sprinkle with kind metaphors and juicy similes and then add the following spices: sweat from the foreheads of *real* authors and saliva from the mouths of literary giants. Eat slowly so you get a taste of what readers experience as they chew on your leathery logic."

"What's for dessert?"

"Humble Pie for your ego, and Joy-Filled Cake for your soul."

"I'll have both."

"Yes, you will."

Where Do We Go From Here?

In the last eight chapters I put forward the evidence and the logic for our dual cognition. I will now leave that evidence for you to ponder. In this chapter my concern is with you, your profession, and your body of knowledge. How does dual-process theory fit with your expertise? Can the differentiation between an allocentric mind and an egocentric mind shape the evolution of your own profession? How can the insights in this book be used in the service of the people you care for and nurture? I pass the baton to you and I wish you well.

In this chapter, I focus on the implications of dual-process theory, especially regarding blindness in special education. Blindness and navigation are my areas of expertise; I can speak with some authority on these subjects. However, during the 33 years that I taught travel skills to blind students, I also taught navigationally disabled children who had other special education labels. Over my career, I worked with deaf and deaf-blind students, with children labeled autistic or emotionally impaired, and with students who had various degrees of physical, emotional, cognitive, auditory, and visual impairments. I could discuss dual-process theory as it relates to any of these special education categories. For example, deafness and autism are two areas where dual-process theory has much to offer. However, I will stick to my field and leave it to other experts to apply dual-process theory to their own professional domains. I cannot presume to understand the wealth of knowledge and experience you bring to the dialogue.

The discussion below about blindness and the implications of dual-process theory is an example of how one professional domain—orientation and mobility—is impacted by the understanding that we have two minds and two kinds of consciousness. Your task, if your expertise is something other than blindness or visual impairment, is to use the example to consider parallels to your own area of expertise.

Blindness

Consider a few important facts as we begin this discussion about blindness, visual impairment, and navigational disability:

- Blindness is rare. Most people labeled "blind" have some degree of vision—they are actually visually impaired. Therefore, the problems that most "blind people" face are complex and variable visual challenges. Even people who are totally blind face auditory-driven image-processing problems that could fall under the heading of "visual" impairment.

- Vision is so complex, and people are so unique, we must be very cautious when we generalize. It is best to consider each person we encounter as highly unique. Blindness and vision impairment are labels that need to be cautiously applied, and only *after* individual uniqueness is acknowledged.

- Children in special education have a higher than normal incidence of vision problems. If we define vision impairment broadly to include all the many varieties of visual anomalies, the number of people in need of orientation and mobility services—especially children in special education programs—is much larger than is usually understood.

- The rule in special education is that most labels are misleading. For example, if we assign the label "visually impaired" to a student, we might miss autistic tendencies, or hearing loss, or cognitive challenges. We might also miss that a particular child is mentally gifted, or exceptionally creative, or has interpersonal skills that make the individual a natural leader or peacekeeper. Assigning a label to a human being can do serious psychological damage to that person.

- Human beings go through physical changes throughout a lifespan. Developmental psychologists have studied these neurophysiological stages and have written extensively about the evolution of human cognition from birth to death. The developmental level of any student is a very significant variable that cannot be ignored. Children do not stay at a developmental level—children constantly evolve, especially as they are nurtured and educated.

- Terminology in special education has always been problematic. For example, when we use the terms "disability," "handicap," "disorder," "anomaly," and "impairment" (for example) we bring along emotional baggage and unintended associations. In other words, careful or not, our language is loaded. I have used the word "disability" in this book, but I am not happy having to use the term, especially because I have so many friends whose many gifts far overshadow their visual challenges.

If your job involves service to blind or visually impaired individuals, you probably agree with the above observations—your experience has taught you the truth of these statements. However, you also have a personal background (education and experience) that makes your viewpoint unique. You might also be blind or visually impaired yourself, in which case you have personal observations to bring to the discussion—you understand in a fundamental way the issues that I am raising here.

Orientation and Mobility

In the dedication to this book, I wrote:

> To the profession of orientation and mobility—in the hope that what is discussed here will become a neurophysiological foundation for the discipline.

I knew when I began to write that dual-process theory—applied to human navigation—was the missing neurophysiological foundation around which the profession of orientation and mobility could develop a richer clinical and theoretical base. However, knowing how the profession might adopt a philosophy—even making the case in this book—is only the beginning. There is a long road ahead convincing the academic and professional communities that this new perspective has long-term value. To see why the profession of orientation and mobility needs a neurological foundation, we need to understand the historical development of the field.

The profession of orientation and mobility did not start with a neurophysiological foundation like other therapeutic professions did. For example, the profession of occupational therapy developed around the research findings and philosophy of Dr. A. Jean Ayres, a psychologist. Dr. Ayres understood that many children in special education suffered from sensory integration problems. From this philosophical and research-based foundation, the field of occupational therapy was able to build a clinical and theoretical practice. In comparison, the profession of orientation and mobility never had such a philosophical foundation; instead, the profession was built around a set of procedures for using a long cane.

Orientation and mobility is a profession that was born after the Second World War. Many service men were blinded during combat, and when they came home they found themselves in a culture ill-equipped to handle so many blind young males. Blinded veterans coming back from battlefields in Europe and Japan were housed at Valley Forge Hospital in Phoenixville, Pennsylvania. It was at Valley Forge where the profession of orientation and mobility began.

In 1944, army sergeant Richard Hoover (1915-1986) assumed a leadership position at Valley Forge. Hoover and his staff developed techniques to help blinded veterans become self-sufficient. Most notably, Hoover developed a cane technique—later called the Hoover technique or the long-cane technique. This simple innovation allowed blind individuals to walk safely through any environment by swinging a long thin cane back and forth in front of them as they moved about. Consequently, the profession of orientation and mobility, the skill set, curriculum, and philosophy, were all designed around Hoover's long-cane technique.

Before Hoover and his staff developed the long-cane technique, veterans had used a short, fat support cane. The traditional short wooden cane was eventually replaced with a lightweight long slender cane cut to a prescription length that matched the specific height of the blinded veteran. The user moved the cane from side to side, with the tip touching the ground ahead of where the cane user was walking. In this manner, the cane hit the ground where the foot was about to step. In effect, the cane was whispering: "It's safe, it's safe, it's safe," as the soldier moved forward.

Where Do We Go From Here?

Eventually, the long-cane technique was adopted by blindness centers across the country. University training programs were established, and eventually certification standards and a code of ethics were established—the profession of orientation and mobility was born. By 1990, more than 2000 mobility instructors would graduate from 15 university training programs. The core of these masters-level university training programs was Hoover's long-cane technique. As I write this in 2017, the emphasis on cane training is still the foundational core of these university programs. The identity of orientation and mobility instructors is still wrapped around the long cane and the Hoover technique.

The field of orientation and mobility remained focused on adults until the early 1960's when the first mobility specialists appeared in public schools. These school-based mobility specialists quickly learned that the Valley Forge curriculum was not adequate for blind children. An adult can learn cane techniques over a few weeks. However, it takes children their entire childhood to grasp the concepts and develop the fine motor skills necessary to duplicate what blind veterans quickly learn. In addition, children in special education settings often have multiple impairments and health problems—it is not easy to apply a curriculum based on adult cane skills to young children.

Children also go through developmental stages. They crawl, then walk, then run. They develop their large skeletal muscles first before they are able to use fine motor control. The brain rapidly and steady evolves over the first 18 years of life—with each passing year, the child is more capable. The skills necessary for efficient navigation slowly evolve but are not fully developed until a student is in high school. The original methods used to teach orientation and mobility to fully developed adults did not consider the rapidly changing abilities of blind children.

When I graduated from Western Michigan University in 1980, I was well-equipped to teach blind adults to use a long cane, to use a sighted guide, and to cross streets. I was also ready with a standard approach to teach blind adults how to navigate through space. However, I soon discovered—as a special education teacher in a public school system—that children did not have the neuromuscular ability to use "proper cane skills" until they were in their late teens. They didn't begin to cross even the simplest of streets until age five (often much later). And most critically of all, they evolved the cognitive skills necessary to understand space very slowly. A serious question arose early in my career: Just what is it that I should be doing with young blind students if they aren't ready to learn cane skills?

Veterans, blinded by war wounds, had been able to see throughout their childhood. They knew what vision was and they had visual memories. However, blind children in public schools are often born without sight or have only a few years of normal vision before going blind. It is far easier to learn orientation skills with a 20-year base of knowledge about visual spatial layouts. In contrast, blind children are not born with a conceptual base of understanding about the world. They must systematically study environments. They have to move through space to develop mental schemas about the world. Teaching a blind child to be oriented in space, and to stay oriented while moving, is the primary challenge for the public school mobility specialist.

It is also true, as I said above, that many blind students in public schools have multiple impairments. A blind child with no hearing loss, no orthopedic impairment, and no cognitive anomaly is a rare student. The mobility curriculum designed to teach blinded veterans does not take into account cognitive challenges, cerebral palsy, congenital hearing loss, or combinations of impairments.

Teaching cane travel to blind veterans could easily be accomplished using one instructor working with one student. In the public schools this one-on-one approach, although useful, is not always the best choice. While the one-on-one practice is still stubbornly adhered to by many mobility specialists who teach children, I found that young children benefit from role models. They also need and respond to their peers. For these reasons, mobility in the public schools should often involve the teaching of groups of blind children. I successfully did group teaching in a program called *Community Travel* in which blind high school students played the role of teacher (with my guidance) for younger blind children.

Veterans could also receive hour-long lessons with specific objectives. This approach does not often work with children. Children need to play. The younger the child, the more the emphasis has to be on having fun, playing a game, anything except a dry lesson. Also, the younger the child, the shorter the attention span; an hour lesson is too long.

The history of the profession defines the mobility teacher as a professional trained to work with individuals who have vision impairments. This definition is appropriate in an adult rehabilitation setting, but it is not broad enough for public education. The blind rehabilitation model fails to address the travel problems of many students in special education besides those who are blind or visually impaired. In Saginaw, Michigan, at the Millet Learning Center, I redefined orientation and mobility as *a profession that addresses navigational disabilities*, whether these disabilities are caused by vision impairment, damage to navigational centers in the brain, lack of travel experience, or because of oculomotor, proprioceptive, or perceptual vision anomalies. The problem student's face in special education is not blindness or visual impairment; it is an inability to navigate accurately and efficiently. This redefinition of mobility teachers as *navigation specialists* is an important shift of emphasis with important consequences.

Valley Forge Hospital was a rehabilitation setting and a medical facility. Consequently, the first mobility specialists followed a medical model for servicing patients. This medical model was based on a doctor-patient relationship. The medical approach was—and still is—an extreme egocentric design. Patients, as passive recipients, are *taken care of* by doctors and nurses. The medical model does something *to the patient*. In contrast, an allocentric model does something *with a fellow human being*. The medical model is based on *ego and other*, doctor and patient. In comparison, the allocentric model is based on relationships, mutual respect, and interconnectivity.

The new neurophysiological foundation for the profession of orientation and mobility is a balance between the medical model—an egocentric approach—and the educational model, an allocentric approach. The mobility specialist becomes an expert in human navigation and serves all children in special education. Cane skills are still taught, but the cane becomes an extension of human perception and the student and teacher are in a nurturing relationship that typically lasts for at least 12 years.

Two Minds, Two Different Teaching Strategies

If dual-process theory is correct, then we have two minds: an egocentric mind and an allocentric mind. We understand our egocentric minds quite well—indeed, we think of this mind as our personality, our ego, who we really are, day in and day out. Consequently and understandably, we designed our educational strategies, our linear curriculums, and our classroom-based temporal-spatial environments to mirror this egocentric perspective. However, as I repeatedly emphasized throughout this book, the ego is blind to the allocentric mind. Consequently, teaching strategies based on an understanding of our allocentric mind are lacking, even denied in public education. It is no wonder then that the profession of orientation and mobility would have an egocentric-based approach to teaching navigation skills to children. However, in both regular and special education, this egocentric approach is a half-developed methodology with only limited success—ego-based strategies ignore one-half of our inherent essence, the allocentric mind.

I will use the differentiation between the *ego*—a product of the egocentric mind—and the *self*, a product of the allocentric mind. Here is a quick review and comparison of the characteristics of the ego and self:

- The ego reduces, the self integrates.
- The ego is dual, the self is non-dual.
- The ego uses logic, the self uses analogy.
- The ego is time-based, the self is spatial.
- The ego processes serially, the self processes in parallel.
- The ego processes *in regard to*, the self processes *in relation with*.
- The ego processes slower than the self.
- The ego freezes the world, the self flows through the world.
- The ego examines, the self experiences.
- The ego judges, the self appreciates and cooperates.
- The ego has beliefs, the self has faith.
- The ego is secular, the self is spiritual.
- The ego measures quantities, the self determines quality.
- The ego is aggressive and competitive; the self is non-aggressive and cooperative.
- The ego use verbal language, the self uses non-verbal communication.
- The ego uses internal dialogue, the self uses silent knowing.
- The ego trains, the self educates.
- The ego evolved intelligence, the self evolved wisdom.
- The ego reacts, the self responds.
- The ego concludes that all men are islands, the self concludes that no man is an island.
- The ego appears to be based in the cortex of the brain, the self is a whole-body phenomenon that is intimately interwoven with the environment.

The egocentric mind ignores, denies, or is ignorant of its twin, the allocentric mind. Since most of us feel that the egocentric mind is all there is to us, it is natural that our institutions would be designed around the egocentric mind. For example, in education we have primarily created an ego-directed system. Orientation and mobility naturally followed this egocentric perspective, so blind children in special education have been *trained* rather than *educated*:

- Blind students have been taught to isolate landmarks, rather than to integrate themselves within spatial gestalts.
- Blind students have been taught to perceive a freeze-frame world, rather than a world that flows and through which they flow.
- Blind students have been taught to dissect the environment, rather than blend and flow with the environment.
- Blind students have been taught that they are separate islands, egos separate from the domain they move through.
- Blind students have not been taught to "make friends" with their surroundings, to participate within the environment.
- Blind students have been lectured and guided, rather than allowed to freely explore and discover.
- Blind students have been taught to *define and examine* the environment, rather than to purely *experience* their surroundings.
- Blind students have been encouraged to be verbal, rather than to deeply listen to their surroundings.

Egocentric-based education has resulted in a cookbook-based, recipe-driven curriculum for teaching blind students. Because long-cane training is the core of the profession today, this linear, recipe-based approach is the main emphasis of the profession. Egocentric attention is based on one-thing-at-a-time perception. There is no learning that is "out of sequence." First principles, first steps, must be mastered before the student can progress to the next section of the curriculum. For example, when learning a route from the classroom to a bathroom, the student must learn a sequence of landmarks. The sequence of landmarks is then reversed on the return trip—which can be quite confusing for a congenitally blind child.

In egocentric mode, the human body is the center of a personal universe that is always invariant, regardless of location. This is the world that blind students easily understand. However, they do not have an allocentric visual gestalt that shows them at-a-glance how landmarks are spatially located relevant to each other. For this reason, I speak of blindness as an allocentric

403

disability. Blind students are at a severe disadvantage when it comes to forming spatial scenes in their mind. Teaching them using a strictly egocentric approach does not directly address this allocentric disability. Blind students do not have a visual sense of "flow," and egocentric training does not address this basic need to experience flow.

The egocentric mind is also language-bound—it is dialogue-dependent. Mobility instructors spend a lot of time explaining about the environment while their students remain immobile. Blind students get very good at explaining, but when it comes time to actually navigate, egocentric-based, language-bound instructions prove inadequate—students struggle to smoothly and efficiently move through space no matter how articulate they are.

Allocentric education is just the opposite of egocentric instruction. Allocentric education is a non-verbal approach that requires the student to be moving and exploring almost all the time—with minimal instructor interference. Flow is essential, and the speed of movement through environments is very important. The core of this strategy is to flow, explore, encounter, and investigate with minimal instructor feedback. Important allocentric concepts are self-discovery, fostering curiosity, self-expression, self-error correction, self-reflection, and relationship building—with the environment and with others. The environment is constantly communicating—there is enough information in this constant flood of information to read and respond to the surroundings.

The ideal approach, of course, is to combine the egocentric with the allocentric. Mobility specialists do combine these approaches in everyday practice, although the main emphasis is on egocentric approaches. The allocentric frame of reference is addressed by O&M specialists but without an understanding of the allocentric mind. Consequently, the emphasis of mobility training is almost all egocentric.

Starting in the early 2000s, the allocentric approach has been exemplified by the philosophy and practice of Daniel Kish and his colleagues at World Access for the Blind. Daniel calls his approach—which he finds to be almost diametrically opposite to the standard O&M curriculum—perceptual navigation.

Perceptual Navigation

Daniel Kish became blind before the age of two. All through his early years, he used echolocation to perceive the world. By the time he got to college, Daniel was a proficient and exceptional navigator because of his ability to use active echolocation. His ability to move about efficiently was in stark contrast to many other blind individuals—even those who had received extensive orientation and mobility training. Daniel eventually realized, as a young adult, that his ability to self-generate echoes to create sound-based images of the world was revolutionary. Essentially, what Daniel had discovered was an allocentric approach to educating blind individuals. As he developed his ideas and his principles, he came up against the egocentric medical model of training that was at the heart of the profession of orientation and mobility. Not only did he have to teach echolocation to others who were blind, he also had to debate his professional colleagues about best practices. Daniel and I became good friends and allies because we both understood this need to use allocentric strategies to redefine orientation and mobility.

Daniel Kish's philosophy, called *perceptual navigation,* is designed around allocentric education. Daniel speaks of *environmental literacy, proxy perceiving,* and *seeing with sound (active echolocation).* His *freedom formula* is a summary of his allocentric teaching philosophy. These conceptualizations—*environmental literacy, proxy perceiving, seeing with sound,* and the *freedom formula*—define perceptual navigation.

Perceptual navigation evolved over time and the concepts were gradually refined by Daniel as he lectured and dialogued about his ideas. His entire approach was originally centered on his expertise in echolocation. In 2016, Daniel and co-author Jo Hook published a comprehensive overview of Daniel's philosophy and practice. The book *Echolocation and FlashSonar* is available from the American Printing House for the Blind.

Daniel has a master's degree in developmental psychology; he is an expert in human perception. He also has a master's degree in blind rehabilitation with a specialty in orientation and mobility. As the founder and CEO of World Access for the Blind, Daniel has become a world authority on human echolocation. He has a robust career lecturing and teaching all across the globe and has presented at numerous TED X talks. His full-length TED talk has over a million hits. Daniel gave blind individuals access to environmental literacy in much the same way that Louis Braille gave blind people access to academic literacy.

Daniel's ideas are too revolutionary to go unexplained, so I will spend a few paragraphs outlining his approach. The most significant observation is that Daniel's strategy directly addresses the allocentric mind. For this reason, his methods are exceptionally important, not only for the profession of orientation and mobility, but also for blind individuals everywhere.

Daniel is exceptionally intelligent and his parents allowed his native intelligence to blossom—most importantly, they did not restrict his movements. They allowed him to explore his environment, and they maintained high expectations for him. In essence, his parents simply treated him like they would any other child. On many occasions, I have heard Daniel express warm appreciation to his parents for their support of his independence.

Early on, before Daniel was two years old, he began to make clicking sounds with his mouth. No one knew why he was doing this, but his parents and teachers did not discourage him. Daniel was a college student before a professor (Dr. David Warren) told Daniel that his clicking was a form of echolocation. This was the beginning of a long journey that has made Daniel a world authority on the use of echolocation for the blind.

I wrote about Daniel's philosophy in the book *Bugs, Blindness, and the Pursuit of Happiness*, so I won't go into great detail here. Daniel's website at World Access for the Blind is the richest resource in the world for those who want detailed information about the use of echolocation for blind navigation. I have personally witnessed Daniel teaching a wide range of people to echolocate, including a three-year old girl and an 80-year old man in Iceland. I have sincerely compared Daniel with Louis Braille and with Helen Keller. Few realize that he is a global treasure and that his insights will go down in history and be revered.

Seeing with Sound
FlashSonar

Blind people who echolocate emit sharp clicks from their mouths. The clicks generate sound waves called sonar. Sound travels outward from the mouth until the waves encounter objects. The objects reflect the sound back to the ears of the echolocator. Daniel calls this process *FlashSonar* because each click generates a flash of perception, a momentary glimpse of a scene. Imagine you are standing in a completely dark space. You have two pieces of flint which you crack together to get a flash of light. The flash only lasts a fraction of a second, but it is enough light to make out forms and the spaces between forms. When learned, FlashSonar is sufficiently powerful to allow for efficient navigation.

Dolphins and bats are the best known echolocators. They navigate at high speed with great accuracy using sound rather than light for perception. Images created in the brain using sound reflections may be equal to visual images that human beings use to generate visual scenes. Compared to bats and dolphins, human echolocators are not able to click fast enough to generate sharp images. Furthermore, the images that are created by human echolocators do not last; the images quickly fade unless followed by additional clicks.

The important concept to understand is that echolocation is both egocentric and allocentric depending on whether the echolocator is *paying attention* egocentrically or *being aware* allocentrically. When Daniel emits a click from his mouth, he is asking two questions of an object: Where are you? What are you? From an egocentric perspective, Daniel is perceptually exploring an object's size and characteristics. From an allocentric perspective, he is locating several objects in a scene and is determining their relative position to each other and to himself.

Daniel does not emit constant clicks. He just occasionally interrogates the environment to update his gestalt. A single click allows Daniel to generate an entire scene. Because he knows the invariants in any given environment, he is able to say what is in a scene. For example, after a single click emitted in a community neighborhood, he can report the location of houses, garages, fences, cars, bikes, bushes, and trees. This is sufficient information to navigate without running into everything.

Daniel gets different kinds of information from the environment depending upon the sharpness of the clicks and their direction. A soft click is more likely to generate a scene (allocentric awareness), while a sharp click is more likely to highlight an object in a scene (egocentric attention). Clicking upward might reveal the canopy of a tree. Clicking along the surface of an object in a street might highlight a small compact car or a larger truck. Clicking at a tree with slight head movements can establish the size of the tree trunk.

This ability for a blind individual *to actually see* using sound reflections is amazing. It is revolutionary. Unfortunately, in the field of orientation and mobility, echolocation has not attained primary status. Perhaps this is because echolocating—actually seeing—undermines a totally egocentric teaching practice. For example, there is no need to trail a wall or memorize landmarks in a sequence when you can just flow down a pathway looking left, right, up, and down as needed. Echolocation, at first glance, threatens the basic egocentric teaching philosophy of orientation and mobility. But it shouldn't have this effect as long as the field incorporates echolocation training into the core curriculum.

Perceptual navigation simply refers to an ability to *actually see using sound*. Blind people can perceive as they move (flow) just as visual people perceive as they move. Perceptual navigation training involves learning how to see using sound. Of course, there is more to the complicated act of navigation. For example, there is a necessary environmental literacy that must accompany seeing with sound.

Environmental Literacy

The ability to *read* an environment is critical for blind individuals. Those people who are environmentally literate know the different kinds of invariants that specific environments contain. Those who are environmentally illiterate are unable to surmise what reflected sounds offer them for accurate navigation. For example, there are basic environmental invariants that define a land mammal's existence. The ground is below, the sky is above. The solid, immovable objects of the world are embedded in the soil and rise up toward the sky. For example, trees are rooted to the earth and rise upward. Buildings are also connected to the earth and rise upward. Some things move across a scene. Cars move along streets, birds fly above, bees and flies buzz around. People walk, run, and ride bikes on sidewalks or streets. The world is actually quite predictable and redundant once you understand a few basic design ideas. But these invariants are not intuitive for a blind child.

Only experience will build a spatial database in the brain. *Telling* a blind child about the wind or about trees that grow from the ground upward, or the movement of the sun as the day changes, and so on, will not help the child build a spatial understanding of environments. Allocentric awareness will develop only by studying and experiencing the real world. For example, to understand trees in the real world, a blind child should climb them, feel the bark and the leaves, pick and eat the fruit of trees, measure the thickness of the trunk, experience the shade of the canopy of deciduous trees, on so on.

Egocentric education relies on explanations, but words do not build spatial gestalts. Only self-movement, exploration, and investigation *through space* can build spatial brain maps. Perception relies on flow, on purposeful movement. Blind children who don't move on their own do not develop adequate spatial brain maps. Therefore, not only will blind children fail to develop environmental literacy if they do not move on their own—if they don't explore on their own—they will also develop an allocentric disability. The allocentric system of the brain will fail to grow the neuronal connections necessary for competent navigation without practicing purposeful movements. The sooner spatial brain maps are laid down during development, the better the chance that an allocentric disability will be avoided.

Unfortunately, egocentric practices can actually cause an allocentric disability. There is great irony here: the more egocentric the approach, the less capable a blind child becomes. Daniel Kish could see this paradox—that egocentric practices can retard a blind child's mental (spatial) development. He called these unfortunate practices *proxy perceiving*.

Proxy Perceiving

From an egocentric point of view, there is no way that a blind person can navigate with the same fluidity as a sighted person. This egocentric perspective believes it is in the best interests of the blind to allow sighted people to assist them with their navigation. In other words, Daniel Kish would say that those who hold an egocentric perspective insist that sighted people must perceive for blind people. Therefore, from the egocentric viewpoint, a blind person needs what Daniel calls a *proxy perceiver* to navigate safely and efficiently.

Even though the shortcomings of the egocentric perspective are now becoming obvious, the field of orientation and mobility, and many well-meaning parents and relatives, still persist with perceiving for a blind person. Here are some ways that the sighted do this proxy perceiving:

- Sighted guide is the most notorious. The blind person takes the arm of a sighted person and is guided through space. The blind person need not pay any attention at all to the surroundings. There is no need for any kind of environmental literacy—so no literacy develops.
- A second approach Daniel calls "puppeting." Well-meaning sighted people push, pull, and turn the blind person. The body of the blind individual is manipulated through space. Again, there is no need for the blind person to monitor and use spatial information for self-movement.
- The third approach Daniel calls "remote control" perceiving. In this approach, a blind person follows the voice commands of a well-meaning sighted helper. This is just another form of proxy perceiving that inhibits self-exploration and the laying down of spatial maps in the brain.

- The fourth method Daniel calls "catering." This simply means that all purposeful movement is done by others. The blind person simply sits and waits for help. The coat is put on and taken off. Food appears and plates disappear. The bed is made in the morning and the clothes are always washed. There is no active participation of blind individuals within the spaces they occupy. There is no chance for environmental literacy to develop.

From an allocentric perspective, there is no self-sufficient flow going on when perceiving is handled by others. Consequently, the brain fails to develop the nerve networks, the brain maps, which are necessary for developing navigation skills. Children treated in such a way are forced into a dependence on others. In Daniel's words, children are forced to give up their freedom when others perceive for them. That is the reason he developed his *freedom formula*—to help students regain self-respect and self-sufficiency as they develop their perceptual capabilities.

The Freedom Formula

Navigation is what defines us at our most basic physiological level: we have two minds and two kinds of consciousness because we move from place to place with a purpose—we are not rooted to the earth for life. Navigation is the foundation for the development of intelligence and wisdom; health and happiness are also inherently linked to navigation. Therefore, allowing a blind child to be immobile is a crime against their humanity; restricting self-initiated movement is a condemnation of the essence of blind children that affects their intelligence, wisdom, health, and happiness.

According to Daniel, orientation and mobility specialists have a moral mandate to defend a blind individual's right to self-discovery—freedom of movement. Daniel believes the O&M profession is in a unique position to directly guide the development of a blind child's cognition and well-being. Of all the helping professions in rehabilitation and special education, orientation and mobility has a special mandate with a supremely relevant responsibility: to ensure that blind individuals are allowed the freedom to navigate independently wherever they are in the world.

Human beings are programmed dually. The mandate of the egocentric mind is to explore, to independently discover, and to self-fulfill. The ego is driven to question, to solve problems, to repair and rebuild, to remain constantly busy with duties and commitments. That is the job of the egocentric mind and it knows no other. The meaning of life for the ego involves going on a quest, a hero's journey to accomplish goals. The ego will wither and die unless it finds a purpose, a mission, something to do in life.

Contrary to the mandate of the ego, the allocentric mind is designed to simply experience life. The allocentric mind only exists within relationships. These relationships can be with friends, family, nature, or pets, but the primary hunger of the allocentric mind is *participation in the act of living*. The allocentric mind will wither and die if it becomes isolated, unable to participate, or unable to share in relationships. The allocentric mind must build and nurture relationships to ensure mental and physical health.

The allocentric mind is where love, peace, joy, wisdom, mindfulness, equanimity, and thankfulness arise and blossom. The allocentric mind has no tasks to accomplish, no duties, and no obligations to fulfill. It exists in the present moment and has no awareness of the past or future.

It is this intuitive understanding of human duality that led Daniel Kish to craft a Freedom Formula. Daniel could see the impact of well-meaning sighted people who unintentionally did harm to blind children by limiting their self-exploration and participation. As a blind person, Daniel felt negative judgments within Western culture during his own life. As a professional, Daniel also witnessed the devastating impact of global ignorance in cultures where blind people are isolated and restricted.

His Freedom Formula has seven principles to counter egocentric-based practices:

Consciousness: A New Slant on an Old Conundrum

1. To immediately stop guiding blind students and blind friends—of course, Daniel is referring to those blind individuals *who are capable* of independent self-discovery. The human perceptual system does not develop if it is not allowed to function and lay down memories. Likewise, executive functioning—the ability to make decisions—is hindered if it is not allowed to make spatial and temporal decisions. There are few opportunities to make decisions when everything, including perception, is done for you. Therefore, allow freedom of movement. Allow and encourage self-discovery.

2. Stop teaching blind children to trail the walls of rooms and hallways. Blind students should not go from landmark to landmark, groping their way through space. There is no need to maintain constant contact with surfaces if the perceptual system is being used effectively.

3. Use a *full-length cane* as a perceptual tool to provide information within walking-space—the area immediately in front of a blind person, the surface area that is about to be traversed. A full-length cane is longer than the field of orientation and mobility currently issues.

4. Blind individuals must engage the environment through the use of echolocation. Daniel says that blind students should be aggressive in their effort to understand the spaces they inhabit or move through. Echolocation is both *passive*—absorbing the constant information that is available in a space—and active, using tongue clicks to interrogate a specific space.

5. Use perceptual technologies to enhance the acuity of images. Perceptual technologies enhance echolocation, just as eyeglasses and hearing aids enhance vision and hearing. There is a long history to the development of perceptual technologies dating back many decades. We are now entering an exciting time when these technologies are maturing and becoming affordable.

6. Educate the fingers and toes to enable greater egocentric perception. Purposeful exploration of the environment is necessary

to build gestalts—to understand the big allocentric picture. However, blind students should also be encouraged to explore—to use "hand-and-eye" coordination (ego-generated images) as they skillfully examine the objects within each environment that they encounter.

7. Understand how the brain and nervous system work so that intelligent decisions can result. Blind students need to understand the distinction between the egocentric mind and the allocentric mind, and they need to develop expertise in both.

Navigational Disability

If it is true that we developed two minds and two kinds of consciousness because we navigate, as I have come to believe, then we can make at least five speculations regarding special education:

1. Any physical impact (insult) on the body—especially on the brain and nervous system—would affect either our ability to accomplish purposeful movements (tasks) or our ability to have experiences. We would expect and look for challenges to either or both of our minds as they attempt to process or develop after damage to the body. We would expect to find evidence that one or the other (or both) of our two minds is in some way affected by physical damage. The extent and complexity of the challenges would be in proportion to the extent and severity of the physical impact. We would also expect to witness adaptation in the two cognitive systems to compensate for the physical insult.

2. The categories of special education—for example, "blindness" or "autism"—would be challenged. Each special education category would have to be re-examined in light of dual-process theory. At the least, we would have to acknowledge two overall categories: allocentric disabilities, and egocentric disabilities. As an example, I applied

this line of thinking to blindness in the discussion above. Because the egocentric mind in humans is primarily served by hearing and speaking, egocentric processing is not as impacted by blindness as is allocentric processing. Therefore, I suggested that we might consider blindness to be an allocentric disability. Using the same logic, we might suggest, for example, that deafness is an egocentric disability. This is a heavy-duty generalization, so I offer the perspective with caution. We are complex individuals; generalizations must always come with personal caveats.

3. We allocate funding in special education based on labeling—that's a major reason we put students into categories—but it is a dubious practice. Labeling can be dangerous and is almost always unfair or limiting. I know of one example where labeling led to a widespread misleading conclusion. When we historically labeled individuals "visually impaired," we used a set of restrictive definitions for "vision impairment." When we did this restrictive labeling, we concluded that "vision impairment" must be a low incidence population and, therefore, does not need as much funding or professional support as autism, for example. This is absurd. The incidence of vision impairment in special education is massive if we take into consideration all the varieties of visual anomalies. If we broaden our understanding of vision, including *all* that can go wrong with visual processing, then just about every child in special education has a visual impairment.

4. Each of us is defined by our experiences—which enable wisdom, or sophistication—and by our ability to use our intellect to conceive of and manipulate patterns. If developmental psychologists are correct that human beings develop cognitive capabilities in successive stages—just as physical maturity progresses through stages of maturation—then we would expect that the egocentric mind and the allocentric mind each mature at a regular and predictable rate. This means that we each evolve through levels of consciousness that more or less parallel physical maturation. We can further surmise that people with lower levels of consciousness are blind to

the capabilities of people with higher levels of consciousness. This is a huge understanding that requires a whole book to explain. I took a long look at these developmental levels of consciousness in my book *The Confusion Caused by Being Your Own Twin*.

5. Technologies would impact the balance between the two minds. I address technology and levels of consciousness in my book *The Confusion Caused by Being Your Own Twin*. I will just make two observations here. First, downloading the human brain into a computer, as many scientists studying artificial intelligence advocate, does not preserve the mind. At best, it would preserve the language processing aspect of the egocentric mind. The brain requires a body to navigate and to have experiences—without the body it came in—an exact duplicate—the brain is useless. Robots are another story because they do offer a body to manipulate— we might preserve some aspect of ourselves if we transported our whole nervous system from our body to an artificial body. More likely, however, we will become cyborgs; there will be a blending of our total essence with molecular and quantum computers. Second, technology almost always addresses consciousness as if it is egocentric, totally ignoring the allocentric mind. What results when a technology is all egocentric is something totally devoid of empathy, love, peaceful intentions, joy, wisdom, mindfulness, equanimity, and appreciation. Consequently, by understanding mind to be just egocentric, we end up building dull-witted, self-centered metallic creatures without souls—indeed, there is no understanding of *soul* in modern technologies.

At one point in my writing, I tried to use dual-process theory—contrasting allocentric and egocentric disabilities—to discuss attention deficit disorder, Parkinson's Disease, bipolar disorder, Asperger's Syndrome, physical disabilities, face blindness, Developmental Topographic Disorder, gender differences, and even day-to-day personality fluctuations. Very quickly, I became overwhelmed. I soon realized that there were many others who were much more knowledgeable and experienced than I am in these domains. I have to leave my observations as a gift—a new slant on an old conundrum.

Each expert will have to explore whether the understanding of allocentric and egocentric processing applies to their discipline. For example, an expert in autism would ask if autism impacted a person allocentrically or egocentrically. Do autistic people perceive gestalts (faces, scenes, relationships) normally—is allocentric processing normal in autism? Do autistic individuals focus intently (obsessively), from an egocentric perspective, on an object of interest? Therefore, could autism be reframed as a disability of allocentric processing, in association with egocentric overcompensation?

When rethinking Parkinson's disease, an expert might ask if this motor disorder was causing allocentric disabilities, disrupting the ease of flow, balance, and proprioceptive ability in the whole body, or if, to the contrary, the egocentric frame of reference was being disturbed. If Parkinson's is a disorder of allocentric processing, how does the egocentric mind adapt and compensate? What therapies might be recommended based on this new perspective?

Could Bipolar Disorder reflect an imbalance between allocentricity and egocentricity? Is the egocentric mind overheated while the allocentric mind is severely depressed in this disorder? Could behavior swing back and forth as the two minds struggled to find harmony? Using dual-process theory, what behavioral strategies might be employed to rebalance a person diagnosed with Bipolar Disorder?

Does hearing loss cause an allocentric processing disability, or are deafness and hearing impairments egocentric disabilities? What implications arise if, for example, we understand deafness to be an egocentric disability? How might a deaf child be impacted if that child did not develop an egocentric language capability?

Every person is unique; often there are multiple impairments and understandable emotional and social variables that impact an individual life.

Where Do We Go From Here?

We must be very cautious when we apply any theory to the circumstances of an individual life—a life that is a process, never a static entity. However, I am confident that dual-process theory can be helpful—a valuable perspective that suggests avenues for remediation and understanding. I am left with this hope that experts in other professions will apply what they have discovered here to the challenges faced by their students, friends, clients, and patients.

Final Comments

If you've enjoyed learning all this, it's because the brain generally likes reading about itself. ~ *Beyond Biocentrism, Robert Lanza, 2016.*

When I began writing this book, I had a simple insight: consciousness is related to navigation. I didn't necessarily believe the supposition to be true on the first day when I picked up my pen and began to write. The connection between consciousness and navigation was just a thought experiment when I started—just a postulate. The idea that consciousness evolved because animals navigated from place to place simply felt plausible and worth exploring.

Early in the researching and writing process, I realized that I had a sufficient knowledge-base to make the case connecting consciousness to navigation. Over my career, I developed a neurophysiological understanding of human navigation. I also had extensive education—in optometry and in blind rehabilitation—which enabled me to comprehend mammalian anatomy and physiology. My knowledge base was practical, not just theoretical.

Furthermore, dual-process theory—as I came to understand it—seemed increasingly important as I did research for the book. What I actually discovered was an extension to dual-process theory. No one, as far as I was able to determine, had made the connection between navigation and our inherent duality.

The main insight that I uncovered—using dual-process theory as a guide—was that navigation requires two processes, one to perceive figures embedded in the ground, and a second system to provide for the ground—a gestalt or scene-generating system. Navigation is not possible unless the figure and ground oscillate. From this observation, I concluded that we were designed by nature to be inherently dual creatures—we eventually evolved two minds and two kinds of consciousness because *navigation dictated that it had to be that way.*

Furthermore, *these two minds had to be mutually exclusive.* They are co-dependent, but one mind cannot manifest unless the sister mind is relatively inhibited. We can either process the figure or we can process the ground, but not both at the same time. This insight—mutually exclusive processing—led me to suggest that our "habit" of always finding opposites, polarizations, and contrasts in the world had an origin in our dual biological makeup.

I began the book by comparing a tree and a human being. A tree has no brain, no legs, no heart, no nervous system, no sensory systems, and no muscles. Only animals have these physical manifestations. Yet the tree is older in evolution than the human being—the ancestry of trees stretches back 385 million years. In contrast, the first modern humans appeared only 200,000 years ago. I made a long and hopefully convincing case that a tree is as magnificent and mysterious a creature as a human being. And yet human beings can write books, make movies, compose symphonies, build space telescopes, heal the sick, and build (or destroy) massive cities. Human beings are wonderful yet tragic creations.

The question that began this journey was "Why do only animals have brains and nervous systems, while plants have none of these attributes?" And the answer is that human beings navigate. Trees spend their entire lifespan rooted to one spot on the earth. There is no need for brains, nervous systems, senses, and muscles if there is no purposeful movement.

Consciousness: A New Slant on an Old Conundrum

As the months of research and writing rolled by, and as the evidence mounted that navigation and consciousness were, indeed, inherently connected, I began to convince myself that dual-process theory was correct. I now feel, as this book ends, that my initial insight has great importance and needs to be an essential part of any dialogue about the evolution and characteristics of consciousness.

I was caught off guard a few times when I found myself writing statements that seemed absurd: space and time are not real, for example. I now know that this is true, both space and time are creations of the mind. The figures of the world, as well as the background for the figures, are simply manufactured inside the human frame. There are no colors in the physical world, no shapes, and no motion—navigation itself is an illusion. This revelation still bewilders me. None of my friends are as they seem. I am not as I seem. Cherry pie is a lie of sorts.

Perhaps the most amazing revelation came about when I was trying to explain the importance of proprioception. Most lay people have no idea what the proprioceptive sense does, or why it has such a key role to play in the evolution of human consciousness. The full scope of the importance of proprioception appeared as I was writing. One morning, I found myself writing: "The eyes enable vision. The ears enable hearing. Proprioception enables consciousness—both of them." I was actually surprised at this revelation, and I struggled to explain how an internal sensory system could give rise to both allocentric and egocentric consciousness. I eventually was able to articulate how this came about, but it was a process that occurred as the book evolved.

I started out with a well-accepted understanding of proprioception as a kind of muscle memory. Doesn't it make sense logically, I surmised, that consciousness—which came about because of navigation—would have muscle memory at its core? Consciousness is a combination: an *awareness of location* combined with an awareness of *change of location*. I won't go into the debate again, but I found this understanding quite profound. Luckily, I could point to others who found this out long ago. My skill is in synthesis and explanation. The pioneers have left their thoughts for people like me to appreciate.

At first, I could only see that proprioception was the reason sophisticated egocentric minds evolved in the human species—this is psychologist Zoltan Torey's great insight. Our egos, our language skills, our internal dialogue, and our personalities are products of proprioception—of muscle memory. Later, however, I realized that the allocentric mind was also a proprioceptive phenomenon dating back to the dawn of life on earth. The breakthrough here was that proprioception was a whole-body system based on cellular communication. Proprioception is also quantum. Our whole body is a brain. Quantum (cellular) communication has its own kind of total-body awareness and memory. Given this logic about egocentric and allocentric proprioception, I was able to conclude that both our minds and our dual consciousness evolved from proprioception.

However, this discovery of the fundamental importance of proprioception was not the end of the journey. It dawned on me one early morning that the earliest complex cellular organisms all moved with wonderful purpose. These cellular organisms seemed to navigate without brains, nervous systems, external senses, and without muscles. This threw a serious wrench into the gears of my logic. I did some patchwork thinking by simply redefining how navigation should be defined. I called everything before the Cambrian Explosion (660 million years ago) the Age of Wayfinding. Only after the Cambrian Age, after animals and the neocortex evolved, could we speak of the Age of Navigation. This bit of emergency logic helped me through the middle of the book. Then, on another fine morning, I awoke thinking about the work of quantum physicist Werner Heisenberg.

Heisenberg was the key for a final insight. I realized that what Heisenberg had done was define allocentricity and egocentricity using the language of mathematics. His famous formula—in layman's terms—simply says that we cannot measure location at the very same time as we measure momentum. In other words, we cannot measure motion at the very same time that we measure no-motion. If you freeze an entity so that you can pinpoint its exact location, you lose all information about velocity—you can't tell where an entity is going. In contrast, you can't measure velocity without losing all the information about location. The important understanding is that Heisenberg gives us a fractal, a fundamental law of the universe. Our two minds and our two kinds of consciousness evolved from this fundamental duality of nature. I did my best to explain this in the text, but, even so, the concepts are bewildering in their complexity.

That is all I will say about this journey. My hope is that you will take these insights and apply them to your life and your profession. There are greater truths hidden in this story, which I leave for you to uncover. I will give Dutch and Surge the final word as we transition from this book to the next *The Confusion Caused by Being Your Own Twin.*

The Rest of This Must Be Said In Silence

"Back so soon, Dutch?"

"I just came back to say goodbye to the reader, and to you Surge—I came to say thank you for all your insights and kindness."

"Are you onto the next book, then?"

"Yes. Other people play golf; I write."

"Well, writing books is like playing golf."

"I don't think so!"

"It's just another kind of game: driving home the point, keeping score, sand pits and water hazards. What's the next book about?"

"The book-fairies want me to write about levels of consciousness. They say that everyone dies on the journey, so goals and passions are left unfulfilled. That's the way it is, the rules of this game, the contract we signed when we checked in. But we can make some personal progress, I guess. Our minds can evolve. Then, if enough people get just a little more compassionate, a tiny bit more empathetic, whole cultures can evolve."

"Just levels of consciousness? That's it?"

"No. I said in this book that the evidence for our duality was everywhere, but I didn't give enough examples. I need to correct that oversight. I need to explore duality in psychology, religion, philosophy, and poetry. You know, do more documenting in my next book."

"What do you call the next book?"

"The Confusion Caused by Being Your Own Twin."

"Well, the title is good, I like it."

"Thanks. I've said enough for now, Surge, so I will wish you well. I am on to my next journey."

"You haven't said goodbye to the reader."

"Ah, yes. Thank you, Surge. Well, my friend, dear reader, our journey has ended. We have shared many words over many hours. I hope I didn't bore you to tears, as they say. My only regret is that I didn't get to know you as much as I had hoped. However, I was able to detect how sophisticated and lovely you are. If only our two minds could merge—what wonderful insights would arise. Thank you for your companionship. Thank you for sharing the journey. May your existence be mindful and filled with joy. May you be free of suffering and may you find happiness in all your days."

"Well, Dutch, we will miss you at the Third-Eye-Watching Vegan Restaurant."

"Yeah, I'll miss you too, Surge. But it's time for me to move on. The train is waiting to depart. Any last thoughts?"

"How about one last Rumi message:

"The rest of this must be said in silence
because of the enormous difference
between light and the words
that try to say light." ~ *"A Cleared Site," A Year with Rumi, 2006.*

Bibliography

A

Adams, Douglas

- *The Hitchhiker's Guide to the Galaxy*, 1995

Ananthaswamy, Anil

- *The Man Who wasn't There*, 2015

Archer, Lawrence M.

- "The Heart of Consciousness," *Journal of Medical and Dental Science Research*; Volume 2, Issue 3, pp:10-20, April, 2015

Ardley, Gavin

- *Aquinas and Kant, the Foundations of the Modern Sciences*, 1950

Armstrong, John

- *Love, Life, Goethe: Lessons of the Imagination from the Great German Poet*, 2007

B

Bakewell, Sarah

- *At The Existentialist Café: Freedom, Being, and Apricot Cocktails*, 2016

Baldwin, Doug

- *Bugs, Blindness, and the Pursuit of Happiness*, 2016
- *Consciousness, A New Slant on an Old Conundrum*, 2016
- *The Confusion Cause by Being Your Own Twin*, 2017 (in process)
- *Knights for the Blind in the Battle Against Darkness*, 2017 (in process)

Bibliography

Barfield, Owen

- *History in English Words,* 1925
- *Poetic Diction, A study in Meaning,* 1928
- *Romanticism Comes of Age,* 1944
- *Saving the Appearances, A Study in Idolatry,* 1957
- *What Coleridge Thought,* 1971
- *The Rediscovery of Meaning and Other Essays,* 1977
- *History, Guilt & Habit,* 1979
- *Owen Barfield on C. S. Lewis,* 1989
- *A Barfield Reader, Selections from the Writings of Owen Barfield,* 1999
- *The Case for Anthroposophy,* 2010

Baron-Cohen, Simon

- *Mindblindness: An Essay on Autism and Theory of Mind,* 1997
- *The Essential Difference: Male And Female Brains And The Truth About Autism,* 2004
- *The Science of Evil: On Empathy and the Origins of Cruelty,* 2012
- *Zero Degrees of Empathy, A New Theory of Human Cruelty and Kindness,* 2012
- *Understanding Other Minds: Perspectives from developmental social neuroscience,* 2013

Batson, Glenna

- *Graviception, F. M. Alexander's Science of Poise,* 2004

Baumeister Roy, and Masicampo, E. J.

- *Cognitive Psychology, a Student Handbook,* 2010

Berendt, Joachim-Ernst and Bredigkeit, Helmut

- *The World Is Sound: Nada Brahma: Music and the Landscape of Consciousness,* 1991

Berthoz, Alain and Weiss, Giselle

- *The Brain's Sense of Movement*, 2002

Bertucci, Paola

- "Somatic and Visceral Nervous Systems, an ancient duality," *BMC Biology*, 2013

Blakeslee, Sandra and Matthew

- *On Intelligence*, 2005
- *The Body has a Mind of its Own*, 2007
- *Sleights of Mind: What the Neuroscience of Magic Reveals about Our Everyday Deceptions*, 2010

Block, Ned

- Co-author of *The Nature of Consciousness: Philosophical Debates*, 1997
- *Consciousness, Function, and Representation*, 2007

Bohm, David

- *Causality and Chance in Modern Physics*, 1971
- *Science, Order, and Creativity: A Dramatic New Look at the Creative Roots of Science and Life*, 1987
- *Changing Consciousness: Exploring the Hidden Source of the Social, Political, and Environmental Crises Facing Our World* (with Mark Edwards), 1991
- *Thought as a System*, 1994
- *The Limits of Thought: Discussions between J. Krishnamurti and David Bohm (with J. Krishnamurti)*, 1999
- *Wholeness and the Implicate Order*, 2002

Bregman, Albert

- *Auditory Scene Analysis: the Perceptual Organization of Sound*, 1990

Burge, Tyler

- "Two Kinds of Consciousness" in the 1997 publication *The Nature of Consciousness: Philosophical Debates*, 1997
- *Truth, thought, reason: essays on Frege*, 2005
- *Foundations of Mind*, 2007
- *Origins of Objectivity*, 2010
- *Cognition Through Understanding: Self-Knowledge, Interlocution, Reasoning*, 2013

Buzsaki, George

- *Rhythms of the Brain*, 2011

C

Cabezón, José Ignacio

- "Buddhism and Science: On the Nature of the Dialogue." In: Wallace, Alan B. (ed.). *Buddhism and Science: Breaking New Ground*, 35-68, 2003

Chopra, Deepak

- *How to Know God: The Soul's Journey into the Mystery of Mysteries*, 2001
- *Life After Death: The Burden of Proof Life After Death*, 2008
- *The Third Jesus: The Christ We Cannot Ignore*, 2009
- *Spiritual Solutions: Answers to Life's Greatest Challenges*, 2012
- *Super Genes: Unlock the Astonishing Power of Your DNA for Optimum Health and Well-Being* (with Rudolph E. Tanzi PhD), 2015
- *The Future of God: A Practical Approach to Spirituality for Our Times*, 2015
- *Quantum Healing (Revised and Updated): Exploring the Frontiers of Mind/Body Medicine*, 2015

D

The 14ᵗʰ Dalai Lama, Tenzin Gyatso. Editors: Francisco Varela, Jeremy W. Hayward

- *Gentle Bridges; Conversations with the Dalai Lama on the Sciences of Mind,*1992

The 14ᵗʰ Dalai Lama:

- *Kindness, Clarity, and Insight: The Fourteenth Dalai Lama, His Holiness Tenzin Gyatso,* 1984
- *The Universe in a Single Atom: The Convergence of Science and Spirituality,* 2006
- *How to See Yourself As You Really Are (with Jeffrey Hopkins),* 2007
- *For the Benefit of All Beings: A Commentary on The Way of the Bodhisattva,* 2011
- "Attitude-Training, Like the Rays of the Sun—His Holiness the Fourteenth Dalai Lama," an essay.

Darwin, Charles

- *Charles Darwin's Notebooks, 1836-1844: Geology, Transmutation of Species, Metaphysical Enquiries,* 1987
- *The Voyage of the Beagle: Charles Darwin's Journal of Researches,* 1989
- *The Descent of Man,* 2004
- *On the Origin of Species: By Means of Natural Selection,* 2006
- *The Expression of the Emotions in Man and Animals,* 2009

Dehaene, Stanislas

- *Consciousness and the Brain: Deciphering How the Brain Codes Our Thoughts,* 2014

Dennett, Daniel C.

- *From Bacteria to Bach and back*, 2017

Depraz, Natalie and Varela, Francisco. Editor: B. Alan Wallace

- "Imaging: Embodiment, Phenomenology, and Transformation," *Buddhism and Science, 2003*

David-Neel, Alexandra and Yongden, Lama

- *The Secret Oral Teachings in Tibetan Buddhist Sects*, 1967

Dudchenko, Paul A.

- *Why People Get Lost; The Psychology and Neuroscience of Spatial Cognition*, 2010

E

Edelman, Gerald

- *Bright Air, Brilliant Fire, on the Matter of the Mind*, 1992
- *wider than the sky; the phenomenal gift of consciousness*, 2004
- *Second nature; brain science and human knowledge*, 2006

Evans, Jonathan and Frankish, Keith

- *The duality of mind: An historical perspective*, http://www.open.ac.uk/Arts/philos/ The_duality_of_mind_preprint.pdf
- *In Two Minds: Dual Processes and Beyond*, 2009

F

Flanagan, Owen

- *Science of the Mind: 2nd Edition, 1991*
- *Varieties of Moral Personality: Ethics and Psychological Realism, 1991*
- *Consciousness Reconsidered, 1992*
- *Self-Expressions: Mind, Morals, and the Meaning of Life, 1996*

- *Dreaming Souls: Sleep, Dreams and the Evolution of the Conscious Mind, 2000*
- *The Problem Of The Soul: Two Visions Of Mind And How To Reconcile Them, 2002*
- *The Really Hard Problem: Meaning in a Material World, 2007*
- *The Bodhisattva's Brain: Buddhism Naturalized, 2013*

Fawcett, Jonathan, Risko, Evan, and Kingstone, Alan, editors

- *The handbook of Attention, 2015*

Fulton James T.

- *Processes in Biological Vision*, an online book
- *Processes in Biological Hearing*, an online book

G

Gallagher, Shaun

- *Hermeneutics and Education, 1992*
- *Hegel, History and Interpretation, 1997*
- *How the Body Shapes the Mind, 2006*
- *Brainstorming: Views and Interviews on the Mind, 2008*
- *The Phenomenological Mind, 2012*
- *A Neurophenomenology of Awe and Wonder: Towards a Non-Reductionist Cognitive Science (New Directions in Philosophy, 2015*

Gibson, James Jerome

- *The Perception of the Visual World* (1950)
- *The Senses Considered as Perceptual Systems* (1966)
- *The Ecological Approach to Visual Perception* (1979)

Goldberg, Elkhonon

- *The Wisdom Paradox: How Your Mind Can Grow Stronger As Your Brain Grows Older, 2006*

- *The New Executive Brain: Frontal Lobes in a Complex World*, 2009

Goldstein, E. B.

- "The ecology of J. J. Gibson's perception," *JSTOR Leonardo*, Vol. 14, No. 3, pp. 191-195, 1981

Goldstein, Joseph

- *Insight Meditation: The Practice of Freedom*, 2003

Graziano, Michael

- *Consciousness and the Social Brain*, 2015

Gurdjieff, George

- *Meetings with Remarkable Men, 1963*

Guzeldere, Guven

- Co-author of *The Nature of Consciousness: Philosophical Debates*, 1997

H

Hanson, Rick

- *Buddha's Brain*, with Richard Mendius, 2009

Hayes, Michael

- *The Hermetic Code in DNA*, 2008

Hawkins, Jeff

- *On Intelligence*, 2005

Heisenberg, Werner

- *The Physical Principles of the Quantum Theory* (with Carl Eckart), 1949
- *The Physicist's Conception of Nature*, 1970

- *Philosophical Problems of Quantum Physics,* 1979
- *Encounters with Einstein,* 1989
- *Physics and Philosophy: The Revolution in Modern Science,* 2007
- *Nuclear Physics,* 2015

Houston, Jean, with Ervin Laszlo, and Larry Dossey

- *What is Consciousness?* 2016

Husserl, Edmund

- *Husserl,* David Woodruff Smith, 2013

J

Jacobs, Lucia

- "Olfactory Orientation and Navigation in Humans," *Journal of The Public Library of Science,* with Jennifer Arter, Amy Cook, and Frank J. Sulloway, June 17, 2015

James, William

- *The Principles of Psychology,* 1950
- *Talks to Teachers on Psychology and to Students on Some of Life's Ideals,* 2001
- *On Some of Life's Ideals: On a Certain Blindness in Human Beings; What Makes a Life Significant,* 2012
- *The Varieties of Religious Experience; A Study in Human Nature,* 2013

Jung, Carl

- *Memories, Dreams, Reflections,* 1963

K

Kahneman, Daniel

- *Thinking, Fast and Slow,* 2011

Karmiloff-Smith, Annette

- *Beyond Modularity: A Developmental Perspective on Cognitive Science,* 1995

Kish, Daniel, and Hook, Jo

- *Echolocation and FlashSonar,* 2016

Kosslyn, Stephen and **Miller, G. Wayne**

- *Top Brain, Bottom Brain: Harnessing the Power of the Four Cognitive Modes,* 2015

L

Lawrence, Krauss

- *A Universe from Nothing,* 2013

Lachman, Gary

- "Meditation and Spiritual Perception," *Classics from the Journal for Anthroposophy,* 2011
- *A Secret History of Consciousness,* with Colin Wilson, 2003
- *Turn Off Your Mind: The Mystic Sixties and the Dark Side of the Age of Aquarius,* 2003
- *A Dark Muse: A History of the Occult,* 2004
- *Rudolf Steiner: An Introduction to His Life and Work,* 2007
- *The Quest for Hermes Trismegistus: From Ancient Egypt to the Modern World,* 2011
- *Swedenborg: An Introduction to His Life and Ideas,* 2012
- *Madame Blavatsky: The Mother of Modern Spirituality,* 2012
- *Jung the Mystic: The Esoteric Dimensions of Carl Jung's Life and Teachings,* 2012
- *Politics and the Occult: The Left, the Right, and the Radically Unseen,* 2012

- *The Caretakers of the Cosmos: Living Responsibly in an Unfinished World,* 2013
- *Aleister Crowley: Magick, Rock and Roll, and the Wickedest Man in the World,* 2014
- *In Search of P. D. Ouspensky: The Genius in the Shadow of Gurdjieff,* 2014
- *Revolutionaries of the Soul: Reflections on Magicians, Philosophers, and Occultists,* 2014
- *The Secret Teachers of the Western World,* 2015
- *Beyond the Robot: The Life and Work of Colin Wilson,* 2016

Lanza, Robert

- *Beyond Biocentrism,* 2016.

Laszlo, Ervin

- *Science and the Akashic Field: An Integral Theory of Everything,* 2007
- *The Immortal Mind: Science and the Continuity of Consciousness beyond the Brain,* 2014
  (with Anthony Peake)
- *The Self-Actualizing Cosmos: The Akasha Revolution in Science and Human Consciousness,* 2014
- *The Systems View of the World: A Holistic Vision for Our Time (Advances in Systems Theory, Complexity, and the Human Sciences)* 2nd Edition, 1996
- *Quantum Shift in the Global Brain: How the New Scientific Reality Can Change Us and Our World,* 2010
- *The Akashic Experience: Science and the Cosmic Memory Field,* 2009
- *The Consciousness Revolution,* 2003

Lindley, David

- *Uncertainty,* 2007

Lusseyran, Jacques

- *What One Sees without Eyes: Selected Writings of Jacques Lusseyran,* 1999
- *And There Was Light: The Extraordinary Memoir of a Blind Hero of the French Resistance in World War II,* 2014
- *Against the Pollution of the I: On the Gifts of Blindness, the Power of Poetry, and the Urgency of Awareness,* 2016

M

Macy, Joanna

- *Mutual Causality in Buddhism and General Systems Theory: The Dharma of Natural Systems,* 1991
- *Spiritual Ecology: The Cry of the Earth,* 2013

Mandelbrot, Benoit B.

- *The Fractal Geometry of Nature,* 1977

Mansfield, Victor

- *Synchronicity, Science, and Soul-Making,* 1995

McTaggart, Lynne

- *The Field: The Quest for the Secret Force of the Universe,* 2008
- *The Intention Experiment: Using Your Thoughts to Change Your Life and the World,* 2008
- *The Bond: How to Fix Your Falling-Down World,* 2012

Miller, George A.

- *Psychology: The science of mental life,* 1973

Milner, David and Goodale, Melvyn

- *The Visual Brain in Action* (1996)
- *Sight Unseen: An Exploration of Conscious and Unconscious Vision* (2004)

Merleau-Ponty, Maurice

- The *Phenomenology of Perception*, 1945

Morris, David

- *The Sense of Space*, 2013

Murphy, Todd

- *Sacred Pathways: The Brains' Role in Religious and Mystic Experiences*, 2015.

N

Neisser, Ulric

- *Cognition and Reality: Principles and Implications of Cognitive Psychology*, 1976

Nietzsche, Friedrich

- *Thus Spoke Zarathustra: A Book for Everyone and No* One, with R. J. Hollingdale, 1961
- *The Will to Power*, with Walter Kaufmann, 1968
- *The Birth of Tragedy: Out of the Spirit of Music*, with Michael Tanner, 1994

Noë, Alva

- *Out of Our Heads; Why you are not your Brain*, 2009

Novak, Peter

- *The Division of Consciousness; The Secret Afterlife of the Human Psyche*, 1997

Bibliography

O

O'Keefe, John

- *The Hippocampus as a Cognitive Map*, John O'Keefe, with Lynn Nadel, 1978

Ouspensky, P. D.

- *The Fourth Way: An Arrangement by Subject of Verbatim Extracts from the Records of Ouspensky's Meetings in London and New York*, 1971
- *The Psychology of Man's Possible Evolution*, 1973
- *A New Model of the Universe*, 1997
- *In Search of the Miraculous*, 2001
- *Tertium Organum*, 2010

P

Paivio, Allan

- *Mind and Its Evolution; a Dual Coding Theoretical Approach*, 2007

Penrose, Roger

- *The Emperor's New Mind: Concerning Computers, Minds, and The Laws of Physics*, 1989
- *Shadows of the Mind, a Search for the Missing Science of Consciousness*, 1994
- *The Road to Reality: A Complete Guide to the Laws of the Universe*, 2004
- *Cycles of Time: An Extraordinary New View of the Universe*, 2010
- *Fashion, Faith, and Fantasy in the New Physics of the Universe*, 2016

Pietsch, Paul

- *Shuffle Brain: The Quest for the Hologramic Mind*, 1981

Ponlop, Dzogchen

- *Rebel Buddha: A Guide to a Revolution of Mind*, 2011

R

Rajneesh (Osho), Bhagwan Shree

- *Book of the Secrets,*1977
- *Book of the Secrets Two,* 1979
- *Book of the Secrets,*1980
- *Hammer on the Rock; a Darshan Diary,* 1980
- *The Great Challenge; the Rajneesh Reader*, 1982
- *Philosophia Ultima,* 1983
- *New Man: The Only Hope for the Future,* 1987
- *Death: The Greatest Fiction,* 1988
- *The Psychology of the Esoteric,* 1989
- *The Art of Dying,* 2007
- *Enlightenment: The Only Revolution Talks on the Great Mystic Ashtavakra,* 2008
- *In Search of the Miraculous,* 2009

Ranck, James B.

- "Head direction cells in the deep layer of dorsal presubiculum in freely moving rats," *Soc Neuroscience Abstracts*, 10:599, J.B. Ranck Jr, 1984

Richard Rohr

- Quotes taken from his Daily Blog

Rūmī, Jalāl ad-Dīn Muhammad

- *A Year with Rumi*, translated by Coleman Barks

S

Sacks, Oliver

- *The Man Who Mistook His Wife For A Hat: And Other Clinical Tales,* 1998
- *Musicophilia: Tales of Music and the Brain,* Revised and Expanded Edition, *2008*
- *The Mind's Eye, 2011*
- *Hallucinations,* 2013
- *On the Move: A Life,* 2016

Samuels, Richard

- "The magic number two, plus or minus: Dual-process theory as a theory of cognitive minds," *Two Minds: Dual Processes and Beyond,* 2009

De Salzmann, Jeanne

- *The Reality of Being: The Fourth Way of Gurdjieff,* 2010

Siegel, Daniel

- *The Developing Mind: Toward a Neurobiology of Interpersonal Experience,* 1999
- *The Mindful Brain: Reflection and Attunement in the Cultivation of Well-Being,* 2007
- *Mindsight: The New Science of Personal Transformation,* 2010
- *Brainstorm: The Power and Purpose of the Teenage Brain,* 2013
- *Developing Mind; Co-author with Tina Payne Bryson,* 2014
- *Mind, A Journey to the heart of Being Human,* 2017

Sheets-Johnstone, Maxine

- *The Primacy of Movement,* 1999

Schneider, W. & Shiffrin, R. M.

- "Controlled and Automatic Human Information Processing: Perceptual Learning, Automatic Attending, and a General Theory." *Psychological Review*, Vol 84, No. 2, March, 1977

Silverman Irwin, and Choi, Jean

- *Buss's Handbook of Evolutionary Psychology,* 2nd edition, 2005

Shlain, Leonard

- *The Alphabet Versus the Goddess: The Conflict Between Word and Image,* 1999
- *Sex, Time and Power: How Women's Sexuality Shaped Human Evolution,* 2003
- *Finding Balance: Reconciling the Masculine/Feminine in Contemporary Art and Culture,* 2006
- *Art & Physics: Parallel Visions in Space, Time, and Light,* 2007
- *Leonardo's Brain, Understanding Da Vinci's Creative Genius,* 2014

Smith, David Woodruff

- *Husserl,* 2013

Stanovich, Keith E.

- *Who Is Rational?: Studies of individual Differences in Reasoning,* 1999
- *Progress in Understanding Reading: Scientific Foundations and New Frontiers,* 2000
- *The Robot's Rebellion: Finding Meaning in the Age of Darwin,* 2004
- *Decision Making and Rationality in the Modern World,* 2009
- *What Intelligence Tests Miss: The Psychology of Rational Thought,* 2009
- *Rationality and the Reflective Mind,* 2011
- *How to Think Straight About Psychology,* 2012
- *The Rationality Quotient: Toward a Test of Rational Thinking,* 2016

Stein, Edith and Stein, Waltraut

- *On the Problem of Empathy: The Collected Works of Edith Stein, 3rd Volume,* 1989
- *Essays On Woman (The Collected Works of Edith Stein),* with Freda Mary Oben, 1996

Stein, Murray

- *Jung's Map of the Soul,* 1998

Steiner, Rudolf

- *How to Know Higher Worlds: A Modern Path of Initiation,* with Christopher Bamford, 1994
- *Theosophy : An Introduction to the Spiritual Processes in Human Life and in the Cosmos,* with Catherine Creege, 1994
- *Intuitive Thinking As a Spiritual Path: A Philosophy of Freedom,* with Michael Lipson, 1995
- *An Outline of Esoteric Science,* with Catherine Creeger, 1997
- *Staying Connected: How to Continue Your Relationships with Those Who Have Died,* with Christopher Bamford, 1999
- *Knowledge of the Higher Worlds and Its Attainment,* 2012
- *Road To Self Knowledge,* 2013
- *The Way of Initiation: How to Attain Knowledge of the Higher Worlds,* with Max Gysi, 2015

T

Taylor, Jill Bolte

- *My Stroke of Insight,* 2008

Thaler, Lore

- "Using Sound to Get Around; Discoveries in Human Echolocation," The Observer Series; Association for Psychological Science, https://

www. psychologicalscience.org /observer/using-sound-to-get-around#. WPy_X9LyvyQ

Thompson, Evan

- *Mind in Life: Biology, Phenomenology, and the Science of Mind*, 2007
- *Waking, Dreaming, Being, 2015*

Torey, Zoltan

- *Out of Darkness: A Memoir*, 2003
- *The Crucible of Consciousness: An Integrated Theory of Mind and Brain*, with Daniel C. Dennett, 2009
- *The Conscious Mind*, 2014

V

Varela, Francisco and Depraz, Natalie. Editor: Wallace, B. Alan

- "Imaging: Embodiment, Phenomenology, and Transformation" *Buddhism and Science, 2003*
- "Radical embodiment: Neural Dynamics and Consciousness," *Trends in Cognitive Sciences* 5:418-425, Evan Thompson and Francisco Varela, 2001

W

Wallace, B. Alan

- *Tibetan Buddhism from the Ground Up: A Practical Approach for Modern Life*, with Steven Wilhelm, 1993
- *The Taboo of Subjectivity: Toward a New Science of Consciousness*, 2004
- *Genuine Happiness: Meditation as the Path to Fulfillment*, 2005
- *The Attention Revolution: Unlocking the Power of the Focused Mind*, with Daniel Goleman, 2006
- *Hidden Dimensions: The Unification of Physics and Consciousness (Columbia Series in Science and Religion)*, 2007

Bibliography

- *The Four Immeasurables: Practices to Open the Heart,* 2010
- *Minding Closely: The Four Applications of Mindfulness,* 2011
- *Dreaming Yourself Awake: Lucid Dreaming and Tibetan Dream Yoga for Insight and Transformation,* with Brian Hodel, 2012

Ward, Anthony

- *Attention, A neuropsychological Approach,* 2004

Watts, Alan

- You Tube presentation, part of a lecture called the *Tao of Philosophy*
- *An Outline of Zen Buddhism,* 1932
- *The Spirit of Zen: A Way of Life, Work and Art in the Far East,* 1936
- *The Legacy of Asia and Western Man,* 1937
- *The Meaning of Happiness,* 1940
- *Theologia Mystica: Being the Treatise of Saint Dionysius, Pseudo-Areopagite, on Mystical Theology, Together with the First and Fifth Epistles,* 1944
- *Behold the Spirit: A Study in the Necessity of Mystical Religion,* 1947
- *Easter: Its Story and Meaning,* 1950
- *The Supreme Identity: An Essay on Oriental Metaphysic and the Christian Religion,* 1950
- *The Wisdom of Insecurity: A Message for an Age of Anxiety,* 1951
- *Myth and Ritual in Christianity,* 1953
- *The Way of Zen,* 1957
- *Nature, Man and Woman,* 1958
- *Beat Zen Square Zen,* 1959
- *This Is It and Other Essays on Zen and Spiritual Experience,* 1960
- *Psychotherapy East and West,* 1961
- *The Joyous Cosmology: Adventures in the Chemistry of Consciousness,* 1962
- *The Two Hands of God: The Myths of Polarity,* 1963
- *Beyond Theology: The Art of Godmanship,* 1964
- *The Book: On the Taboo Against Knowing Who You Are,* 1966

- *Nonsense, illustrations by Greg Irons (a collection of literary nonsense)*, 1967
- *Does It Matter?: Essays on Man's Relation to Materiality*, 1970
- *The Temple of Konarak: Erotic Spirituality, with photographs by Eliot Elisofon*, 1971
- *The Art of Contemplation: A Facsimile Manuscript with Doodles*, 1972
- *In My Own Way: An Autobiography 1915–1965*, 1972
- *Cloud-hidden, Whereabouts Unknown: A Mountain Journal*, 1973

Wheeler, John

- "Law without Law," *Quantum Theory and Measurement*, 1983

Wilber, Ken

- *The Spectrum of Consciousness*, 1977
- *The Holographic Paradigm*, 1982
- *Integral Psychology: Consciousness, Spirit, Psychology, Therapy*, 2000
- *The Integral Vision: A Very Short Introduction to the Revolutionary Integral Approach to Life, God, the Universe*, 2007
- *Integral Meditation: Mindfulness as a Way to Grow Up, Wake Up, and Show Up in Your Life*, 2016

Wilson, Colin

- *The Essential Colin Wilson*, 1986
- *The Strange Life of P.D. Ouspensky*, 1993
- *Poetry & Mysticism*, 2001
- *Super Consciousness: The Quest for the Peak Experience*, 2009
- *The Philosopher's Stone*, with Colin Stanley, 2013

Wittgenstein, Ludwig

- *Tractatus Logico-Philosophicus*, http:// archive.org/ stream/ tractatus-logicop 05740gut/tloph10.txt

Wu, Wayne

- *Attention*, 2014

I love **The Great Courses** lectures and I have used three as references in my recent books.

- A Great Courses lecture entitled *Philosophy, Religion, and the Meaning of Life*, by Georgetown University Professor Francis Ambrosio.
- Robotics
- Consciousness

TED Talks:

Daniel Wolpert

Daniel Kish

Jill Bolte Taylor

Index

Index

Index

Index